The Apollo Guidance Computer
Architecture and Operation

Frank O'Brien

The Apollo Guidance Computer

Architecture and Operation

 Springer

Published in association with
Praxis Publishing
Chichester, UK

Mr Frank O'Brien
West Windsor
New Jersey
USA

SPRINGER–PRAXIS BOOKS IN SPACE EXPLORATION
SUBJECT *ADVISORY EDITOR*: John Mason, M.B.E., B.Sc., M.Sc., Ph.D.

ISBN 978-1-4419-0876-6 Springer Berlin Heidelberg New York

Springer is a part of Springer Science + Business Media (*springer.com*)

Library of Congress Control Number: 2009936113

Cover design: Jim Wilkie
Copy Editor: David M. Harland
Typesetting: BookEns Ltd, Royston, Herts., UK

Printed in Germany on acid-free paper

Contents

To my mother and father

*Mom is no longer with us, but I know that future
explorers will use her star in Heaven to guide
them safely home.*

List of figures

Author's preface

Much of the story of mankind's first voyage to the moon centers on its engineering achievements. The sheer physical size of the vehicles and facilities not only dwarfed the previous generation of manned spacecraft, but still represents one of the greatest engineering achievements in history. But size alone is hardly the most valid metric for such an endeavor. Almost every conceivable figure of merit, whether pressures, temperatures, flow rates or stresses (and the list goes on) pushed the limits of the possible. But the magnitude of raw power overshadows equally impressive engineering at the other end of the spectrum. This book focuses on very different technologies, each with a unique set of attributes: fantastically small, breathtakingly precise, and in the most impressive feat of all, completely invisible to the eye. The Saturn V, with its 3,300 tons of structure and propellant would be helpless without a brain to give it a purpose. That brain, in the form of a small computer, guided the spacecraft and its precious human cargo to a new world, and home again.

In the 1960s, the Apollo Guidance Computer defined the state of the art in digital electronics, reliability and real-time computer control. However, the journey from a laboratory bench to a flight-ready system can only be described as tortuous, as the evolving requirements for a lunar mission forced nearly continuous redesigns. With so much time and effort already invested in an existing design, starting over with a new "clean sheet of paper" architecture was never an option. As a result, the AGC grew like a newlyweds' house. Over time, the happy couple's family grows, and so too must their house. Never willing to tear it down and begin anew, the house becomes a collection of architectural contradictions, with a blind hallway or two and plumbing that makes little sense. Still, the house is quite functional, with each iteration representing the best ideas at the time.

Some aspects of the AGC were truly revolutionary for their time. The exclusive use of integrated circuits in the processor ushered in a new era of computing, the novel memory design stored large amounts of data in a small space, and the human interface allowed real-time interaction with software. Other characteristics were like the newlyweds' house; the accumulated changes made the design awkward and inefficient. Finally, the severe lack of memory perpetually challenged the developers during the entire life of the project.

The *Apollo Guidance Computer: Architecture and Operation* presents the three major organizational levels of the AGC. First, the programmers' view of the AGC, known as the hardware architecture, is described in detail. This section discusses the key processor components and how they interact with each other. Importantly, the balance between power, flexibility and the limitations of the hardware is made evident by the designers' struggle to extend the capabilities of the system. The final design is full of exceptions and special cases, but is especially notable for its elegant solutions for optimizing every bit in memory. Understanding these constraints sheds much light on the decisions affecting the other parts of the system, from the Executive software to the operating procedures.

Efficiently running all the programs that are necessary for a complex spacecraft like Apollo demands a sophisticated operating system. The second section is devoted to the operating system software, known as the Executive, and a powerful execution environment called the Interpreter. All of the computer resources used by the mission programs were managed by the Executive. The Executive oversaw all of the processing necessary for an interactive, real-time computing environment: namely scheduling mission programs for execution,[1] accepting commands from the crew and displaying its results, and recovering from any errors that occur. Complex programming logic used in the mission programs required equally sophisticated capabilities that were unavailable in the hardware. A novel and compact language was designed which, by reflecting the problem to be solved, greatly reduced the workload on the programmer. This language was processed by a program called the Interpreter, which created the illusion of an entirely new and highly capable computer architecture. Slow, but rich with features, the Interpreter simplified the programming effort needed for the mission programs and greatly reduced their memory requirements.

The third and final section leaves the hardware and Executive behind to enter the world of mission software and its operation. Each mission phase, whether leaving Earth, landing on the Moon or performing a navigating sighting requires a variety of programs in the AGC. In turn, the execution of these programs must be coordinated with the operation of other spacecraft systems to satisfy a mission requirement. Properly covering each mission phase requires placing the AGC's operation into a larger perspective, by including the operation of selected spacecraft systems in the discussion. Each Apollo mission built on the experiences of its predecessors, so there is no "typical" mission to use as an example. Procedures and techniques reached their highest level of refinement in the latter missions – Apollo 15, 16 and 17 – so many of the descriptions will reflect these flights.

It is also worth noting what is *not* included in this book. With few exceptions,

---

[1] When discussing the wide variety of software used in computer systems, it is useful to separate their function into two distinct areas: "Executive software" or "operating system software", manages the resources and services of the computer. "Applications software" or "mission programs" focus on solving a particular problem for the user, whether it is calculating a payroll or the trajectory to the Moon.

there is no description of the underlying circuitry and chip technology, as the internal architecture of the AGC is invisible to both the programmer and the astronauts. To discuss both computer science and spaceflight in a single volume is a bold step in itself; to expand it to include electrical engineering would run the danger of overwhelming the reader. Also, despite its extensive treatment of the AGC hardware, Executive and Interpreter, this book is not intended to be an AGC programming tutorial. Including material on programming techniques, especially necessary for novice readers, followed by the obligatory problem exercises would move this book far from its intended scope. Finally, descriptions of the architectural evolution of the hardware and software are consciously avoided. Specifically, the book reflects the Block II configuration of the AGC, which was used in all manned Apollo flights. The compressed timeline to achieve a manned lunar landing meant there were several concurrent development projects at any one time, and a description of every variant would result in a hopelessly complex account.

Apollo Guidance Computer: Architecture and Operation is intended to be the essential reference for both spaceflight historians and those with computer or engineering backgrounds. Bringing those disciplines together into a single work is not just an interesting academic exercise; it offers a new way to understand the mission design process. Both the design decisions and the compromises made during the AGC's development directly influenced the overall operation of the Apollo spacecraft. Once defined, the spacecraft's operational capabilities became the principal driver in creating mission objectives and procedures. By understanding the AGC's capabilities and constraints, the reader will gain important insights into how the lunar missions assumed their final form.

Acknowledgments

Award ceremonies in the entertainment industry are often criticized for their endless series of thank-yous, often without recognizing that memorable works are rarely created in a vacuum. Writing a book is little different, and I drew upon the knowledge and guidance of a several people to whom I'll always be indebted:

Eric Jones and David Woods, the editors of the *Apollo Lunar Surface Journal* and the *Apollo Flight Journal*. Their work has resulted in the canonical reference on the Apollo missions for historians and researchers. While the encyclopedic documentation and annotation of the missions is impressive, even more so was their use of a fledgling Internet to assemble a group of highly motivated people from all over the world to contribute their expertise. With assignments ranging from researching obscure spacecraft systems to the essential but mundane task of proofreading, the Apollo Journals are a true collaborative effort. Gentlemen, it has been a privilege working with both of you.

David Harland has chronicled mankind's greatest adventure in a series of highly readable books. Much to my surprise, he felt that writing a book on the Apollo Guidance Computer was a worthy idea. Though I was initially skeptical, both he and David Woods felt that my work on the Apollo Journals could be expanded considerably. In addition to planting the seed for this work in my mind, David Harland also undertook the role of editing my manuscript. His suggestions and comments on my manuscript, especially in the early versions, have made it a far better work.

John O'Neill, retired Director of Space Operations at NASA, who lavished a young Frank O'Brien with all things Apollo. I devoured every scrap of information he passed along to me, using it to follow along with the missions as they were happening. It is not an exaggeration to say that this book is a direct result of that generosity.

Joshua Stoff of the Cradle of Aviation Museum, Uniondale, NY, and Paul Fjeld, for allowing me to work with and learn from them. What they have created at the Cradle of Aviation will bring tears of joy to your eyes.

Fred Carl of the Infoage Science/History Learning Center, Wall, NJ, who supported my work in creating their AGC exhibit and using it as part of their community outreach.

I had the special opportunity of spending time with Eldon Hall, chief engineer for the AGC at the MIT Instrumentation Laboratory, and Hugh Blair-Smith, one of the principal architects of the AGC. Their explanations provided much needed insights, and were given with admirable patience and wit.

Independently, Julian Webb and Ron Burkey have created AGC emulators that run actual Lunar Module software. Ron has extended the scope of his work to include the Lunar Module's Abort Guidance Computer and an extensive collection of documentation. Their emulators and the archived references were especially valuable for my research.

I especially have to thank Clive Horwood and his staff at Praxis for their help in bringing this book to life.

Finally, none of this could have been accomplished without the support of family and friends, especially my wonderful wife Stacey.

Frank O'Brien
January 2010

0

The state of the art

FROM WHENCE WE CAME: EARLY COMPUTING

The dizzying pace of technological change is so ingrained into our consciousness, it is impossible to remember what was considered "the state of the art" as recently as a few decades ago.

Consider the first practical computer – the room sized, vacuum tube ENIAC. In the years since its creation in 1946, the power of perhaps a million ENIACs has shrunk to an attractive package that comfortably fits in a briefcase. In such a 21st century laptop, with its multiple processors and gigabytes of memory, there is rarely any thought given to its internal complexity. It is by most definitions, an appliance – a black box, and few need to know or care about the principles that make its magic happen. Creating such a sophisticated device demands a standing army of thousands of engineers and technicians, each using sophisticated tools to design the chips, disks and displays. Manufacturing these components demands such cleanliness and precision that humans cannot touch the parts until their fabrication is complete. Software developers, using formalized methods, automated tools and debuggers also create products with formidable complexity. Software reliability is an important goal of course, but so is the need for standardized interfaces, security, lush graphical interfaces and a myriad of other factors necessary for a product to be successful in the marketplace. In the end, engineers and programmers manage to create systems that operate seamlessly with each other (at least, most of the time).

Such images of hardware and software development are so prevalent that the reader might be excused for believing that comparable tools and processes existed in the early 1960s to create the AGC. This, of course, cannot be further from reality. Most of the hardware design tools for the AGC were paper and pencils, oscilloscopes and slide rules. Automated design tools did not exist, so engineers tested and proved designs by actually *building* the circuits on the laboratory bench, and ultimately, full-scale prototypes with thousands of hand-wired components. Software development environments, with elaborate debuggers and other productivity enhancing tools also would not exist until the 1980s. Written and reviewed by hand, each revision of the program code was laboriously checked for any logic errors that might have slipped in. Only when there was a reasonable expectation that the program's logic was correct was it ready for execution on actual computer hardware. The program, carefully written on standardized paper forms, was then sent to a pool of workers who typed each line of code using a keypunch machine. The keypunch did not enter the program or its data into the computer; it only punched holes into a card, with each set of holes corresponding to a single typed character. The completed program, now in a form readable by a computer, often required hundreds if not thousands of cards. Loaded into the computer's memory, the program began executing and referenced data from additional cards, or magnetic media such as disk or tape. Output was sent to a line printer, where it was printed on lengths of fan-fold paper.

Lights and switches on the computer console were not for dramatic effect, but indicated the various states of the processor and were essential for debugging the program code. Each light indicated whether a specific bit was a 1 or 0, and switches allowed the programmer to start and stop the program, or execute it one step at a time if necessary. All the while the program was executing, the programmer could look at memory in search of hidden logic errors. Time on a computer was scarce, and the opportunity to run a program was often limited to only a few attempts per day. Computing in this era had a surprisingly large manual component, and the computer operation departments organized the day's work according to strict schedules. Large amounts of data were stored on tapes and removable disks which had to be collected before running the program. The hardware for the processor itself was large, easily the size of dozens of filing cabinets, but was hardly the only device in the computer room. Equipment for reading and writing tapes, printing output and communicating over telephone lines often dominated the remaining space in the room. With this environment in mind, it should be no surprise that computing was restricted to specialized computer rooms, with elaborate power and cooling facilities.

OUTSIDE THE COMPUTER ROOM: EARLY COMPUTING IN AVIATION AND SPACE

Digital computers did not exist in aviation in any meaningful way in the early 1960s. Autopilots and navigation electronics were exclusively analog devices and frequently used vacuum tubes in their construction. Even the military's most advanced radar and weapon fire control systems were analog, although digital uplinks were being integrated. Given the size, weight and power requirements, digital devices did not present significant advantages over the analog systems that had been refined over the years. Successfully integrating computers into the cockpit required an intuitive interface for pilots to enter data and interpret the results, especially in a combat environment. This was a difficult task, in large part because the concept of a small, interactive display was still evolving. An even more practical problem was finding space for such an interface in a fighter aircraft cockpit, already packed with flight instrumentation, aircraft system controls and radios.

In spaceflight, the first generation of Earth orbiting satellites simply broadcasted their data "in the blind", without having any idea of where a tracking station might be to receive its signals. Such a design was quite reasonable, as placing an object in orbit – *any* orbit – presented huge challenges in the 1950s. With more accurate rockets and smaller, lighter and more reliable electronics, later spacecraft added the ability to accept commands from the ground. Rather than continuously broadcasting, a tape recorder would save the data and a command from the tracking station would request its playback. The need to command a series of instructions led to the development of programmable sequencers. Vastly more flexible than the hardwired logic of early spacecraft, sequencers were a highly successful technology on the evolutionary road to onboard general purpose computers. Sequencers, as their name implies, are designed to perform operations on a set of commands stored in memory. For example, a command sequence might include the following operations:

- Wait 908 seconds
- Rewind the data tape recorder
- Pitch down to 14 degrees
- Take 27 pictures.

Each command directs a piece of hardware to perform a specific task. Sequencers usually lacked the decision making and computational operations normally found in a computer, but this simplified their design at a time when digital hardware was heavy and complex. Although not considered a true computer, sequencers did require memory, a rudimentary processor to fetch and interpret commands, and a basic arithmetic capability. Still, the above example illustrates several important capabilities introduced with programmable sequencers. Once the sequence is stored, the spacecraft can perform operations outside the range of tracking stations. This allowed the spacecraft to operate autonomously and collect data only when it was over areas of interest. Complex operations are made possible by storing multiple commands, each scheduled to execute at specific times. Command sequences can be arranged in any order (within reason), and can be updated as necessary. Finally, a key to flexibility is allowing variable data in sequencer commands. A camera, for example, can be commanded to take any number of pictures, not just one.

COMPUTING IN MANNED SPACECRAFT

The American Mercury and its Soviet counterpart, the Vostok/Voshkod, were the first generation of manned spacecraft. Both began development in the late 1950s when computing was in its infancy and the conversion away from vacuum-tube computers was still underway. Mission objectives for the spacecraft were also similar: to explore the effects of the space environment on both man and machine. On these early missions, there was little need for onboard computing, as ground-based systems performed the guidance and navigation calculations, and routed the results to the flight controllers. As a result, astronauts and cosmonauts had little situational awareness of their orbit. A model globe of the Earth was mounted on the instrument panel, and crosshairs identified the approximate location of the capsule as it traveled around the globe. Attitude control onboard the spacecraft was a combination of manual inputs and analog electronics, using gyros to sense changes in attitude. A few critical events were controlled by automatic sequencers, but these relied on analog electronics. The digital revolution had not yet arrived in spaceflight.

The ambitious objectives of the next generation of American spacecraft, called Gemini, created the need for real-time, onboard computing. For this new spacecraft to rendezvous with another vehicle, a computer must process a continuous flow of navigation and radar data and present its solutions to the crew. With an orbital path that was frequently out of range of tracking stations, Gemini demonstrated that it is possible to perform all the calculations and maneuvers independently from ground support. Operating the Gemini computer was straightforward, reflecting the task-oriented design of its interface. A rotary switch selected which program to run, such as Ascent, Rendezvous and Reentry. The user interface was a numeric keyboard to enter and read back data, and the display consisted of the address and contents of a single word of memory, using mechanical, not electronic digits. Data was entered by modifying the contents of a location in storage, and, likewise, retrieving data was performed by entering a memory location and requesting that the data be read out. The Gemini computer also presented velocity information on specialized displays located on the instrument panel, and directed the "fly-to" needles on the crew's attitude indicators. The result was the first spacecraft to have a computer fully integrated into the guidance and navigations system.

DEFINING COMPUTER "POWER"

Over the course of this book the discussion will center on the capabilities of a small computer whose complexity will quickly become apparent. The first reaction is often, "All this capability simply can't fit in a machine that size!" But is this really true? The current worldview of computers and software is usually from a person's experience with their personal computer. Desktop computer systems are equipped with a high performance processor, a few billion bytes of memory and upwards of a trillion bytes of disk space. What makes these systems require so many resources, and how can the AGC perform all of its capabilities with so little? To begin, it is important to dismiss the oft-repeated claim: "My watch/calculator/microwave has more processing power than the Apollo Guidance Computer!" But does it really? There is certainly no debate that modern electronics are far smaller and faster than those built in the 1960s. Still, how is the "power" of a computer measured without falling into arcane technical arguments? Much of this comparison is based on the most common and misleading metric for computer performance, its internal clock speed. The job of a computer clock is not to tell time, but to act as a metronome, sending a pulse through the hardware to orchestrate the processing of data. Clock speeds in 21st-century computers use the gigahertz as a unit of measurement – billions of clock pulses per second. In this area, the AGC with its internal clock speed of 2 megahertz (millions of pulses per second), simply does not stand a chance in this comparison. But this does not tell the entire story.

To illustrate the futility of this argument, consider another subject of frequent comparison: automobile engine displacement. The engine in a Honda Civic is roughly 350 times smaller than the total displacement of the Crawler-Transporter engines used to carry the Saturn V and the Space Shuttle to the launch pad. A Civic can drive 100 times faster than the Crawler and with much better gas mileage, but would be unable to pull millions of pounds up an incline. Using a simplistic metric such as clock speed does not provide a valid insight to the overall performance of a computer. So, what other factors make a computer "powerful"?

A more practical measure is the ability of a computer to manage a large number of dissimilar devices and the speed at which data flows in and out of them. Consider a modern laptop. Virtually every moving part, every light, every connection, and even the display exists to bring data into the processor or to send data out in some form. A small embedded controller of the type employed in a microwave or a calculator has only a limited interface with the outside world, and performs a very limited range of functions with very few devices. Designed to achieve a singular purpose, there is no need for multitasking, elaborate timing control or a large catalog of software. With such a limited range of memory and I/O capability, dedicated processors such as those found in simple consumer devices are clearly inappropriate for real-time computing. If forced to name a 21st-century example that might be a valid comparison against the AGC, it would be fair to suggest a high-end cell phone. Running several programs at once for mail, Internet, texting and photos as well as basic telephone functions, the software must present a simple user interface and be highly reliable. The phone contains a wide variety of internal devices that the software must manage, such as the antenna, battery, camera and touch screen. A user can select any program at any time, so software must be stored using a rugged and reliable memory technology. Finally, the requirements of small size and low power are as important in a cell phone as in the AGC.

It is also instructive to understand what makes today's PC software so large and complex. This book is being written on a current model IBM ThinkPad, which contains hardware and software to play a movie, a wireless connection to the Internet providing streaming music, and an overflowing E-mail inbox. The feature-rich word processor makes writing and editing (relatively) painless, and protects the reader from having to endure my generally horrific spelling. Programs to prevent malevolent software from attacking my machine are running in the background. Everything running in its cleverly designed window is isolated from the other windows, which in turn are isolated from the operating system. If something crashes (which is rare), the other software keeps on running, generally unaffected by the troubles elsewhere.

This description is only highlighting what a common consumer product is capable of, yet these are only a fraction of all the available features. Is it any wonder why the operating system and applications are so resource intensive? The difference, of course, is that the first generations of computers were far simpler devices, reflecting the limits of technology at the time. Disks had just finished their emergence from the laboratory, and paper, in the form of cards and printed forms, was an important source for input and output. Application programs had little memory or processor power to execute with, and, as a result, were limited to only the most essential functions. The cost and scarcity of computer resources limited their use to only those business or scientific problems that could justify such an investment. The computing world focused inwards, and was defined by what a single machine could accomplish. Even the works of science fiction did not imagine technologies such as the Internet, multimedia and gaming with virtual worlds. For those still unconvinced at how much was possible with these "limited" machines, here are some points to ponder:

- 1952: The UNIVAC 1 was able to predict the outcome of the United States' presidential election, with only 1,000 words of memory and a processor that executed fewer than 2,000 instructions per second.
- 1959: The SAGE (Semi-Automatic Ground Environment) computers were the key to NORAD's control of fighters sent to intercept Soviet bombers. In operation for over 20 years, the last of the 24 vacuum-tube systems was finally decommissioned in 1981. Each site occupied a building the size of a city block, yet each had only 64K words of memory.
- 1972: Early versions of the Unix operating system for the PDP/7 required about 12K of memory.
- 1975: The breakthrough program for the Altair 8800 was an implementation of the BASIC language, written by an obscure startup called "Micro-Soft". It ran in 8K.

Ultimately, the problem with contemporary computers is that their capabilities are so vast, any characteristic is usually expressed using numbers that are far beyond everyday experience. Even in the smallest computers, billions of bytes of main memory is the norm, and the speed of the processor is often expressed in billions of instructions per second. Disks have long been measured in gigabytes, and terabyte drives are now common and inexpensive. To impart a sense of scale, equally impressive comparisons are drawn, such as the ability to save the name, address and other vital data of every living person on Earth using a only home computer. In a modern laptop, it is a trivial matter to write a program which requests 100 megabytes of memory – no, better yet, how about a full gigabyte? The system responds that yes, there are 1,000,000,000 bytes of real, honest-to-god memory available, and ready to be used. The abundance of powerful, inexpensive semiconductor hardware makes this a trivial and forgettable exercise. To engineers in 1961, this task would have been as awe-inspiring as stepping onboard a jumbo jet would have been to the Wright brothers at the beginning of the 20th century.[1]

At the other end of the spectrum is the AGC. Without external storage on devices

---

[1]   Actually, even before Moore's Law was suggested, most engineers of the day would have bet that such a machine was not impossible. But they would have been stunned that such a computer could be slung in a teenager's backpack, connected to a network of similarly capable machines all over the world, where one could order yet another computer (using a simple piece of plastic representing your promise to pay), and have it delivered to your doorstep the very next morning by a courier guided by a constellation of navigation satellites. They would have regarded *this* scenario as bad science fiction.

such as disk or tape, memory capacity is one of the few available figures of merit. The 38K words of erasable and fixed memory are only enough to create a thumbnail image on a web site. More impressive still, is the fact that all of the Command Module and Lunar Module software for all six lunar landings will fit on a single floppy disk, with enough room to include Apollo 8. Even storing the binary code for the AGC on punched cards, the media that defines obsolete technology, requires a deck less than 8 inches thick. Truly appreciating its capabilities requires understanding how much was possible with the small amount of resources available at the time.

THE EVOLUTION OF COMPUTING

Beginning in the 1950s and continuing into the 1980s, operating system software was written in a language called assembly, using individual, hand-coded machine instructions. Assembly is the fastest and most memory efficient computer language, since it allows the programmer full control of the computer's hardware without incurring the overhead of more advanced languages. Direct manipulation of the processor, memory and I/O devices is a primary task of an operating system, and assembly language performs these tasks with ease. Unfortunately, all of this flexibility comes at a significant price. By their very nature, the basic instructions in a computer are primitive, unstructured commands that demand an intimate knowledge of a processor's operation. The resulting code can be tedious to write and difficult to read. Programming in the first decades of computing was regarded as a craft, not a science, and often did not undergo the rigorous reviews or provide the comprehensive documentation considered essential today. Unconventional coding styles (alternately called elegant or unreadable, depending on whether or not you are the programmer) were commonplace, making the software especially difficult to understand. Such practices are acceptable when trying to fit a large amount of capability into a resource-starved machine. In the case of the AGC, where the amount of software strained the available memory capacity, the use of high level languages was never an option. Programmers struggled to eliminate even single words of memory in the effort to fit all of the software into the computer.

In the years between the early AGC design of 1961 and the final Apollo flight in 1975, the state of the art in electronics exploded far beyond anyone's expectations. This era began with the vacuum tube declining as the dominant switching and amplifying device, and even the humble circuit board was not a universal part of electronic equipment. The novelty of small transistor radios was waning, but many models still displayed the transistor count on their faceplates! A single transistor was expensive and difficult to manufacture, but the promise of reliability and low power consumption more than offset its limitations. Over time, manufacturers mastered the art of creating a transistor on a silicon substrate, and were then ready for the next technological leap: placing *two* transistors on a single substrate.

This is hardly a flippant statement, as the creation of the next generation of digital components was a huge challenge. Placing two or more transistors, or a mix of transistors and other components on a single substrate required new manufacturing processes, clean room environments and automated fabrication techniques. Even reinventing the device's packaging technology was necessary to address new requirements for mounting, heat dissipation and connecting the ever-increasing number of wires. All of this effort paid off handsomely, as progressively more complex circuits found their way onto chips. By 1975, commercially available processor chips with thousands of transistors were common. By every reasonable standard, the technology used in the AGC was obsolete. Far more transistors were being squeezed onto a single chip, shrinking the size of electronics and cutting their power consumption.

So, with such fantastic progress in technology in such a short time, why did the AGC stay locked in its original hardware and software rather than being allowed to take advantage of the latest in microprocessor design?

TECHNOLOGY ACQUISITION: CONSUMERS VS THE AEROSPACE INDUSTRY

The history of computing is defined by its breathtaking rate of change. Every year, revolutionary new technologies are introduced and integrated into products at a pace that would astound engineers and consumers only a decade prior. Not surprisingly, many of these new technologies just as quickly find themselves in the rubbish bin, as even faster, cheaper and more capable devices are introduced. In consumer electronics, a product frequently exists in the marketplace for only a year or two before becoming obsolete, even though it is performing its job perfectly well. Manufacturers are only too happy to capitalize on this perception, as they prepare to introduce "The Next Big Thing" to waiting buyers. As a result, it is not uncommon to find products in perfect working order stored in closets, never to be used again. The consumer is not completely at fault here, as the cost to upgrade is small, especially considering the new features available in "this year's product". The fear of obsolescence is real, as owners of 8-track tapes and Betamax VCRs can attest.

Technology procurement in aerospace and the military is vastly different. Rather than trying to anticipate future customer demands, government contracts present specifications that the contractor must accommodate, often in exacting detail. This is especially the case in contracts for a limited number of high value products that must meet extraordinary performance and reliability requirements. Military aircraft procurement presents an excellent example of managing advanced technology, long lifecycle projects. A fighter jet might take 10 years and billions of dollars to develop, test and integrate into the operational inventory. In turn, the aircraft are expected to remain in service for 20 years or more.[2] The long lead-time from initial concept to deployment means that at some point, development must be frozen, allowing the

---

[2]   An extraordinary example is the B-52, which is scheduled to remain in service 85 years after its first flight, with the youngest aircraft fully 75 years old. It is no longer newsworthy to hear that a pilot's father flew his particular aircraft!

work of manufacturing and test to begin. Newer technologies inevitably arrive, along with a long list of potential improvements, but by then the customer and contractor have committed to a particular design. A fighter jet may take a year or more to manufacture, and that model will remain essentially unchanged during its decade-long production run. There are many practical considerations at stake. Testing and evaluating any new technology is tremendously expensive and time consuming. Great fanfare and promise surround every announcement of a faster computer chip, a lighter alloy or more efficient engine, but at this point the technology is immature and unproven. Only after passing many cycles of tortuous evaluations are they certified as ready to survive the harsh environments found in air and spacecraft.

1

The AGC hardware

INTRODUCTION

In our technological society, we define the concept of computer "processing" by the task we need to perform. Producing a video, optimizing an engine's performance or fulfilling an online book order all fall under the generic definition of "processing", but say nothing about how that magic is performed. Stripped of their elaborate interfaces, clever optimization tricks and marketing hype, all computers contain the same three major components: a central processing unit or CPU, memory, and input and output devices. The processor performs all the numerical and logical operations of the computer, and is the "traffic cop" that coordinates the movement of data within the system. Drawing from a surprisingly limited collection of trivial operations, called an instruction set, a programmer writes programs to perform specific sets of tasks. Many programs are themselves components of even larger programs, which in their final form may process a payroll or save the Universe from virtual space aliens.

To help organize the work in a processor requires a number of dedicated, high-speed storage areas called registers. Constructed as part of the processor and not formally defined as a subset of the main memory area, registers perform specialized operations on other parts of the processor or memory. In particular, a register called the accumulator performs the arithmetic and logical operations in the processor. Other registers track the location where the program is currently executing, or help to organize memory by breaking it down into manageable pieces. A clock, more properly thought of as a metronome than a watch, sets the pace of the system, with each tick marking the start of a basic operation in the processor.

Memory, often called storage, has the simplest organization of all the computer components. Memory is divided into distinct cells or containers, each holding a single item of data, but the large number of cells requires some means of uniquely identifying each one. A unique, sequential number known as an address is assigned to each memory cell, in the same way a house is known by its address on the street. Data contained in each memory location takes the form of a single binary number.

Memory itself is indifferent to what its contents represent; the bits in each location can represent data or instructions. During the execution of a program, the processor retrieves data from a memory location, and interprets that data as a machine instruction. How the processor chooses the next memory location to obtain an instruction is dependent on the logic of the program and the results of previous instruction executions. The next memory location might be used, or an instruction might direct the program to continue its execution in a different area of storage.

The remaining computer component allows interaction between the processor and the outside world. Appropriately called the Input/Output, or I/O system, this type of hardware takes commands from the processor and moves data to and fro between the external devices and the processor or memory. Components such as keyboards, displays and disks are collectively called peripherals. Data brought into the computer might be input to a program, or a status bit from a piece of hardware. Likewise, the processor can send data to a display or send commands to external hardware.

Having identified all the major computer building blocks, it is time to return to our most basic question of just how does a computer process information? As the internal clock ticks, data might be read or written from memory, two numbers might be added together, or a command sent to display a character on a monitor screen. To illustrate how all of these components work together, consider an example that is familiar to many: a competition on a reality television show. In these competitions, or "challenges", participants often perform a variety of tasks – the first might require scavenging for clues, then assembling a puzzle, followed by building a fire. Key to winning a challenge is performing each step exactly as required before moving onto the next step. Only when all the events are completed is the challenge officially over. Our "Pretend You Are a Computer" challenge operates in the same way. We start with a row of numbered "memory locations", each containing some sort of "data" that can be interpreted as a task for the contestant, or possibly have no intrinsic meaning at all. A "processor" sits to one side, with an "accumulator" where the data is stored, retrieved, and operated on. Various "I/O devices" complete the challenge landscape, each with its own rules of operation. A contestant is assigned a memory location to start at, and fetches the data stored there. Taking it to the processor station, our contestant matches the "instruction code" against a list of numbered tasks posted on the processor. Some of the tasks might be trivial: take the data from another memory location and place it in the accumulator. Others are more complex: first, look at the data in the accumulator and check if it is greater than zero; if so, subtract one from the accumulator; if not greater than zero, do nothing. After each task is completed, the contestant advances one memory location and the process begins anew. The object of the challenge is not a race to the end of memory; indeed some of the tasks will tell the challenger to move to another location ahead or behind, and continue from there. Only when a contestant reaches a location whose instruction says "Stop" does the challenge end.

This reality television example helps us reinforce an essential fact in understanding how computers function. We have long heard that computers blindly perform computations and have no capability for independent operation. Even this description of a computer and its capability is far too generous. Our silicon savants, with their incredible technology and countless transistors are even more astonishing when we realize what they are actually capable of – which is not very much at all! In fact, even the best 21st-century processors are capable of only a few basic operations:

- Simple arithmetic and logical operations
- Moving data between memory and the processor
- Testing data against another value, and conditionally changing the execution sequence of the program
- Inputting and outputting data
- Memory references that are based on previous calculations.

Although the processor of a high-end laptop operates with vastly more speed and efficiency than the early vacuum tube monoliths, the basic capabilities of the two are very much the same.

By themselves, each of the basic tasks available in the processor has little practical use. With the operation and sequencing concepts of computing behind us, the next task is to combine instructions into a program to perform useful work. The easiest example is writing the instructions of a common business process, such as preparing a payroll. Reduced to its most basic form, the steps are:

1. Open up the personnel register, and locate the first employee record.
2. Q: Is this person a salaried worker?
   If yes, look up their Gross Pay.
   If no, Q: Did the person work 40 hours or less?
   If yes, Gross Pay is Hours Worked times Pay Rate.
   If no, Gross Pay is Pay Rate times 40 plus
   Overtime Hours times Pay Rate times 1.5.
3. Compute Federal Tax as Gross Pay times Tax Rate.
4. Compute Net Pay as Gross Pay minus Federal Tax.
5. Print the payroll check for employee.
6. Q: Are there any more employees to pay?
   If yes, get the next employees record and go to step 2.
   If no, Stop.

All of these steps are familiar to those working in a payroll department, whether they use a computer or not. The procedures developed here represent a well-defined sequence of tasks, each performing an essential part of the overall payroll process. Looking closer, we realize that each step is equivalent to a basic computer task, such as addition, subtraction and multiplication. Several tests determine an employee's pay: first based on whether the employee is salaried, and if not, whether any overtime pay is due. Entering employee data and printing paychecks are the means used to move information into and out of the system. From this, it is possible to see that by carefully describing our problem, it can be stated in terms of operations that are implemented in the computer's hardware.

Perhaps the most overused and poorly defined term in computing is "data". Defining data as "information" (a common mistake) only further obscures the definition, as defining "information" is also a struggle because of the vagueness of the word. Unfortunately, the dilemma is that both of these terms are overly broad. Data can take the form of a singular entity, such as the color of a single car, or a collection of records containing the color of each car manufactured that year. Much of its meaning is defined within the discussion at hand, if the conversation centers on moving a single binary digit, or "bit", it is acceptable to understand "data" as referring to that singular bit. Larger and more complex entities, known as data structures, are seen in digital photographs or spreadsheets. In the most general sense, data is any information that we intend to process within the computer, regardless of its structure or source. In this introductory computer architecture discussion, the focus is on the data in memory and its representation. When discussing memory, one of the more useful properties is the size of each memory location, measured by the number of bits it contains. For example, memory used in personal computers store eight bits per location. These bits are not saved in some jumbled and unstructured form, but are interpreted as a single binary number. But what good is just a number? How do we leap from, for example, a memory location containing the value 42 (00101010 in binary) to a payroll calculation?

Here lies the elegance of computing – a number in memory is able to represent anything one wishes, subject to the requirement that everyone must agree with the same interpretation. Consider a case where the number 42 is saved in a memory location. In one context, the processor can interpret that value as one of its basic instruction codes, for example, the command to subtract one number from another. In a different context, a programmer might interpret 42 as an employee number or the number of hours worked. In both cases, there is no confusion about the how to interpret that value.

In reality, this is not as easy as it might appear. A properly executing program demands that the programmer use each memory location as intended, according to the rules and logic defined by the problem. Even with the best of intentions, errors in program logic occur all too frequently. A program for a retail store that applies an incorrect discount during a sale is fundamentally flawed, even if the discount is applied faithfully and with no other indication of a problem. Using the incorrect memory location for data, such as selecting the storage location containing an employee's middle initial rather than their birth date, is not uncommon when programming directly using the basic instructions of the processor. Errors of this type are maddeningly difficult to identify and, if undetected, lead to unpredictable errors and corrupted data in memory.

Creating reasonable bounds for the range of data is also necessary. A company operating in normal economic times might report its income in millions or perhaps even billions of dollars. Very reasonably, accounting programs used by the company would never expect to see revenues change by several orders of magnitude, at least not in a short period. However, if the economy experiences hyperinflation, it would be no surprise to see revenues in the trillions of dollars, reflecting the decreasing value of the currency. If unprepared for such numbers, a program would fail despite previously error-free operation on smaller values. Programmers must anticipate problems such as these, and gracefully handle the possible errors that will result.

OVERVIEW OF CHAPTER 1

In this chapter, the underlying design and organization of the AGC's hardware is discussed. Often called the low level or machine architecture, its circuits perform logical and arithmetic operations. By itself, the hardware is comparable to an empty vessel, with little purpose and unable to perform the most basic tasks. Only when software is loaded into its memory and the processor is directed to execute the code will this collection of chips perform any useful work. Computer organization at this level is mostly invisible to the users and programmers of 21st-century computers, as layers of software exist to mask the complexities of the underlying hardware. Unlike their modern-day counterparts, AGC programmers did not have these layers of software to protect them from the nuances of their machine. While such layers are beneficial in mature systems by allowing significant productivity improvements, in the case of the AGC they would have imposed unacceptable costs in operational flexibility and resource consumption. When designing a real-time computer like the AGC, directly accessing the hardware environment is not only desirable, it is essential. Much of the AGC's programming effort focused on managing the complexity of the computer and how it interacted with the external systems. In essence, this view of the hardware was the canvas on which the programmers painted their masterpieces.

A key aspect in understanding the AGC is to become familiar with the environment the computer operates in, especially the language used to create the software. The "language" used for a substantial portion of its programming is called assembler, or sometimes assembly, which is a shorthand notation for coding the lowest-level operations. There is no denying that programs written in assembler are daunting. Much of the struggle in understanding assembler programs is due to the need to break down the process into such elementary steps that it becomes easy to lose sight of the original problem. Reading programs require patience, keen attention to details and an understanding of the program's overall objectives.

The first section presents the AGC hardware organization, from the basics of the number systems to the sophistication of how memory is managed. Describing a computer as only a machine that manipulates numbers may be technically correct, but gives little insight into how that process works. This section begins with an introduction to the binary number systems used in the AGC. Going beyond the obligatory review of binary arithmetic, the discussion emphasizes the properties and notations of the numbers, and how they contribute to the overall architectural framework. Two capabilities in the AGC are especially important. Spaceflight guidance and navigation demands complex, highly precise calculations, yet the computer is only capable of integer arithmetic. Using a novel representation called fractional notation, and combining multiple words to improve precision, the AGC is able to perform the mathematics necessary for a mission to the Moon.

The next two sections describe the core of the AGC machine organization. An instruction is the basic unit of computer operation, and programs use their instructions to solve a problem. Creating a rich set of instructions requires more than the three bits nominally allotted in the instruction word format. Elegant solutions are used to reinterpret bits in the instruction word, making a comprehensive and powerful set of operations possible. The final design fully exploits every possible bit sequence and reassigns several instructions, creating far more operations than would otherwise be possible. Manipulating the instructions in unexpected ways would be impossible without a complex scheme to manage the AGC's memory. Storage for programs and data was always in critically short supply, and the short instruction word made it exceptionally difficult to increase the total amount of memory in the computer. By breaking storage down into divisions called "memory banks" and limiting access to only a few banks at any one time, the limited range of memory addressing becomes tolerable. As a result, the instruction set implementation and the memory banking scheme are not separable into distinct topics; each complements and is dependent on the other. For those readers new to discussions on computer architectures, the notion of memory banking may be particularly esoteric. As banking is so integral to the AGC's design, an imaginary computer design called FIGMENT is offered to help introduce memory management concepts.

At this point, the description of concepts such as data, instruction formats and memory gives a rather static view of computer operation. The next topic brings the computer alive with the components that perform the real work in the system: the arithmetic operations, program sequencing and Input/Output. The dynamics of these elements, called the central and special registers, are responsible for performing all of the operations on programs and data. Located at the low end of storage, instructions reference these registers as they would any other storage location, simplifying both the design and programming of the computer. Of these registers, those for arithmetic and control operations are seen only by the software and cannot be sensed outside of the physical processor. For a computer to interact with external devices there must be a means for the programming to exchange data with hardware that is external to the processor. Some of the special registers send and receive data from the platform of the Inertial Measurement Unit, with another group sending signals to a wide variety of spacecraft hardware and receiving data and status information.

A description of the memory, registers and number systems leads inevitably to the instruction set. Early in the discussions of the AGC hardware, descriptions of the computer's instructions were necessarily vague to prevent overwhelming the reader with detail. With that background firmly established, the final part of this chapter introduces the AGC instruction set. The form and use of many of the instructions are the direct result of the AGC's evolution, and often present unique solutions to addressing, control and data manipulation. Interrupt processing, a function necessary for suspending one program to permit another to run, is also introduced. Interrupts are used extensively in the AGC for controlling Input/Output, timing of events and ensuring that no one task monopolizes the processor.

PHYSICAL CHARACTERISTICS OF THE AGC

During the era of Apollo, ground-based computers were a dream only Hollywood could imagine. Usually filling large rooms and demanding elaborate power and environmental requirements, the consoles were covered with hundreds of lights flashing at a maddening rate. Rows of refrigerator-sized machines, each dedicated to reading a single magnetic tape, provided a sense of motion as the reels spun forwards and backwards. Printers chattered as paper flowed through them. And other machines read the ubiquitous media of the day, the punched card. Service engineers from the manufacturer occasionally wore lab coats, not so much to establish a sense of authority, but as a practical matter to keep their mandatory ties and jackets clean and away from any moving parts. The computer room of the day was a busy and surprisingly noisy place, never letting a visitor forget that something important was taking place inside.

With this image of 1960's computing firmly in mind, the appearance of the Apollo Guidance Computer is at best underwhelming. Utterly silent and lacking any indication of its purpose, the AGC can easily be overlooked during an inspection of the spacecraft interior. Crammed into the lower equipment bay of the Command Module, the AGC occupies the space below the Inertial Measurement Unit and the optics system. In the Lunar Module, the AGC is located on the aft wall of the cabin. In both spacecraft, the AGC is mounted on a cold plate, where a water-glycol solution is circulated to carry away the heat generated by the system. The physical characteristics of the AGC are equally unassuming, measuring approximately 21.5 inches (55 cm) long, by 13 inches wide (33 cm) and 6 inches (15 cm) high, totaling one cubic foot (0.028 cubic meters). Designed to survive the rigors of spaceflight, the computer is a heavy 70 pounds (32 kg), owing to the sturdy chassis and extensive potting of all components. Power consumption is only 55 watts, which is on par with most 21st century laptops.

The computer itself is completely sealed, and might well carry the familiar notice found on consumer electronics: "No user serviceable parts inside." Early designs

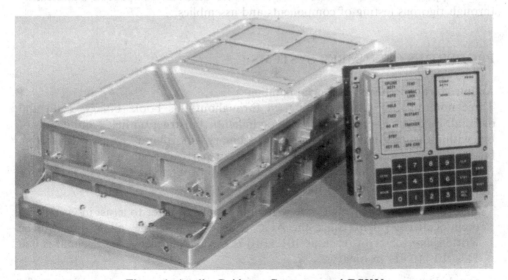

Figure 1: Apollo Guidance Computer and DSKY

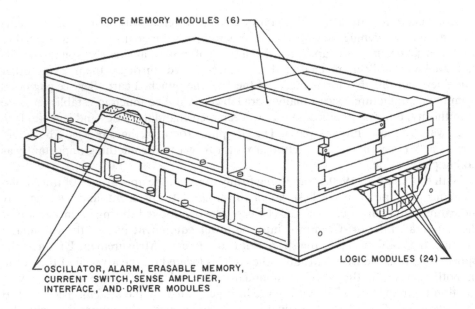

ROPE MEMORY MODULES (6)

LOGIC MODULES (24)

OSCILLATOR, ALARM, ERASABLE MEMORY,
CURRENT SWITCH, SENSE AMPLIFIER,
INTERFACE, AND DRIVER MODULES

Figure 2: AGC component locations

anticipated that the new digital technology might not survive the flight without failing, and engineers created designs where trays of components could be replaced with some effort. In creating a design that allowed servicing, the hardware was open to the environment, and could easily succumb to variations in air pressure, temperature and humidity. In the final designs, the requirement for in-flight repairs was dropped, and at the same time, hardware reliability improved significantly through rigorous testing of components and assemblies.

PROPERTIES OF NUMBER SYSTEMS

An introduction to binary and octal notation
Three notations for numbers are used throughout this book: binary, octal and decimal. Binary is essential for describing individual instruction fields and details of bits within registers and storage. Octal, as a binary "shorthand", is used where long bit strings are impractical to write and read; in particular, in memory addresses. Decimal is used everywhere else, especially when not referring to physical machine representations. The rules for the three notations are:

Binary: Bits, or binary digits, are written in groups of three to assist in visualizing their octal equivalent. Since the AGC is a 15-bit machine, all 15 bits will be written. No subscript (such as $101\ 110_2$) is used. Example: 3,763 decimal is 000 111 010 110 011 in binary.

Octal: Data is represented in base 8, using digits 0 through 7. Octal is a useful notation, as there are exactly five octal digits in the 15-bit AGC word, and groups of three binary digits form one octal digit. All five of the octal digits used for representing an AGC word are written without commas, include all leading zeros where appropriate, and a subscript "8" is appended to the number. Example: 3,763 decimal is written 07263_8 in octal.

Decimal: Data is represented using base 10, with no leading zeros or subscripts indicating the base. Commas are used where necessary. All fractional numbers use decimal notation, unless otherwise noted.

Base 8 is a number system whose digits are multiples of powers of eight. In decimal the powers of 10 are 1, 10, 100 and so forth. From this, the decimal number 123 is the sum of one 100, two 10s and three 1s. Another, more formal way to state this is to have one 10^2, two 10^1, and three 10^0. In base 8, the first few digits are powers of 8 with the decimal values 1, 8, 64, 512 and on and on. An octal value of 123_8 is the sum of one 8^2, two 8^1 and three 8^0, or (1*64) + (2*8) + (3*1), or 83 decimal. Similarly, the bit pattern 000 011 010 100 101 is converted to decimal by adding together one 2^{10}, one 2^9, zero 2^8, one 2^7 and so forth, until we have added the final value, one 2^0, to arrive at the decimal value of 1,701.

An important shorthand in computing is found when making decimal approximations of large binary numbers. Numbers in computing, especially when used to describe an amount of memory, quickly grow from thousands, then to millions, and in 21st century computers to billions and beyond. A binary number of 10 bits can represent 1,024 unique values, allowing the approximation of "1K" for 1 kilobyte as a shorthand. Strictly speaking, in computer notation the K refers to a multiplier of 1,024, not the 1,000 that the notation might imply. Likewise, the M (megabyte) notation implies a multiplier of 1,048,576, or 1,024 times 1,024, and not 1,000,000.

The length of a word of data directly affects the complexity of the machine and the calculations it performs. A long word length of 32 bits allows for a rich instruction set and a large range of data values, but much of this capability is wasted on a small machine such as the AGC. Small word sizes of 12 bits or less are unlikely to have the ability to reference the amount of storage necessary for a sophisticated control computer, nor have a wide variety of basic operations to process the data. A compromise comes from the requirements of the data that will be processed. In a first level approximation, less than one decimal digit of precision is added for every three bits used to represent the number. Therefore, a debate of small differences in word size is not productive. It quickly becomes apparent that without the need for a large instruction set and memory requirements, a small word size is more efficient despite requiring multiple words to establish sufficient numerical precision. An analysis of the required precision for navigation concluded that 28 bits, or 9 decimal digits should be sufficient.[1] Creating these 28-bit values requires two 16-bit words,

---

[1] On a lunar mission, 9 bits of precision translates to distances on the order of 1 foot and velocities of a fraction of a foot/sec.

Numeric Representation

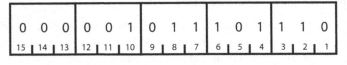

Example: Storing the value 750

0	0	0	0	0	1	0	1	1	1	0	1	1	1	0
15	14	13	12	11	10	9	8	7	6	5	4	3	2	1

$$512 + 0 + 128 + 64 + 32 + 0 + 8 + 4 + 2 + 0 = 750$$

Figure 3: Numeric representation in the AGC word

each providing 14 bits of the 28-bit quantity, plus one bit to define the sign (positive or negative) of the number and a second bit to specify parity.

Used to detect errors in the data, parity bits are invisible to the programmer. They are never seen in data and cannot be manipulated by machine instructions. In generating an odd parity bit, the other 15 bits of the word are scanned and the number of '1' bits are counted. If this results in an even number, the party bit is set to one. Otherwise, the parity bit is set to zero. Each data transfer to or from memory is automatically checked by the hardware to ensure that the parity setting is correct. From the entire 16-bit word, if the total number of bits are even, this is a clear indication that the data word is corrupt. Using only one bit for parity checking is imperfect, as multiple errors can easily fool this scheme, and it is not possible to determine which bits are incorrectly set. Even with these limitations, parity checking is a powerful way to ensure that the data is valid.

Having established a word size of 16 bits (15 visible to the programmer), attention turns to how numerical data is represented. Two types of data are used in the AGC. Simple integers are necessary for counters, indexes and whole-number variables. Generally, the numbers performing these operations tend to be small and easily represented in the 15 bit word of the AGC. Real numbers, those that have fractional values such as 3.14159, are also required and are used extensively throughout the software. Two different notations for data are necessary in order to handle the processing requirements. The challenge is to manage both types with the least amount of hardware and special processing requirements.

One's complement notation
Integer representation is the familiar binary format, with 14 bits for the number and the uppermost bit for the sign. A 0 in bit 15 indicates that the number is positive. The first few positive numbers are:

0	000 000 000 000 000
1	000 000 000 000 001
2	000 000 000 000 010
3	000 000 000 000 011
4	000 000 000 000 100

... and so forth, until

16,387	011 111 111 111 111

Representing a negative value is achieved by simply flipping the bits – setting all the 1s to 0, and all the 0s to 1 in a process called a negation or bitwise complement. With 1 coded as 000 000 000 000 001, flipping the bits to create -1 results in 111 111 111 111 110. Note that the leftmost bit, the sign, conveniently changes to 1 from 0, indicating a negative number. This representation is "one's complement" notation. This makes addition and subtraction a straightforward process, with no special consideration for the sign.

A list of a few negative numbers:

-1	111 111 111 111 110
-2	111 111 111 111 101
-3	111 111 111 111 100
-4	111 111 111 111 011
-5	111 111 111 111 010

... and so forth, until

-16,387	100 000 000 000 000

Basic arithmetic is straightforward. For example, adding 2 to -5 will result in -3:

2:	000 000 000 000 010
-5:	111 111 111 111 010
Add:	111 111 111 111 100 (= -3)

The addition of negative numbers is more problematic. Consider the case of adding -2 and -5, with the answer of -7. In binary notation, the problem is expressed as:

-2	111 111 111 111 101
-5	111 111 111 111 010
Add	111 111 111 110 111 (-8, oops!)

Adding the two negative numbers together in the same manner as the positive numbers results in a value of -8, not the expected -7. At the same time, there is the dilemma of the "carry bit" generated in the leftmost addition, which appears to simply "fall off". One's complement neatly solves both of these issues by introducing the "end-around carry". The rule for addition is to check for a carry out of the high order bit. If a carry exists, add that bit to the sum just created. The example above now becomes:

```
-2          111 111 111 111 101
-5          111 111 111 111 010
Add         111 111 111 110 111 (-8, but generates a carry)
Add carry                      1
            111 111 111 111 000 (-7)
```

The value range available using 14 bits is 16,383 to -16,383. There is a problem when the value 000 000 000 000 000 is complemented. Flipping all of the bits results in a value of 111 111 111 111 111, where the sign for zero is now negative! By definition, mathematics does not define zero with a sign, but one's complement creates positive and negative zeros as an artifact of its notation. The two numbers are not the same, which creates an exasperating situation for software developers when testing for non-positive, non-negative numbers. Surprisingly, the "end-around carry" technique relieves the programmer from the concern about "positive" and "negative" zeros in computations.

Arithmetic in any number system that imposes a finite level of precision must deal with the issue of overflow. For example, if a number is limited to only two decimal digits, what happens when adding 57 and 82 together? The result (139) is more than the number of digits allotted, and is called an overflow condition. An overflow condition produces invalid data, and any computations using an overflowed result are necessarily worthless. Alerting the programmer to an overflow condition is essential, but the task is to create a means to test for the condition that does not impose an excessive programming overhead. Terminating the program when any overflow occurs is a drastic measure, and the AGC finds a solution in a variation on the one's complement notation. This new method, called a modified one's complement notation is the AGC's mechanism for detecting and handling overflows. It is a key part of the accumulator's operation, and is discussed in more detail in that section.

Floating point numbers and fractional notation

When used in computers, real numbers take on an additional name, and are called floating point numbers. Real numbers, as opposed to integers, are the workhorses in scientific processing. All throughout mathematics, science and engineering, numbers can range from the very small, such as the size of an atom, to the very large numbers used when counting the number of cells in the human body. To handle this wide range in magnitudes, computations are performed using a mathematical shorthand known as scientific notation. It is easy to see why. In everyday experiences with numbers such as adding up a grocery bill, the numbers fall within a reasonably small range where they are not excessively large or small. A chocolate bar is not expected to cost thousands of dollars, nor is rice sold by the single grain and priced at a tiny fraction of a cent. Adding the bill may be tedious, but is not especially difficult. Now consider the tasks in physics or engineering, where computations frequently require numbers that are scaled in the millions, and others that are equally small.

In scientific notation, a number is expressed as a fractional value multiplied by an exponential power of 10. The decimal number 123.987 is written as $1.23987 * 10^2$. In this notation, 10^2 is the shorthand for 100, which multiplied by 1.23987 returns with the initial value of 123.987. Specifying much smaller numbers, such as 0.0006587, uses a similar process, and the notation $6.587 * 10^{-4}$ is used. Some representations of scientific notation presume the absolute value *of* the fractional part as less than 1. In this representation, the numbers stated above are $0.123987 * 10^3$ and $0.6587 * 10^{-5}$ respectively. Unfortunately, there is no native floating point capability in the AGC, as implementing such hardware on a computer is very costly in terms of the hardware required and memory necessary to store the numbers. Another notation to represent fractional data uses a small, fixed word size, and existing integer arithmetic hardware. The solution for implementing floating point numbers is surprisingly simple, at the cost of some mental housekeeping.

Continuing with using the decimal representation, take a fractional number such as 0.4375. This number is defined by summing four times 10^{-1}, three times 10^{-2}, seven times 10^{-3} and five times 10^{-4}. The concept of a fractional notation is not limited to decimal numbers. In this example, the number system is in base 10, and uses exponentially smaller values of 10. Fractional numbers expressed in binary use a very similar methodology, but rather than using powers of 10, a binary fractional number uses powers of 2. Taking the uppermost digit as 2^{-1} (or 1/2 in decimal), the next as 2^{-2} (1/4), then 2^{-3} (1/8) all the way down to 2^{-14} (or 0.0000610). Using this notation 0.4375 is represented as 001 110 000 000 000, having zero 2^{-1}, one 2^{-2}, one 2^{-3} and one 2^{-4}, as can be verified by reference to Figure 4.

As with all notations, there is no way to precisely represent irrational numbers such as 1/3, and only a reasonable approximation is possible. In the AGC, 1/3 is represented as 001 010 101 010 101. Summing together the values of one 2^{-2}, one 2^{-4}, one 2^{-6}, one 2^{-8}, and so forth, down to 2^{-14}, the result is only 0.33331298828125. When doubling the number of bits used to represent 1/3, the accuracy doubles to 8 decimal digits. Even greater precision requires still more bits, but eventually a compromise is necessary between the need for additional precision, the amount of memory required, and the additional CPU cycles required to process the larger number of bits.

2^{-1}	0.5
2^{-2}	0.25
2^{-3}	0.125
2^{-4}	0.0625
2^{-5}	0.03125
2^{-6}	0.015625
2^{-7}	0.0078125
2^{-8}	0.00390625
2^{-9}	0.001953125
2^{-10}	0.0009765625
2^{-11}	0.00048828125
2^{-12}	0.000244140625
2^{-13}	0.0001220703125
2^{-14}	0.00006103515625

Figure 4: Fractional powers of 2

Integer Representation

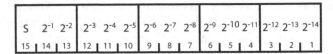

S	2^{13}	2^{12}	2^{11}	2^{10}	2^9	2^8	2^7	2^6	2^5	2^4	2^3	2^2	2^1	2^0
15	14	13	12	11	10	9	8	7	6	5	4	3	2	1

Fractional Representation

S	2^{-1}	2^{-2}	2^{-3}	2^{-4}	2^{-5}	2^{-6}	2^{-7}	2^{-8}	2^{-9}	2^{-10}	2^{-11}	2^{-12}	2^{-13}	2^{-14}
15	14	13	12	11	10	9	8	7	6	5	4	3	2	1

Figure 5: Integer vs fractional representation

This representation of data is not complex, and appears to function well. But the system of numbers that represents integers yields a range of only -16,383 to 16,383, and fractional numbers from 0 to approximately 0.9999. In most applications, data will consist of far larger numbers, such as spacecraft weight (the combined weight of a fully loaded CSM and LM is over 100,000 pounds), and fractional numbers greater than 1 (Earth's gravitational constant is 32.2 feet/sec^2). Using an additional word to represent an integer number only postpones the problem of representing larger numbers, and at a cost of memory and processing cycles. Unfortunately, this still does not completely address the problem of fractional numbers.

In 21st-century computers, floating point numbers are usually implemented in a format that echoes scientific notation by using two components to fully define the value. In creating a floating point representation in binary, a word in memory is divided into two fields: one for the fractional value and the other for the exponent. To illustrate this, recall that at the moment of Apollo 11's launch, the distance to the Moon was 218,096 nautical miles, which can be expressed in scientific notation as $2.18096*10^5$. The fractional part of the number, 2.18096, is known as the mantissa, and the value of 10^5 is the exponent. Both components of the scientific notation are essential for completely defining the number. However, the AGC only uses the fractional part of the number. Isn't this incomplete? How is it possible to describe completely such a number, given the notation we have adopted?

We take our solution from the ubiquitous predecessor of the computer: the slide rule. Seriously!

Despite being maligned as the canonical example of how primitive man calculated before the advent of silicon-based processing, slide rules are fantastically adept at performing most mathematical functions. In skilled hands, calculations are made quickly, or at least, as quickly as a manual process can be. Precision is not the slide rule's strongest attribute – in some calculations it is difficult to get more than three digits of accuracy. Still, using the logarithmic scales on the rules, the user can perform multiplication, division, trigonometric functions, squares, square roots and more. If there is a quirk to using a slide rule, it is that numbers are only mantissas, and the user is responsible for keeping track of the magnitude of the calculation. By managing the magnitudes of the numbers separately, a slide rule can operate with an unlimited range of exponents. For example, consider the problem of multiplying together three numbers of various magnitudes:

$$(2.3*10^{-6}) * (5*10^{1}) * (1.8*10^{2})$$

On a slide rule, the mantissa of the solution is calculated as $2.3*5*1.8 = 20.7$, and the magnitude is obtained by manually adding the exponents (which sum to -3), giving a final solution of $20.7*10^{-3}$ or 0.0207.

Now, if the items used in the calculations are expected to fall within a narrow range of values, the need for knowing the magnitude diminishes. For example, in a discussion about room temperature, the range of values is not expected to vary by several orders of magnitude, and therefore mentioning the magnitude is redundant. To say that the temperature is 72 degrees Fahrenheit, the magnitude is understood to be 10^{0} (1), and not 10^{2} or 10^{-2}. Even in another scale, this type of assumption is acceptable. Where room temperature is 295 degrees Kelvin, a reasonable assumption is that the magnitude is 10^{0}. Switching to a fractional notation from integers, a room temperature is expressed as 0.295 degrees Kelvin, and the implied magnitude is 10^{3}. As the temperature falls to 50 degrees Kelvin, the notation is robust enough to correctly express the value as 0.050 using the same implied magnitude.

Here then, is the secret to handling floating point numbers in the AGC. We have neatly eliminated the complexities and cost of representing a new and complex floating point data type in the computer. However, using the fractional notation technique demands a rigorous understanding of the data. Most importantly, the programmer must ensure that his equations will not generate values that over- or underflow the fractional number. This is especially necessary in a computer with such constrained resources as the AGC. In most cases, data values used in its programs are assumed to be valid, and checking of the data to assure its validity is not done. This places a huge burden on the programmer to identify any problems with data scaling by thoroughly testing the software using a wide range of data.

Up until now, magnitude has been assumed to be a factor of 10, but there is actually no requirement for this. If a factor in an expression is multiplied by a common constant, such as pi, we can skip the multiplication, saving memory and processor cycles. Our value is now expressed as order pi. Providing that other calculations are anticipating such values, there is no problem using this methodology.

Scaling, precision and accuracy

Accuracy and precision are often used interchangeably, but are actually very different concepts. Precision refers to the number of digits used to represent a value, while accuracy defines how close the number is to its true value. Consider a scale in the grocery store. When buying a piece of cheese, the scale might report that it weighs 1.2 pounds. The precision is stated to the tenth of a pound, but there is no indication of its accuracy. How is it possible to know whether some unscrupulous clerk has not placed a bit of lead on the scale? Even if the scale reads 1.2000000 pounds, the level of precision is certainly impressive but the accuracy may still be in question. Conversely, a government agency may have certified the scale to be accurate, but this says nothing about the precision of its display. A one pound piece of cheese registers as one pound on the scale, but if it only registers in whole pounds (no fractions), its precision is certainly lacking. A scale operating only in whole pounds is forced to round its answer so that a 1.2 pound piece of cheese might register only 1.0 pounds, yet the scale might show 2.0 pounds for a 1.7 pound cheese wedge.

Clearly, computing for spaceflight demands a high degree of both precision and accuracy, but the realities of the physical hardware force a compromise between the two. When representing data in a computer, the number of bits used in a floating point or fractional number defines the level of precision. Accuracy is a more complex concept to quantify, as it is strongly dependent on the quality of input data and the mathematical algorithms used to process that data. Earlier, the notion of fractional representation of numeric data was introduced. The presentation of the arithmetic group of instructions will show how this notation works in practice, along with its strengths and weaknesses.

Descriptions of the AGC consistently state that only integer arithmetic is supported; that is, only operations on whole numbers are possible. More correctly, it should be said that the AGC does not have a native capability to perform operations using "floating point" or "real" datatypes. Unfortunately, without a floating point datatype, the hardware is pressed to do duty for both integer and floating point data. Thus, the "fractional" data representation requires that the programmer assume the responsibility for managing the magnitude, or scale of the number. At the same time, accuracy in floating point data representation is limited to the number of bits used to represent the number. These two issues work together to complicate the decision on how to represent a number in storage. As an example, the value of pi is expressed as $0.31415926*10^1$. Just as accurately, the notation $0.00000031415926*10^7$ is feasible. Now, assume there is a restriction that limits the fractional component to eight digits. In this case, $0.31415926*10^1$ remains a reasonable approximation of the value of pi. However, combining the eight digit restriction when using a different scaling can produce a value of $0.00000003*10^7$, creating a serious loss of accuracy. A large part of the AGC program design effort is devoted to this issue, adjusting the data to ensure that accuracy is not lost. This involves making sensible decisions on the scaling of the data.

As scientific notation is based on powers of ten, our numeric vocabulary is expressed in tens, hundreds, thousands and the like. Certainly, data in the previous example was adjusted by orders of magnitude to make notation easier, but there is no fundamental reason to choose any specific factor other than convenience. Managing scaling is often a matter of striking a balance between accuracy and overflow protection. In a fixed-precision system, scaling down by a decimal order of magnitude results in the loss of at least three bits of accuracy, which may be an unacceptably large amount. Often, there are situations where data may only need to be reduced by a factor of two or four (one or two bits) to prevent an overflow. Further, the idea of scaling can be extended to the elimination of redundant constants.

As with irrational numbers, virtually all floating point data is an approximation. With a finite number of bits available in a computer, it is not possible to represent most values exactly. Occasionally a case exists where a number is a fractional power of two (such as 1/4 or 1/2), but the vast majority of fractional data is irrational. Returning to the problem of representing 1/3 as a floating point or fractional number, its value is 0.33333333..., with the sequence never terminating. If only two digits of precision are available, only a value of 0.33 is possible, but this number suffers from being somewhat imprecise. Ten digits is a great improvement, but still only results in 0.3333333333, and even millions of digits cannot produce the exact value. Consequently, some loss of precision is inevitable. During the process of defining a program, a level of precision must be agreed upon. This is invariably a compromise between the required accuracy and the cost of memory and processing power. In the case of the AGC, eight decimal digits of precision are sufficient for its most critical role of guidance and navigation. In fractional notation, 28 bits are necessary to provide this level of precision. Concatenating two adjacent words into a "double word" creates a datatype that supplies the necessary number of bits. Efficiently handling this datatype requires special instructions that are discussed later in this chapter.

Yet another problem exists with fixed precision numbers. Errors from rounding and truncation invariably accumulate when performing extensive floating point calculations, and cannot be eliminated. Double and triple precision can limit this effect, but cannot completely eliminate it. Additional work is often necessary to characterize the magnitude of the resulting error. Again, a compromise is necessary to balance numerical accuracy with the available computational resources.

DOUBLE PRECISION NUMBERS

Up to this point, only datatypes using a single word of storage have been discussed. Appropriately categorized as single precision datatypes, they are the workhorses of the AGC in that they provide both fractional and integer data. However, as its name implies, a single word provides the minimum amount of precision. But there are applications in the AGC that require greater precision in the data, and create the need for more bits to represent these numbers. Of the 15 bits in the word, the sign occupies one bit and contributes nothing to the precision of the data. The remaining 14 bits translate to approximately 4 decimal digits of precision, which is sufficient for a wide range of uses. Solving guidance and navigation equations is much more demanding, and requires a higher level of precision to accurately solve the complex equations of motion. After much analysis, mission designers and AGC engineers agreed that 28 bits, equivalent to 8 decimal digits of precision, were acceptable. To contain this amount of data, the double precision datatype was created using two adjacent words in memory to represent a single data value. Double and triple precision quantities are limited to fractional numbers as the AGC does not support integers larger than a single word.

Double precision is implemented as a bit string spanning two consecutive words in storage. This can yield some surprises in its structure. For example, it is possible (and legal) for a double precision value to have a positive value in one word, and a negative value in the other. At first glance, this seems illogical if not impossible. But is it? Consider a simple, two digit decimal number, say 53, which is defined as (5 * 10) + (3*1). However, 53 is also correctly defined as (6*10) + (-7*1). While this is unconventional, the mixing of positive and negative values in this alternative version is quite reasonable.

Mixing words with unequal signs during arithmetic operations is not an irrational task. An operation (add, subtract, etc) is performed against one word at a time, rather than the entire doubleword at once. With the problem broken down in this manner, the traditional rules for arithmetic operations apply as before. An important exception is in handling overflow on the low order word. An operation that results in an overflow must carry over into the high order word, an action called an "interflow". Interflows operate as a carry between words in a double precision datatype, taking the overflow from the lower order word into the upper word.

The exceptional case of two's complement
Two's complement is used universally in 21st-century computers, and has the essential attribute of a singular value for zero. In a quest for simplicity some early processors implemented serial adders, where the digits are summed one-by-one. While not particularly fast, serial adders require very little hardware to implement, greatly simplifying the hardware design. Key to the simplicity of such adders is the requirement that numbers employ one's complement for their representation. As computer architectures moved from serial to parallel adders, the necessity for one's complement disappeared, and with it the troublesome issue of managing two values for zero. For positive numbers, notation in two's complement is identical to one's complement. The sign of the number is contained in the leftmost bit, and like one's complement is zero for positive numbers. The remaining bits define the number as a binary value. Negative numbers begin with a bitwise complement (the same as one's complement), but a one is added to the complement.

One's Complement		Two's Complement	
29:	000 000 000 011 101	29:	000 000 000 011 101
Complement	111 111 111 100 010	Complement	111 111 111 100 010
		Add 1	000 000 000 000 001
-29:	111 111 111 100 010	-29:	111 111 111 100 011

Figure 6: Negation of one and two's complement

In two's complement, there is never the problem of two values for zero; only the binary value of 000 000 ... 000 is defined as zero. In the 15-bit word found in the AGC, the range of numbers is 16,383 to -16,384.

Internal and external systems of units
Mission programs in the AGC maintain their internal measurement system in metric, recognizing its importance as the standard notation in science. Crews, however, desired English units in all the displays. This does not present as much overhead for the AGC

as it first might appear. Fractional notation, used extensively for data representation, requires routines to convert numbers to human-readable values. Additionally, during the conversion from a fractional notation to displayable digits, a scaling factor must be used. Conversion routines, already required to adjust for the effects of scaling, simultaneously incorporated the difference between metric-to-English units.

FIGMENT

Discussions of computer architectures are necessarily complex, and it is easy to get lost in the minutiae of a particular system. Unfortunately, the AGC presents a number of special cases that make it a poor choice for introducing important concepts. To avoid unnecessary diversions, we will introduce an idealized computer to illustrate architectural fundamentals, known as FIGMENT (First Imaginary Guidance Machine for Education aNd Training[2]). Although intentionally similar to the AGC, the intent of FIGMENT is to create "breadboard" examples to build up and tear down as we see fit, to introduce the more complex concepts of the AGC. By conveniently ignoring some practical issues, it is possible to demonstrate alternative architectures without burdening ourselves with inconvenient realities. Certainly, there is no assumption that a particular configuration under discussion applies in later examples. In general, any discussion involving FIGMENT will be quite separate from any reference to the AGC, in order to minimize any confusion between the two. In its simplest form, the structure of FIGMENT is quite conventional. Many of the basic architectural elements are the same as in the AGC, including the instruction format and a word size of 15 bits. A number of special registers for the accumulator, program counter and return addresses are included. While FIGMENT includes all of these components, it will also evolve and introduce new features to illustrate how design decisions are made.

With the introduction of FIGMENT, we can explore the instruction set and memory organization of the AGC.

INSTRUCTIONS: THE BASIC UNITS OF COMPUTER OPERATION

In computing, the most recognizable unit of operation is the instruction. Various more fundamental processes occur deep within the hardware, but the processor's instruction set reflects the computer's basic capabilities. In the early generations of computers, programmers used this level of machine organization directly to perform the operations necessary to create software for both the operating system and the mission-oriented applications. The sets of instructions found in processors evolve over time as newer systems are developed, building upon and extending the previous generation's design.

---

[2]   OK, it is a bit of a stretch.

An instruction is a word in memory whose bits are assigned specific meanings by the processing unit. The purpose of an instruction is twofold. First, the instruction directs the processor to perform a specific operation, such as adding two numbers or comparing a value against zero. Secondly, the instruction also specifies the data (if any) that the instruction operates on. Thus, an instruction word is composed of two groups of bits. First, the operation code, or opcode, is a unique sequence of bits that define the instruction's particular function. The address, also known as the operand field, is the second group of bits and contains the memory location of the data used by the instruction. Some instructions change the execution sequence of a program, and use the address field as the location at which processing will continue.

The conventional approach for discussing the AGC's architecture is to break the instruction word down into its constituent bit fields, and discuss each field (opcode and addressing bits) independently. While ideal in theory, the interpretation of the bit fields in the AGC's instruction word varies considerably depending on the operation code and memory address. This understandably creates a large amount of confusion in those expecting a fixed, well-defined instruction format. If the instruction set and memory schemes seem somewhat byzantine, it is important to understand that the design engineers did not set out to create an instruction set full of exceptions and special cases! The original AGC designs, reflecting the then-current design requirements, were far simpler. Early iterations specified only eight instructions and 4K words of memory. The relentless demands of new system requirements continually challenged the AGC engineers, and required even more sophisticated instructions and additional memory. The basic format is seen in Figure 7. Over time, however, the format of an AGC instruction was extended several times to accommodate the need for more instructions and ever larger amounts storage.

In actuality, only about one quarter of the AGC instructions use this basic format. As the computer architecture evolved during the early and mid-1960s, new operation codes became necessary. Adding additional operation codes requires additional bits to define them. With the word length fixed, any additional bits must be scavenged from another source. But from where? After accounting for the bits used for the operation code, the only remaining bits in the word were those allocated to addressing. The solution is to take a close look at how the instructions address their data, and in particular the type of storage they are referencing. Before this discussion on instruction sets continues, it is necessary to consider how data is accessed in memory.

Figure 7: AGC instruction format

MEMORY MANAGEMENT

The AGC has a total of 38K words of memory,[3] beginning with 2K (2,048) 16-bit words of standard coincident-current core for erasable, or read/write memory; the rest of its storage is fixed, or read-only storage, consisting of 36K (36,864) words using a novel implementation of storage named core rope. The first problem that arises in referencing even this meager amount of memory is that the 12 bits in the addressing field can only reference 4K of storage. This is simply not enough to address all 38K of memory, which requires at least 16 bits. Virtual storage, with its extensive hardware requirements and dependence on secondary storage, was still in its infancy in the 1960s and was only beginning to be a practical solution for large commercial computers. For the AGC, a solution is found in an early cousin of virtual storage, called "memory banking".

Sophisticated memory management is the key to a computer's ability to process large amounts of data. When considering a computer architecture and the amount of memory physically installed on the machine, there are usually two different but complementary problems. The first occurs if the architecture supports far more memory than it is practical to include in a computer. This is the case in 21st-century computer systems, where 64 bits may be available to address memory, yet a computer with millions of gigabytes of memory is not expected until later in the century. More relevant to the AGC is the issue of managing more physical memory than the number of addressing bits in the instruction would normally allow; a common situation in the early history of computers where memory was a rare and precious commodity. As the introduction of more advanced storage technology dramatically reduced its cost, early computer architects were faced with a dilemma. Computer memory was getting smaller and cheaper by the 1960s, and large amounts of memory for small computers quickly became a reality. As the amount of available memory increases, so does the number of bits necessary for referencing each storage location. Unfortunately, each new computer design cannot blithely add additional bits to its instruction format in order to access all the memory included in the hardware, otherwise instruction words would quickly grow to unmanageable sizes. Most importantly, making changes in the instruction format, a structure that is fundamental to the computer's overall architecture, makes the new machine incompatible with its predecessors.

Although the AGC's 12 addressing bits are insufficient to reference all 38K words of storage installed in the computer, there is no reason to look at all of memory at once.

Before presuming the previous statement to be patently absurd, a practical observation on how programs reference storage is pertinent. A large program, regardless of whether it is a guidance targeting program or a word processor, is never written as one large, monolithic piece of software. In breaking a program down into far smaller subprograms (subroutines), the task of designing, coding and testing is

---

[3] Although a 15-bit word is almost twice the size of the now-common 8-bit byte, storage in the AGC is never expressed in bytes.

hugely simplified. Each subroutine is usually designed to perform one well-defined function on a specific set of data. It is also common practice to arrange the data accessed by the subroutine within a small range of addresses, for no other reason than to make it easier for the programmer to read. As the code executes, only a few areas of memory are actively referenced at any one time: the instructions that are being read and executed, and the data that is being acted upon. Exploiting this phenomena, called locality of reference, creates a technique for accessing large amounts of physical memory using only a few addressing bits. The solution is no longer to battle the addressing limitations, but to embrace them. Even if the instructions could reference all of the memory in the computer at once, the software still only operates on small sections of program and data during any given time interval. The limited addressing range of the instruction allows us to formalize a technique to exploit locality of reference.

Special hardware registers are necessary to implement memory banking. The bank register contains the high order bits of the complete physical address, and in doing so, defines the banks in specific locations and sizes. Generating the complete physical address is simple; concatenating the contents of the base register to the beginning of the instruction address field yields the physical memory address. Bank registers allow programmers the ability to address a large amount of physical memory in a machine with a restrictive instruction format. There is no artificial limit to the amount of physical memory, but the complexity of managing large bank registers places a practical limitation on its size.

It is now time to explore the details of memory banking, using the imaginary computer called FIGMENT as a much simplified AGC. By design, FIGMENT implements a 15-bit instruction as a 3-bit operation code and a 12-bit addressing field, which is the same as the basic instruction format in the AGC. Memory on this particular FIGMENT system is 4K words, and the 12-bit addressing field matches the memory requirements perfectly, as 12 bits allow for accessing all 4,096 words of storage. Next, assume the specification is upgraded, and now requires addressing 64K words of memory. This is impossible with only the 12 bits that exist in the available addressing field, which lacks the additional 4 bits necessary to reference all 64K. The solution is to implement a new hardware register, appropriately called a banking register, to store the additional 4 bits. These 4 bits in the banking register are concatenated with the 12 address bits of the instruction, yielding the 16 bits necessary to reference all 64K of memory. These 16 bits define the "physical address" of the word in memory. FIGMENT maintains the 16-bit physical address as an internal register, which is not accessible to the programmer.

With a banking register defined as 4 bits long, FIGMENT now has the ability to reference 16 memory banks, each of 4K. There are no special attributes associated with each bank or limitations in the type of data stored in the memory. All memory locations can contain instructions or data, may be written to as well as read, and have the same 15-bit length. The bank register has 16 values, from 00_8 (for physical memory in the address range 000000_8 to 007777_8), then 01_8 (010000_8 to 017777_8), and all the way to 17_8 (170000_8 to 177777_8). However, banking registers introduce an important new limitation. All memory references are limited to the single 4K area of memory specified by the banking register. At startup, FIGMENT begins with the bank register set to 00_8, forcing execution to start in the lowest 4K memory bank. All is well until we need to cross a bank boundary, as all other banks are outside of our addressing range and are effectively invisible. We now recognize the need for special instructions allowing us to enter and return from one bank to another.

Instructions used to manage banking registers are often complex, not because the operations are particularly difficult to implement, but because they must not be interrupted. All computers include a hardware feature to abruptly stop, or interrupt, the current work from running and switch to another task. During the time that a program is modifying the banking registers, it is especially sensitive to such an interruption. If an executing program were interrupted when only partially finished with its banking register update, the program will most likely go awry in an entirely unpredictable way. Consider the case where a program wishes to transfer control to a memory location in another bank. The sequence to transfer control requires saving the physical address of the calling program (both bank number and address within the bank), loading the new bank number, and using that information to transfer control to the new location. All of these operations must occur within the course of a single instruction, as data may be lost and the program fail if it were to be interrupted. FIGMENT lacks a common memory area that is accessible to all programs, so two new registers are necessary to save the calling program's bank register and the return address within the bank. Next, transferring control requires the simultaneous changing of the contents of the bank register and concatenating a 12-bit address to it. Changing just the bank register in one instruction, and then trying to transfer control to the subroutine address using a second instruction will not work because once the contents of the banking register is altered, the next instruction will be taken from the new bank without being directed to the intended subroutine location.

The need to move freely through memory creates all the ingredients for a basic computer operation called "Transfer Into New Bank". The purpose of this instruction would be to combine all the operations necessary to transfer control from one bank to another, all in a single step. Because the 16-bit physical address of the destination will not fit in the instruction, FIGMENT will expect to find this information in two consecutive memory locations in the current bank. The first word contains the destination bank number, and the second word contains the offset into the bank where the subroutine begins. With these new registers defined, it is possible to completely specify the steps that "Transfer into New Bank" (TNB) must perform:

- Copy the current Bank register into the "Saved Bank" register
- Copy the 12-bit instruction address following the TNB instruction into the "Return Address" register
- Load the Bank register with the new bank number, and the program counter with the bank offset with the values pointed to by the TNB instruction.

Upon entry, the called subroutine must store the contents of the Saved Bank Number register and the Saved Bank Return Address register. Later, another TNB instruction would use these values to return to the calling program. Therefore, to implement banking, we now need several new registers:

- Bank Number
- Saved Bank Number
- Saved Bank Return Address.

What started as a simple little system has now become quite complex!

Memory banking in the AGC

This exercise with an imaginary system has built the necessary foundation for discussing memory banking in the AGC. Unfortunately, as the AGC evolved, the need to increase both the instruction set and storage size was limited by the fixed instruction length. The architecture that resulted is a collection of compromises that, on the surface, appear impossibly complex. But on closer inspection many of the memory management elements are surprisingly straightforward, if not intuitive.

Memory in the AGC is divided broadly into two types: 2K (2,048) words of erasable memory (i.e. read/write) and 36K (36,864) words of fixed (i.e. read-only) memory. Memory addresses refer only to whole 15-bit words, which are not divided into smaller units such as bytes. The range of memory addresses comprises one contiguous range, starting at 0 and continuing up to the highest number of words in the system (38,912). Of course, operating on such a large number is not possible using the 15 bits available in a word, necessitating the use of memory banking to reference most erasable and fixed storage areas. Although both types of storage can hold either instructions or data, essential programs are loaded into the fixed memory, along with a number of constants. Implementing the programming in fixed memory has huge implications in software reliability. A common and often insidious software bug occurs when accidentally overwriting a piece of programming code. Program logic that is tested and accepted as error-free can fail in unexpected ways if overlaid by data. By placing code into unchangeable memory, a large class of problems is neatly eliminated. An errant memory reference cannot destroy essential code, thus preserving the integrity of the software. Debugging is simplified, since any attempt to store data in fixed memory is rejected.

Up to this time, we have made the assumption that any data referenced by a program is expected to change, creating the requirement that data resides in erasable memory. This is not always true. Much of the critical data in the AGC is stored as fixed values, such as navigation constants or star locations. Saving this data in fixed storage also protects it from corruption in case of errors. Efficient usage of fixed and erasable memory follows a basic philosophy that anything that must not change should be implemented in fixed memory, with everything else saved in erasable. Despite the advantages of insisting that instructions are stored only in fixed storage, with data only in erasable, the AGC does not make this convention compulsory. If the AGC enforced the requirement that instructions are stored only in fixed memory, the ability to execute software in erasable memory is lost. At first, this might appear to be a reasonable restriction. After all, the rationale of placing instructions in fixed memory is to protect the software from accidental overwriting. The only requirement, quite sensibly, is that any data that is expected to change must reside in erasable memory, and instructions that store data must reference only erasable

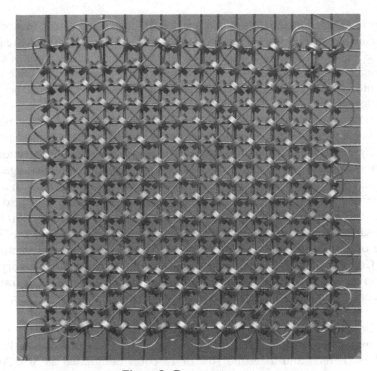

Figure 8: Core memory

memory. As expected, extensive testing of AGC software occurs before releasing the software for manufacturing into the memory module. Invariably, there are occasions where a programming error or invalid data has escaped the testing process. Without any means to change the contents of fixed memory, the only option is the expensive and time consuming process of manufacturing a new rope memory module. Entering software updates into erasable memory is possible, and are normally typed by hand into the DSKY. Manually entering and running a program is necessary only in urgent situations, but the capability exists in order to address specific problems. This class of programs, called Erasable Memory Programs, is discussed in a later section.

Memory technology in the AGC

Magnetic core memory was the dominant storage technology from the late 1950s until the early 1970s. Two characteristics of core memory make it ideal for spacecraft use: it is immune to most radiation-induced errors, and it retains its contents after the power has been removed.[4] Cores are small ferrite rings, typically only a few millimeters (or less) in diameter. When magnetized in one "direction", they are interpreted as storing

---

[4]   Indeed, the advantages of core technology endured long after their exit from the commercial data processing marketplace. Core memories were used in the Space Shuttle computers until 1990, when they were replaced by solid-state memory.

a binary '1'. When magnetized in the other direction, the core is said to store a binary '0'. The core is threaded with several very fine wires, used to select the core itself, read (or "sense") its contents and change that content if desired. Thousands of cores, each storing one bit, are used in constructing storage for a computer. Erasable memory is implemented using coincident-current magnetic cores, which were the prevailing memory technology of the 1960s. Fixed storage uses a novel technique called "core rope" whose storage density is considerably greater than single core planes. The premise behind core ropes is brilliantly simple. A core will hold a magnetic field indefinitely, and is magnetized one way for a '1', and in the opposite direction for a '0'. Conceptually, erasable storage is designed as a set of magnetic cores arranged in a two-dimensional plane. Each core can be considered as a point on a grid, with a unique x- and y-coordinate obtained from the physical address bits. Importantly, the direction of magnetization is the sole definition of that core's state, and is fundamental to the idea of an erasable, or read/write memory. But what happens if we do not need to change the state of a core; that is, we want to ensure that the contents of memory do not change? Core rope, as contrasted with conventional read/write core memory, does not depend on the magnetization of the cores. Rather, the ferrite rings act as transformers, and the contents of a word are dependent on the wiring scheme of the memory and not on any particular magnetic state.

The construction of core rope is also fundamentally different from its erasable cousin. Imagine constructing a fixed memory so that there is a row of 15 cores, all assigned as '1' bits, and a second row of 15 cores designated as '0's. Again, it is important to remember that these cores are simply torodial shaped transformers, and are not themselves magnetized. Figure 9 depicts what this arrangement looks like. To represent the binary string 101 001 010 111 110, the first wire is threaded through the core magnetized to a '1', the second wire through a '0' core, the third back to a '1' core, continuing this process until the entire binary string is wired. At

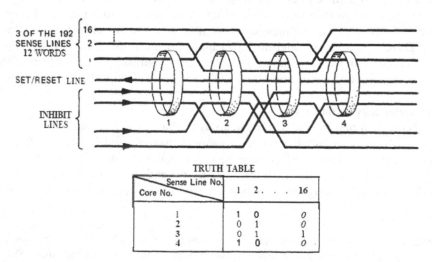

Figure 9: Schematic of core rope

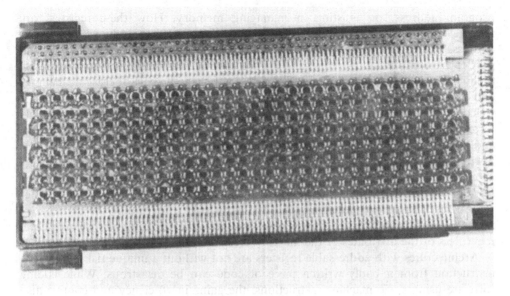

Figure 10: Flight version of core rope

first, this appears to be a wasteful exercise – only half of the cores are used to define a single word! Creating a second word in fixed memory using this technique again requires twice as many cores as necessary. But what if the original two rows of 15 cores are reused? There is no reason why we cannot thread the wire through the two rows of cores for our second word of storage, using the same process as with the first storage word. Creating a word with the binary value 000 101 100 001 010, the first three '0' cores are threaded with a wire, then a '1' core, etc. Adding a second word of storage is the break-even point in this implementation of core rope. Two 15-bit words require 30 cores, the same number required for two words of conventional storage. Implementing four words of storage in core rope now requires one-half of the cores needed in a conventional core plane. As more words reuse the original cores, the savings is even greater. Eventually, a practical limit is reached for how many wires can pass through a single core. For the AGC, this limit is 24 sets of wires – that is, each core is reused an astonishing 24 times!

Before leaving this topic, there is still need to address the etymology of the term "core rope". In the original implementation, the cores were strung out along the wires and not neatly arranged in a grid. Eventually, the design took on the appearance of a long bundle of cores and wire, looking like a rope with cores along its length.

A TOUR OF LOW CORE AND THE CENTRAL REGISTERS

The actual processing of data in a computer demands a large number of dedicated areas where computation and control occur. Collectively referred to as "registers", each of these areas performs a specialized function, such as computation, saving a program address, or assisting in managing memory. How these registers are referenced is one of the defining factors in a computer's architecture. "Non-addressable" central registers cannot be accessed as part of storage, yet are a vital part of the computer's operation and explicit instructions are needed to manipulate their contents. "Addressable" central registers appear to be part of erasable memory and are referenced in the same manner as all other memory locations, creating a welcome level of flexibility. To the programmer, addressable registers are simply locations in memory, typically in the lowest addresses of storage. While arithmetic and logical operations still require special instructions dedicated to a specific operation, simple data movement instructions can interact with the central registers as easily as they do with other storage areas. For example, the same "compare" instruction used to test a memory location is also usable with the accumulator, other central registers or a counter register. An instruction to exchange the contents of memory locations now provides a general mechanism for loading and storing data, regardless of the attributes of the memory location.

Architectures with addressable registers are not without a unique risk, as a rogue instruction from a badly written piece of code can be disastrous. While coding software using basic machine instructions, the gains in efficiency and control also demand a near-fanatical attention to detail. A common error is to save data in an unintended location, usually from a miscalculated memory address. Erroneously saving data anywhere in the central register area produces unpredictable results and places the stability of the overall system in danger. The guilty code is often difficult to identify and debugging is usually a long and tedious process. Code from compiled high level languages is usually of sufficiently high quality that the programmer need not worry about unintended overlays of storage. However, the use of high level languages is not a panacea; memory can still be corrupted in unexpected ways, as many a C programmer can attest.

The central registers are implemented in the lowest storage addresses in the AGC, making them accessible to all programs. The first 48, or 60_8 words of physical memory are reserved for the central registers such as the accumulator, program counter and return address register. Huge efficiencies are realized by placing these registers within the addressable memory space.

In explaining the central and special registers and their relationships, a diagram helps to visualize this hardware. Appendix C shows the lowest part of storage in the AGC, with the central and special registers labeled. The first, and perhaps the most familiar register is the accumulator, which is in location 00000_8. Arithmetic functions are performed in the accumulator, as well as logical operations such as binary OR, AND, and NOT. Many instructions compare the contents of the accumulator against data elsewhere in memory to decide what processing should be performed next. With so many operations using the accumulator, its unique properties play an important role in the AGC.

The center of the action: the accumulator – location 00000_8
Most of the arithmetic and logical processing in the AGC requires the accumulator, either as the source or destination of its operations. Part of the flexibility of the

accumulator is its multiple personalities. As with all the special and central registers, the accumulator is mapped to erasable storage. Assigned to storage location 00000_8, AGC instructions may reference the accumulator as they would any other location in erasable storage. Similarly, the accumulator is also interpreted as an I/O register. Designation as an I/O register is somewhat misleading, as it implies that there are connections to devices external to the AGC that are controllable by placing data into the accumulator. Rather, this interpretation is intended to allow the accumulator to enjoy the use of special logical instructions that are available only to I/O registers.

Overflow and modified one's complement

The one's complement technique described earlier explains how the AGC performs arithmetic operations. Indeed, from a programmer's perspective, the central registers operate only on 15 bits of data. To address the problem of arithmetic overflows, the AGC implements a 16-bit accumulator and a variant called the modified one's complement system. A 16-bit wide accumulator is one bit wider than any data used in the AGC. The extra bit is used to preserve a copy of the sign bit, which is then used to detect overflows from arithmetic operations. When data is loaded in the accumulator, all 15 bits of data and the sign move into the lower end of the register. Bit 15 contains the sign, and is referenced as S_1 in Figure 11. The modification of the standard one's complement occurs when copying the S_1 bit into bit 16, referred to as S_2, creating two copies of the sign bit. At this point, the duplication of the sign bit is invisible to the programmer. It becomes relevant only when an overflow occurs. An important property of S_2 is that it remains static during the arithmetic operations; any overflow that propagates to the S_1 bit does not continue to propagate into S_2. The overflow is sensed when the S_1 bit differs from the S_2 bit.

The effects of arithmetic operations on the two different representations of one's complement are shown in Figure 11. In this simplified example, the "accumulator" is only 4 bits wide with the second sign bit S_2 present in the modified one's complement. In the trivial case of adding (or subtracting or another arithmetic operation) two numbers where an overflow does not occur, the contents of the S_1 and S_2 bits are unchanged, and no special handling is required by the computer or programmer. When adding two numbers that do produce an overflow, the S_1 bit changes as a result of an overflow, as expected. In a standard one's complement system, the overflow problem becomes apparent when adding two large positive numbers unexpectedly creates a negative number.

In overflow situations, the function of the S_2 bit becomes apparent. The contents of S_1 and S_2 are no longer equal, and this inequality is the indication that an overflow has indeed occurred. As seen in the example, an overflow exists, but how can the programmer determine this state? At first glance, it would seem reasonable to test the accumulator after the arithmetic operation to determine if the S_1 and S_2 bits have changed, but the S_2 bit, which is necessary for identifying whether an overflow exists, is inaccessible to the programmer. Although there is no means to test the S_2 bit directly, an overflow does alter AGC processing in two ways.

First, when the accumulator contains an overflow, interrupts are inhibited in the AGC. Interrupts, as briefly discussed before, halt the normal processing of a

	Standard						Modified					
	S_1	2^3	2^2	2^1	2^0		S_2	S_1	2^3	2^2	2^1	2^0
Both Positive, No Overflow	0	0	0	0	1		0	0	0	0	0	1
	0	0	0	1	1		0	0	0	0	1	1
	0	0	1	0	0		0	0	0	1	0	0
Both Positive, Overflow	0	1	0	0	1		0	0	1	0	0	1
	0	1	0	1	1		0	0	1	0	1	1
	1	0	1	0	0		0	1	0	1	0	0
Both Negative, No Overflow	1	1	1	1	0		1	1	1	1	1	0
	1	1	1	0	0		1	1	1	1	0	0
	0	1	0	1	0		1	1	1	0	1	0
	+				1 Carry		+					1 Carry
	0	1	0	1	1		1	1	1	0	1	1
Both Negative, Overflow	1	0	1	1	0		1	1	0	1	1	0
	1	0	1	0	0		1	1	0	1	0	0
	0	1	0	1	0		1	0	1	0	1	0
	+				1 Carry		+					1 Carry
	0	1	0	1	1		1	0	1	0	1	1

Figure 11: Arithmetic examples in standard and modified one's complement

computer to switch to another, more important task. Inhibiting interrupts prevents this switching from occurring. This is necessary because the S_2 bit is not preserved during interrupt processing. Only after clearing the overflow condition will interrupt processing resume. It is essential to resolve the overflow condition quickly, since withholding interrupts for any length of time prevents normal AGC processing. Loading a new value into the accumulator, or clearing its contents is sufficient to remove the overflow condition. A more direct way of testing if an overflow exists is through an AGC instruction named Transfer to Storage (TS). TS solves the overflow detection dilemma by saving the accumulator contents only if there is no overflow present. If a positive or negative overflow occurs, a +1 or -1 replaces the value remaining in the accumulator. TS also skips to the next instruction in the program when an overflow exists, on the presumption that the programmer has provided code to handle the overflow. A single instruction now provides an easily tested means for indicating a positive or negative overflow.

Overflow correction
When saving the contents of the accumulator, the lower 14 bits and the S_2 bit (but not the S_1 bit as might be expected) are saved. In cases where there is no overflow present, either S_1 or S_2 may be saved, as they are identical. In cases where an overflow exists, what value remains in the accumulator? In the following example, two numbers, 14,908 (35074_8) and 8,265 (20111_8) are added together. In decimal, the result is 23,173 (55205_8) and is beyond the maximum value that the accumulator can hold. Overflow has occurred, since S_1 and S_2 are now set to different values. The positive sign is taken from the S_2 bit, along with bits 14 through 1 of the overflowed value, resulting in 6,789 (15205_8) being stored in memory. This technique, known as "overflow correction", preserves the correct sign of the overflowed accumulator value. Importantly, overflow detection and correction extends beyond the case when performing operations on numbers that are one word in length. In double and triple precision values, discussed later, an overflow in an operation indicates that a carry into the next word is necessary. Because TS responds to an overflow by replacing the accumulator contents with $\pm$ 1, it is not appropriate for saving an overflow corrected value. Other instructions that do not modify an overflowed accumulator are used instead.

The L register – the low order accumulator – location 00001_8
Using a single word for the calculations is the most straightforward and efficient means of processing data when dealing with a fractional value. But it limits numerical precision to only four decimal digits, which is insufficient for many calculations. Double precision, a technique that combines two contiguous words to create a single value with a higher level of precision is widely used in the AGC. The L register, or Low Order Accumulator, is paired with the accumulator for multiplication and division operations, which use double precision values as an operand or result. Although the L register is used as part of these two arithmetic operations, it is not capable of performing arithmetic operations on its contents. When not used for multiplication or division operations, the L register is most frequently used as a temporary storage location to hold intermediate data values. Although incapable of the logical and arithmetic operations of the accumulator, the L register assumes many different characteristics. Assigned to the erasable storage location 00001_8, all instructions have access to the L register. Additionally, it behaves as an I/O register, allowing the full complement of logical operations on the data it contains.

Storing the return address – the Q register – location 00002_8
Implementing subroutines in software is a necessary part of ensuring readable, structured code. Subroutines are usually small pieces of programming that are part of a much larger program, and perform a well-defined function. Subroutines are essential when a particular operation is executed frequently, or when several programs need to use the function. A typical example of a subroutine is a program to

find the square root of a number. Imbedding the square root code in the rest of the program is a poor option, since storage is wasted if other programs include their own versions of square root routines. Additionally, while the programmer might be expert at the nuances of a navigation routine, he might not be particularly knowledgeable on how to optimize the Newton-Raphson method of finding roots. Here, a programmer with expertise in numerical methods will write a single copy of the square root routine, and make it available to all other programs. A subroutine can be called by a program that is anywhere in the computer, and must return to that program once the subroutine has finished processing. Therefore, the program calling the subroutine must also save the memory address of where the program should resume once the subroutine has completed processing. The Q register contains the 12-bit memory address of the instruction the subroutine will return to, and relies on the banking registers to generate the appropriate storage address.

The banking registers
Banking is the heart and soul of addressing in the AGC. While the FIGMENT system is a useful exercise, it is time to look at the real-world implementation of banking in the AGC, which is significantly more complex. Here, the fundamental problem of memory management in the AGC is apparent. Simply, there are fewer bits in the address field of the instruction than are required to access all the physical memory installed in the computer. Recalling the earlier description of the AGC instruction layout, the 15-bit field is generally divided into two unique fields. A 3-bit operation code defines what action the instruction is to perform, and the remaining 12 bits specify the operand. However, 12 bits are only sufficient to reference 4,096 unique locations in memory, far less than the 38,192 words physically installed in the computer. Both erasable and fixed storage use banking, and dedicate three registers to this purpose. Two of these, named EBANK and FBANK, perform the expected duties for the erasable and fixed memories respectively. A third register, BBANK, reflects the contents of both these registers.

Managing memory in the AGC is a complex topic, and requires more detail than is appropriate in this tour of the central registers. For now, only the essential characteristics are presented. A full description of how the banking registers and instruction formats are used together to manage memory in the AGC will be given in a later section.

EBANK: the erasable storage banking register – location 00003_8
Erasable memory contains 2,048 words of read/write storage, divided into eight banks of 256 words. Together, they occupy the lowest address range in memory. Within any 256-word bank, only eight bits are needed to uniquely reference any word within the bank. An additional three bits are necessary for selecting which bank to use. These are contained in the Erasable Storage Bank, or EBANK register.

FBANK: the fixed storage banking register – location 00004_8
Fixed storage is handled in a manner very similar to erasable storage. At 1,024 words, the size of a fixed memory bank is much larger than that of its erasable

counterpart and 10 bits are needed to reference every word in a bank. Thirty-six banks of 1,024 words of fixed storage are available in the AGC and contains most of its programs and constants. Addressing such a large amount of storage requires 16 bits, exceeding the length of any AGC register or memory word and making banking essential for any memory reference. In the same manner as the erasable memory banks, the Fixed Storage Bank, or FBANK, contains five bits. This is sufficient to reference 32 banks. Another 10 bits are taken from the addressing field of the instruction, making it possible to reference 32K of fixed storage. Nevertheless, this is not sufficient, as the AGC contains 36K words of fixed storage. A third element in the banking scheme, the Fixed Extension Bit, or Superbank Bit, is used in conjunction with the Fixed Storage Bank to reference the final 4K words of fixed storage.

BBANK: the both-banks register – location 00006_8
A program's instructions and data are tightly intertwined with one another, making it essential to manipulate the FBANK and EBANK registers together. When transferring control to another program that requires an entirely different set of banking register values than the currently running program, both the EBANK and the FBANK registers must be changed simultaneously. As its name implies, the BBANK (Both Banks) register contains the contents of both the FBANK and the EBANK. Whenever either of the two individual registers is modified, that update is immediately reflected in the contents of the BBANK. Similarly, a change to the contents of the BBANK automatically updates the FBANK and EBANK registers. Using BBANK to manage the fixed and erasable banking registers is for more than its convenience. Transferring control to another program requires simultaneously changing the banking registers and the program counter. Without properly setting up the new banking environment, the memory switching process will fail. This register makes it possible to perform the operation in a single instruction, and the section on interbank transfer of control discusses this topic in more detail.

The program counter – the Z register – location 00005_8
Execution of a program is a step-by-step process, with the state of the program defined by the contents of storage, the registers, and the point in the program that is currently executing. The Z register is also known by its more common name of the program counter. As a program is simply a sequence of steps, the Z register points to the next instruction to be fetched from memory and executed. After the instruction is processed, the Z register increments by one. This process continues until an instruction or interrupt explicitly changes the order of execution. As the Z register is 12-bits long, it, like all other memory references, must use the banking registers to resolve the memory address. The Z register uses the same memory banking technique as machine instructions, making all storage references consistent and programs easier to write and read.

Programs rarely execute in a linear fashion, at least not for long stretches of time. Eventually, a decision is made to execute one section of programming code rather than another. For example, in calculating a payroll, timesheets are checked to see if an employee worked on a holiday, which is calculated at a higher pay rate. If the employee had worked on a holiday, a "jump" or "branch" instruction transfers to the holiday pay processing by placing the memory address of the pay routine in the Z register. In the next instruction cycle, execution resumes at the start of the holiday pay routine, and continues from that point.

Other events change the Z register, such as an interrupt caused by a timer expiring or an I/O device demanding attention. Hardware in the AGC senses the particular type of interrupt and loads the Z register with the predefined address of the routine assigned to handle the interrupt. Implicit in interrupt processing is the need to save the Z register when the interrupt strikes, and restore it again when normal processing is resumed. As we will see below, a storage location, named ZRUPT, is reserved for just this purpose.

A source of zeros – location 00007_8
The value of zero is used extensively in programming, most frequently for initializing data or in comparisons against another value. When defining a new calculation, placing a zero in the register or storage location clears the previous contents, and new calculations are possible without the danger of including old or unrelated data. Location 00007_8 is permanently set to the value of zero, and is addressable by all instructions, effectively acting as a fixed constant in the erasable storage address range.

Interrupt save areas – locations 00010_8 through 00012_8
Interrupts are an integral part of the computer system hardware, and provide an important means to bring the current process to an orderly halt, to schedule new work in the processor. As its name implies, an interrupt abruptly stops the program when the hardware recognizes that there is other work pending. Once interrupted, a dilemma quickly appears. Although processing has stopped, immediately starting a new program will destroy the contents of many of the central registers, making the resumption of the program impossible. When processing an interrupt, an essential part of housekeeping is to ensure that the contents of critical registers are preserved. In essence, interrupts are based on the concept that the processing flow can be diverted from the current program, yet that program will have no knowledge that it has been disturbed. Two registers are automatically saved during an interrupt. The Z register (the program counter) and B (an internal register containing the next instruction to be executed) are saved in the ZRUPT and BRUPT locations. While processing an interrupt, the AGC inhibits further interrupts until an instruction named RESUME is executed. The interrupt routine itself is not interruptible. During the interrupt processing, it is very likely that other registers will be used, especially the accumulator. Four other storage locations are set aside for saving the remaining central registers: the accumulator, L, Q and BB (or BBANK), appropriately named ARUPT, LRUPT, QRUPT and BANKRUPT. Saving these registers is required only if the programmer will be altering their contents while processing the interrupt. The A, L, Q and BB registers are saved manually, and must be restored manually prior to executing the RESUME instruction to return to the interrupted program.

Much of the earlier discussion of the accumulator focused on its 16-bit length, whose 16th bit is used for overflow detection and is inaccessible to the programmer.

Yet, the ARUPT register has only 15 bits. At first, it would appear that there is the danger of losing information if the accumulator contains an overflow. But the ARUPT register does not need to have a 16th bit! Any time the accumulator overflows, interrupts are disabled until the overflow condition is cleared. As a result, saving and loading the accumulator are possible without worrying about overflow. It is tempting to think of these registers as convenient places to save the intermediate values of a computation. Resisting this enticement is easy, as an interrupt can occur at any time without any capability for saving or restoring these registers across the interrupt. From the perspective of the programmer whose code is invisibly interrupted, the values in the ARUPT, LRUPT, QRUPT and BANKRUPT registers tend to change randomly and unpredictably.[5]

Editing registers – locations 00020₈ through 00023₈

Not all manipulation of data is in the form of arithmetic operations. It is frequently useful to consider data as a string of bits that have their own intrinsic meaning, rather than as a discrete number. Representing data in this way is very efficient, especially if there is a large amount of data with a limited range of values. For example, assume it is necessary to define the day of the week when a particular widget is scheduled for manufacture. One method uses only seven bits to represent the manufacturing schedule by assigning bit 0 for Sunday, bit 1 for Monday, and so forth. The binary string 011110 now represents the Monday through Friday manufacturing schedule of our widgets. Encoding the data as individual bits saves several words of storage by combining the production schedule into one word, albeit at the cost of the extra effort to encode and extract the bits. In the AGC, where memory is perhaps the greatest constraint, representing data in this way is a useful technique for minimizing the amount of storage used.

Shifting data is not a basic logical operation, and can take on many forms. The simplest, a Logical Shift, moves the bits left or right in the register without regard to the presence of a sign bit. Shifting the contents of the register one bit to the right places a zero in the leftmost bit location and discards the rightmost bit.[6] Conversely, a left-shift pads the rightmost bit with a zero and the upper, leftmost bit falls off into the "bit bucket". Finally, we have the Cycling Shift. Bits move within the word left or right as in the Logical Shift, but instead of having a bit drop off, it is "cycled" to the opposite end of the register. In a Cycle Shift, none of the bits are lost. Figure 12 shows these operations more clearly.

---

[5] This would not be the case if the program has already inhibited interrupts, or while there was overflow in the accumulator. Still, it is not good programming practice to use these save areas casually.

[6] Known colloquially as the "bit bucket", it was the source of torment to neophyte computer operators. In addition to the usual menial tasks all new employees are subject to, a manager will make a casual request to empty the "bit bucket" to prevent some grave consequence to the system. Our naïve hero, trying desperately to please, sets out on a fruitless search for the imaginary "bit bucket", much to the amusement of the rest of the staff.

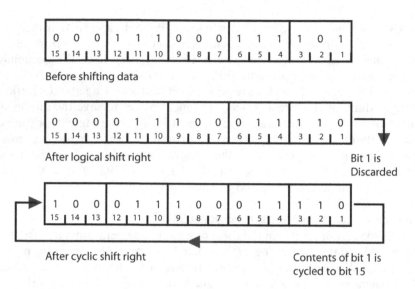

Figure 12: Logical and cyclical shifts

Ideally, instructions to implement shifting operations should allow shifts of a variable number of bits in an arbitrary manner. With a limited instruction set and a potentially large number of opcodes to implement, the AGC needs an alternative mechanism to implement shifting operations. By limiting the flexibility of the instructions required for shifting data somewhat, it is possible to eliminate the need for dedicated instructions altogether. The solution is to create special, single function registers as a part of lower storage, called editing registers. Three shifting registers exist in the AGC, each performing a different function: Cycle Right, Shift Right and Cycle Left. Rather than using dedicated instructions and precious opcodes to shift data, the data is stored into one of these special registers and the shift operation occurs simultaneously. For an example, assume the accumulator has the value 000 000 110 010 111. Saving the accumulator into the Shift Right editing register produces 000 000 011 001 011 as the result. Only one shift operation is performed each time data is saved in the register. Performing multiple shifts on a word of data in a single operation is not possible. Multiple shifts require executing the same shifting process multiple times, copying the data out of the shift register and saving it back into the register again.

Remembering that a word in memory is only a binary number, we see the types of applications facilitated by shifting data. Cycling to the left or right is useful when operating with numbers using multiple words; i.e. double or triple precision. A logical shift to the right has the effect of dividing by 2, and repetitive shifting allows us to divide by 4, 8, 16 and so on. In the same manner, shifting a binary value to the left is a quick way to multiply by two. The need to divide by 2 is common, and it is far more efficient to use a right shift for this division than to use the divide instruction itself. When shifting right to divide, any remainder that exists is lost, but in many cases this is not a concern. Shifting left to multiply, however, runs into the problem of overflowing the register. Shift registers do not trap overflows, as this is recognized only when operations are performed in the accumulator.

The EDOP register (EDit Interpretive OPcode) completes the set of editing registers. Interpretive instructions are implemented in software and use 7-bit opcodes that are stored two per word. The lower opcode is extracted through masking, but isolating the upper opcode requires shifting the word seven bits to the right. Saving a word in the EDOP register shifts the contents seven bits to the right as a single action, leaving the interpretive operation code in bits 7 to 1 of the register.[7] Executing a right shift instruction seven times would achieve the same result, but at the cost of longer execution times and several instruction words.

Care must be taken when using editing registers. Because the AGC has no means to save and reload data in the editing registers, they are effectively "off limits" during interrupt processing. Normally, an interrupt routine will save a register before using it, and then restore it just prior to returning to the calling program. Saving the values in the editing registers is straightforward, but when the editing registers are reloaded the data is re-edited and an entirely new value appears in the register! As a result, using editing registers during interrupt processing might cause the interrupted program to fail.

KEEPING TIME IN THE AGC: TIMERS AND CLOCKS

Timing circuits in a computer system do far more than simply maintain the time of day. In addition to performing standard clock functions, several specialized timers exist to expire at well-defined intervals. Each timer is associated with a specialized task such as scheduling work in the computer, or checking the status of I/O devices. By assigning a discrete function to each timer when they expire, it is possible to manage several different functions at once. In the AGC, there is no absolute date or time, nor does the clock display time in terms of day, month or year. Most computer clocks define time since the beginning of some arbitrary epoch. The Unix operating system, for example, defines "time zero" as January 1, 1970. For the AGC, the epoch, or time zero of the clock is defined as a point several hours before liftoff. All references to events are expressed in terms of "Mission Elapsed Time"; the hours, minutes and seconds since liftoff. In addition to the obvious need to keep mission time, the clock provides critical timing information for mission events, restart and recovery, navigation and tracking.

The AGC has a central clock occupying two words in low memory, followed by

---

[7] In keeping with the notation in the AGC literature, bit positions are noted in diagrams with the highest numbered bit in the leftmost position. This leads to specifying the bits of embedded fields in descending order (e.g. bits 7 to 1) rather than the more intuitive ascending order of everyday sequences.

four timers of one word apiece. Two 14-bit words are used to hold the central clock, and each "tick" of the clock occurs every 10 ms, the coarsest granularity of timing in the spacecraft. The lower word, TIME1, is incremented every 10 ms and after approximately 164 seconds it overflows into the upper word of the clock TIME2, incrementing that in turn. TIME2 is sized to last approximately 31 days, which is more than enough for a two week lunar mission.

Summary of AGC Timers

TIME2	00024_8	T2: Elapsed time	Scaled at 10 ms
TIME1	00025_8	T1: Elapsed time	Scaled at 10 ms
TIME3	00026_8	T3: Wait list	Scaled at 10 ms. Offset to T4RUPT by 5 ms
TIME4	00027_8	T4: T4RUPT	Scaled at 10 ms
TIME5	00030_8	Digital Autopilot	Scaled at 100 ms
TIME6	00031_8	Fine scale clocking	Scaled at 1/1,600 second (0.625 ms)

Timers: TIME3, TIME4, TIME5 AND TIME6
Some of the most important "clocks" in computers do not concern themselves with times in the sense of hours and minutes of mission elapsed time. Four timers exist for scheduling critical events at specific times or intervals. Consider the array of tasks under the control of the AGC – firing and shutting down engines, rendezvous tracking, or scheduling work running in the computer. To properly control many of these operations requires precise timing, often down to a precision of 0.625 ms. Three of these timers (TIME3, TIME4 AND TIME6) are programmable, allowing control over the time before they expire. A fourth timer, TIME5, is set to expire every 10 ms. All four operate in a similar fashion, with the only exception being that TIME6 operates at a much finer scale than the other three (0.625 ms vs 10 ms).

Our first timer, TIME3, controls routines in the waitlist. The waitlist is a list of small routines that run with interrupts inhibited and perform short duration, high priority work. Up to seven routines can be queued in the waitlist, each of which is scheduled to start in a predefined interval. When the TIME3 timer expires, a T3RUPT interrupt is raised and control is transferred to waitlist scheduling. The next routine in the Waitlist starts. When this completes, the next task in the list moves to the top and the TIME3 timer is set for executing the next task. Inhibiting interrupts is essential during waitlist programs to ensure only one program runs at a time, and that it completes quickly.

TIME4 is used to schedule many of the critical periodic checks in the computer. Set with intervals of between 20 and 120 ms, TIME4 interrupts schedule the T4RUPT process to manage the DSKY output, IMU status and the optics, and also to interrogate several instrument panel switch positions. The T4RUPT routines are normally entered every 120 ms, but under certain circumstances much shorter intervals are scheduled. In scheduling these short intervals, the routine adjusts the next entry time for T4RUPT to maintain the 120 ms execution cycle. An important aspect of TIME3 and TIME4 is that while they are both counting at 10 ms intervals, the hardware enforces an offset of 5 ms between them. Skewing the times in this fashion eliminates the inevitable conflict that arises when two timers expire at the same time, especially since waitlist tasks run in interrupt inhibited mode. Waitlist programs must complete within 4 ms to ensure that they complete before the TIME4 interrupt occurs. In the AGC, this interval allows for an average of only 160 instructions.

The Digital Autopilot (DAP) code is one of the largest single software components of the AGC. During its processing cycle, the DAP evaluates the current and desired attitudes, and schedules Reaction Control System (RCS) jet firings to adjust the attitude. Real-time control of the spacecraft demands frequent updates, but the processing requirements quickly become excessive. A compromise of ten passes through the DAP code per second provides acceptable control response without requiring an unacceptable amount of processing time. DAP software, responding to an interrupt generated every one-tenth of a second, executes the main autopilot routine. Phases within the DAP code also use TIME5 for scheduling work in intervals of 20 to 30 ms, but the 100 ms major loop schedule is not affected. Ensuring precise control of the spacecraft requires accurate timing of the RCS jets. Using mathematical rules that define how to maneuver the spacecraft, the autopilot issues commands to the thrusters to maneuver the vehicle or maintain a fixed attitude. Over- or under-control from inaccurate thruster burns will not only waste fuel, it can create serious and potentially dangerous handling issues. Accurate maneuvering requires that the duty cycles of the RCS jets be continuously variable from their minimum impulse time of 14 ms up to steady-state, "always on" operation. A timer with centisecond resolution does not have sufficient granularity for such fine control. TIME6 provides the ability to time events at a resolution of 0.625 ms, which is adequate to satisfy the DAP control requirements. When selecting a thruster to maneuver the vehicle, the TIME6 timer is loaded with the jet's firing duration. By setting a bit in the output channel, the thruster receives its firing command and TIME6 begins counting. At the end of the interval, the timer expires and the T6RUPT interrupt routine will reset the bit in the output channel to shut off the thruster.

Timers are not set to expire at an absolute time, such as 125 hours 11 minutes and 14.39 seconds. Rather, they are defined as an interval between the current time and when the timer is to expire. In controlling events that involve only a short duration (in comparison to the time scale of human perception) timers are often more flexible and require less overhead, as explicit times do not have to be computed.[8] Timers are set by taking the largest positive value for the word, adding 1, and subtracting the timer interval in 10 ms increments. Data is expressed in increments of 10 ms (or 0.625 ms for TIME6), so a timer setting for 120 ms is:

---

[8] We use this case in everyday life. Waking up in the morning, groggy and without that essential first cup of coffee, we want to boil an egg for three minutes. It doesn't matter what time the egg is to finish cooking, so we set our egg timer for three minutes, heedless of the absolute time of day the egg is ready to eat.

$$
\begin{array}{lll}
 & 011\ 111\ 111\ 111\ 111 & 2^{15} - 1 \\
+ & 000\ 000\ 000\ 000\ 001 & \text{Add 1 (forcing an overflow)} \\
- & 000\ 000\ 000\ 001\ 100 & \text{120 milliseconds} \\
= & 111\ 111\ 111\ 110\ 100 & \text{Contents of timer}
\end{array}
$$

For the three timers scaled at 10 ms resolution, we can specify a maximum interval of about 2 minutes 43 seconds. TIME6, at 0.625 ms resolution, will count for a maximum of 10.2 seconds before overflowing. Once the timer is loaded with its value, it begins incrementing with every 10 ms tick of the clock. Eventually, the timer reaches the maximum value (32,767), and on the next clock tick the timer overflows. This overflow triggers an interrupt, and control switches to the address associated with that timer. Each timer has its own processing routine that determines the action to perform next, such as waitlist or DAP processing.

The Inertial Measurement Unit

The primary instrument for all attitude and acceleration data in the Apollo spacecraft is the Inertial Measurement Unit, or IMU. A gyroscopically stabilized assembly, called an inertial platform, is mounted within three concentric rings that allow the platform three degrees of rotational freedom within the IMU case. Each ring contains bearings to connect itself to the next ring. The ring and bearing assembly is jointly known as the gimbal assembly. Because the gimbal rings allow rotations in all directions, the stable member of the IMU will maintain its orientation in space regardless of the attitude of the spacecraft.

A single gimbal allows rotation around only one axis. Therefore, three gimbals are necessary for rotations in all three axes. The first, inner, gimbal attaches to the platform, allowing rotations along the Y-axis. When adding a second, middle, gimbal to the assembly, the bearings cannot mount directly to the platform, as this would interfere with the motion of the inner gimbal. The solution is to mount the middle gimbal to the ring supporting the inner gimbal, offset by 90 degrees from the inner bearings. In this design, the Z-axis is assigned to the middle gimbal. In a similar manner, a third, outer, gimbal is attached for rotations along the X-axis. This is attached to the spherical IMU case, which in turn is mounted on the spacecraft. As the spacecraft rotates, a sensor called a resolver in the pivot bearing senses the angle between the gimbal and the platform, and sends it to the Coupling Data Unit (CDU). The CDU, essentially an analog to digital converter, takes data from the resolver, converts it to digital pulses and sends them to the AGC. Each pulse represents 39.55 seconds of arc, with a complete rotation producing 32,768 pulses. A CDU is dedicated to each of the three rotational axes of the IMU, plus the shaft and trunnion of the sextant or the rendezvous radar. Because the sextant is unique to the Command Module, and the rendezvous radar exists only in the Lunar Module, they are able to use the same counter locations in the AGC.

The inertial platform contains three accelerometers, one for each spacecraft axis, and these sense the lateral motions of the spacecraft. Determining the accumulated velocity from those motions involves integrating these accelerations over a known time. The appropriately named Pulsed Integrating Pendulous Accelerometers

Figure 13: Inertial measurement unit

(PIPAs) perform this velocity calculation and send the pulses to the AGC. Although the Command Module and Lunar Module IMUs are identical, their velocity scaling is different. Velocity changes sensed by the CSM during launch, translunar injection and atmospheric entry are large, ranging from zero to 36,000 ft/sec (11,000 m/sec), while the Lunar Module will see less than 5,500 ft/sec (1,700 m/sec) in velocity changes on a typical mission. The difference in overall velocity changes between the two vehicles is reflected in the magnitude represented by one counter pulse. A single count in the CSM represents 0.191 feet/sec (5.85 cm/sec) velocity change, while in the LM it represents 0.033 feet/sec (1 cm/sec). With the LM's lower overall velocity changes, higher precision measurements become possible for the lunar landing and the subsequent rendezvous.

COUNTERS – CDUS (X, Y, Z, OPTS, OPTT) AND PIPAS (X, Y, Z)

Spaceflight is a tremendously dynamic endeavor. From the moment of launch to when the crew returns safely to the Earth, the vehicle undergoes tremendous accelerations and attitude changes. Much of the AGC's raison d'être is to track these changes and incorporate them in equations for guidance, navigation and control.

Sensing changes in attitude and acceleration is the role of the Inertial Measurement Unit, not the AGC itself. With its stable platform containing highly accurate accelerometers and gyroscopes mounted within three precision gimbals, the IMU is the most important I/O device for the computer. This presents one of the more important design decisions for computer architects. For any I/O device to communicate with a computer, what does the data look like? A single bit appears as a simple on/off or yes/no flag, with only a few machine instructions necessary to isolate a flag's value. But how is rapidly changing data described to the computer, especially when a high degree of precision is necessary? Much of the data originates with analog devices, such as accelerometers and angle resolvers. Before the computer can digest this data, it must be converted into a digital format. At first inspection, the solution is to build analog to digital converters for each device, and then take measurements of the device and pass all of this data to the computer at a high rate.

While this solution might appear workable, it quickly becomes impractical to implement. In this arrangement, data sent to the computer must be able to represent the full range of the mission requirements, which may span up to six orders of magnitude and require a large number of bits to represent. This large number of input data lines, multiplied by the degrees of freedom provided by the IMU, would require a tremendous number of wires and other hardware. Most of the time, the spacecraft is in coasting flight with acceleration and attitude changing very little, yet this same data changes at a rapid pace during powered flight. Solving this problem becomes an exercise in extreme simplification. By eliminating the use of the fractional data representation, much of the interface hardware and the scores of wires that come with it are also eliminated. Rather than representing such data as a full numeric value, the computer assumes each input from the IMU or other device was initialized to a known state, such as zero velocity or an offset of zero degrees. Each device connected to the computer then defines a *change in state* at a certain level of granularity. For example, an accelerometer (integrating the acceleration over a known time) might report velocity changes in increments of 0.1 m/sec, or an angle sensor detecting changes of 0.01 degrees. Here, whenever the accelerometer senses a change in velocity of 0.1 m/sec, it notifies the computer of that event by sending a single binary pulse.

For recording the changes in velocity or angles, special registers exist in low memory that count the pulses sent to the AGC. Each time a device signals the AGC that a change has occurred, it also supplies the direction of that movement. Changes in both the positive or negative direction are possible, such as a velocity change forward or backward, or a radar rotating right or left. A pulse representing motion in the positive direction adds one to the counter, while a pulse in the negative direction decrements the counter. Over time, the counter retains the net value of all the pulses it receives as a measure of the change in its state since it was initialized.

Counter interrupts and cycle stealing
Each pulse sent to the AGC carries an essential update about the state of the vehicle or other hardware. If the data represented by each pulse is small enough to ensure a high level of precision, it necessarily must represent a small number. Large changes

in velocity or an angle therefore demand a correspondingly large number of counter pulses. To notify the processor that there is data from an external source (such as the IMU) waiting to be incorporated, an interrupt might be reasonably used to process the data. Interrupts, however, introduce a significant processing overhead, as processes are suspended, registers are saved and restored, and the actual interrupt routine is executed. Regardless of the efficiency of the actual interrupt routine, halting and restoring the processing of a job still takes considerable resources. Interrupts arriving at a sufficiently high rate can monopolize the processor to the detriment of other work. In moving to an architecture where every small change sensed by the IMU interrupts the AGC, only the briefest disruption in processing is tolerable. In an extreme but realistic scenario such as a 10g launch abort, up to 1,600 velocity update interrupts per second would occur, and consume a significant amount of resources in the traditional interrupt processing scheme.

Once again, the solution lies in simplicity. If the overhead of interrupting a process (saving and restoring registers, and returning to the interrupted task) is unacceptable, then the answer is to eliminate all of this housekeeping. A likely first reaction to such a patently absurd statement is that the housekeeping is essential, for without it we introduce the potential to violate the fundamental requirement of interrupts; that is, they cannot corrupt the data of a currently running process. The solution in the AGC is to take interrupts back to their most primitive definition.

Several different types of interrupt are usually found in computers, and the AGC is no exception. In defining distinct classes of interrupts, we can direct the transfer of control to the routine specially designed to handle that function. An I/O handling routine processes an interrupt from a keystroke, the dispatcher processes a timer expiring, and hardware recovery routines handle parity errors. In each case, the type of interrupt defines the pre-defined storage location to which control should be transferred. Rather than being defined as the target address where control is transferred, counters are defined as storage locations and are *themselves* the object of the interrupt. For example, a signal line from the X-axis accelerometer in the IMU is associated with location 00037_8, while location 00033_8 contains the counter for the middle gimbal (Z-axis) angle. This technique is often called "cycle stealing", as it consumes clock cycles that would otherwise be used in executing machine instructions.

Processing in counter interrupts does not exist in software routines, since each counter pulse received by the AGC does not "interrupt" processing in the traditional sense. Most importantly, no saving of registers or transfer of control occurs! After the AGC completes the execution of each instruction, the hardware checks for any pending counter interrupts. When a counter interrupt is detected, the execution of the next instruction is suspended and the AGC increments or decrements the counter. After the counter update is complete, the program resumes, unaware that an update occurred. With no transfer of control, nor modification of central registers, there is nothing to restore before continuing with the program's execution. More than just a clever solution to the interrupt overhead problem, counters provide a huge improvement in architectural efficiency.

None of this is without measurable overhead. During coasting flight, the number of counter interrupts might be hardly noticeable, perhaps only a few per second.

During powered flight, interrupts can arrive by the hundreds, if not thousands, consuming a measurable percentage of AGC processing capacity. An extremely high rate of counter interrupts is probably the result of an error. To prevent this runaway activity, sometimes called a "hot I/O", from overwhelming the computer, a safety feature in the hardware triggers a restart in an effort to recover from the problem. Design limits for the CDUs and PIPAs far exceed anything expected during the flight. CDUs can send pulses at a maximum of 6,400 pulses per second, which translates to about 12 rotations of the spacecraft per second, and PIPAs can send 3,200 pulses per second (a load of 19.1g in the Command Module). These are extreme conditions that ought never to occur in routine flight.

Eight counter registers in the AGC record velocity and angular data. Six of these counters are supplied with data from the Inertial Measurement Unit, to provide gimbal angles and velocity information. The remaining two counter registers record the angles of the sextant in the Command Module or the rendezvous radar in the Lunar Module. In summary, counters are sufficiently different from other interrupts for them to be considered a distinct architectural feature. The table below illustrates their major differences:

INTERRUPT	COUNTER
Interrupt address defines type	Counter address defines type
Interrupt transfers control	No transfer of control (cycle stealing)
Registers saved, restored	No registers changed

The problem with one's complement and counters
Counters expose an interesting oddity in the AGC architecture. In all of the discussions of numeric representation in the AGC the one's complement system is used. Of particular importance, one's complement is plagued with the existence of "positive" and "negative" zero, each of which is a distinct value. Use of the one's complement system with its multiple values of zero becomes a problem for the counters that define gimbal angles. The simplest example is when the crew places the spacecraft in a slow continuous roll to evenly distribute solar heating on the spacecraft. Each time the counters pass through the "first" zero, the next count continues to the "second" zero. Since the CDU counters should report the actual gimbal angles, we now have the problem where we cannot uniquely define zero degrees. While the few seconds of arc for each additional counter pulse may by itself be considered acceptable, the error will accumulate over time. Clearly, the use of one's complement is simply the wrong solution for defining gimbal angles; it requires a system where there is only one value for zero.

In the only case where the one's complement system is not used in the AGC, the two's complement number system is used to count the pulses from the gimbal, sextant and radar CDUs. With a single value for zero, all angles are uniquely defined, creating a simpler design for handling gimbal angles. Such a special case is not without a cost. A new instruction is introduced that performs the conversion from the unsigned two's complement used in the CDU counter numbers to one's complement. On a practical level, the use of one's complement is not a problem for velocity measurements. Velocity changes, computed from the accelerometers, are usually the result of a powered maneuver. By their very nature, these velocity changes usually occur continuously and in one direction for an extended period from milliseconds to several minutes. Even in the rare case where the spacecraft reverses direction once or twice, the accumulated error is minor, usually a few tenths of a foot per second, and a one's complement notation is acceptable.

Controlling gimbals through counter outputs
Having discussed how the AGC receives input from gimbals, it is natural to wonder how to command the IMU gimbals, optics or radar to move (velocity counters are input only). We have already seen that the gimbals transmit changes in their positions through pulses that generate counter interrupts. Commanding the gimbals to move uses a similar technique, but operating in reverse. The CDUs operate bi-directionally, and can command the motion of a device as well as report its movement. For example, to move the LM's rendezvous radar on its trunnion axis, the AGC sends pulses to the CDU associated with the trunnion. Beginning with the radar's known current position, it calculates the number of pulses necessary to move the radar to its desired position. This number is placed in the counter output register, and the trunnion CDU is commanded to drive the radar to this position. The CDU converts the number of pulses to the analog signals necessary to drive the torque motors integrated in the gimbal assembly. Although the radar is moving under the computer's command, it is not relieved of the responsibility for reporting its position. As the radar moves, it continues to send pulses to update the counters in the AGC.

There are three counter output registers: CDUXCMD, CDUYCMD and CDUZCMD, one each for the X-, Y- and Z-axis gimbals in the IMU. Two additional counter output registers, CDUSCMD and CDUTCMD, connect to the sextant in the Command Module or the rendezvous radar in the Lunar Module; one each for the shaft and trunnion axis. In a special case for the Command Module, storage areas in the AGC are reused for two distinct purposes. Quite reasonably, the Command Module crew does not use the optics system during the time the Service Propulsion System engine is firing. Here, the CDUSCMD and CDYTCMD special registers are pressed into service to command the SPS gimbals.

RADAR, ENGINE AND CREW INTERFACES

Rotational hand controllers (LM only)
In the Lunar Module, the Digital Autopilot maintains attitude control during normal operations. The primary mode, rate command, permits maneuvering similar to controlling an aircraft. Like the control stick of an airplane, moving the hand controller (formally called the Attitude Controller Assembly – ACA) sends a signal to the computer that is proportional to the amount of the controller's deflection. This signal translates to a rotational rate where, the larger the deflection, the higher the rate the DAP commands the LM to maneuver. Each axis of the hand controller sends a signal to a dedicated location in the computer, P-RHCCTR, Q-RHCCTR

and R-RHCCTR.[9] During the autopilot processing cycle, these storage locations are checked and the appropriate thrusters are fired to produce the specified rotational rate. These registers, plus the counters from the IMU are the interface necessary for the computer to implement the "Fly-by-Wire" digital autopilot.

Radar and ranging data
Radars on the Lunar Module and the VHF ranging equipment on the CSM provide essential range and range rate information to the computer for updating targeting solutions or providing altitude and velocity data during a lunar landing. During rendezvous, the Command Module assumes a passive but vital role in tracking the Lunar Module in its return from the surface. Ever ready to perform a rescue should trouble occur, the Command Module Pilot uses the sextant to take regular position fixes on the LM, while the VHF ranging equipment onboard the Command Module provides the distance between the two spacecraft. The Lunar Module has the more complex job of both landing and performing the active role during rendezvous, requiring two separate radars. Mounted underneath the descent stage, the landing radar radiates four signals, a three-beam Doppler velocity system (one for each axis), and a radar altimeter. On top of the ascent stage, the rendezvous radar provides range and range rate data in addition to the orientation of the radar. By knowing the orientation of the radar and the attitude of the LM, the relative position of the Command Module can be calculated.

   All radar ranging and rate data sent to the computer share the same register, RNRAD, and two different mechanisms exist to retrieve data from the radars and ranging equipment. In the Command Module, VHF ranging uses only one data source, the RNRAD register. But what about the Lunar Module, which has six radar data sources? As only one register is available for the several sources of radar data, the solution is to make several requests, one for each data source. First, a regularly scheduled housekeeping program, the Servicer, selects the radar source by setting bits in Output Channel 13 to one of the values shown in Figure 14. Next, bit 4 in Output Channel 13 is set to begin the read process. A few milliseconds later, the selected source places the data in the RNRAD register and generates an interrupt that saves that data for later processing. When controlling the landing radar, this sequence is repeated for all three velocity axes and then for the range data.

INLINK (telemetry uplink)
Data telemetered to the Earth or received as input to the computer may be lumped together as two sides of a complementary topic, but these operations use completely different processes and data formats. The AGC formats and transmits a wide variety of data for review and evaluation on the ground. In this tour of special registers, the focus is on the uplink register, INLINK, which contains the data received by the spacecraft from ground controllers. Ground controllers send data directly to the

---

[9]   To simplify the LM Digital Autopilot, we define the P, Q and R rotational axes as parallel to the X-, Y- and Z-axis, respectively.

a	b	c	Function
0	0	0	-
0	0	1	Rendezvous Radar range
0	1	0	Rendezvous Radar range rate
0	1	1	-
1	0	0	Landing Radar X velocity
1	0	1	Landing Radar Y velocity
1	1	0	Landing Radar Z velocity
1	1	1	Landing Radar range

Figure 14: Codes for radar data source selection

computer on a regular basis, usually to update the state vectors or other navigation data. While an astronaut could perform the update himself, updating the computer remotely eliminates the tedious and error-prone task of copying down a stream of numbers and entering them into the DSKY. The format of the data sent to the computer is the same as used by the DSKY. In effect, controllers are remotely typing data into the computer, using the same key sequences as would an astronaut. Data uplinked to the spacecraft is deposited in the INLINK register and generates the same interrupt as a manual key input.

Engine throttle control
During the lunar landing, the Lunar Module computer is continually calculating the optimal thrust setting for the descent engine. Early in the descent, the objective is to slow the Lunar Module as fast and efficiently as practical. This requires the engine, after a short period of calibration, to fire at full thrust against the velocity vector of the vehicle. As the LM continues the descent, the computer compares its guidance and targeting requirements against its current position and velocity, and modulates the engine thrust to achieve those targets. During the final phase of the landing, the LM is hovering or descending at a slow rate. At the same time, the computer is continuously adjusting the descent engine thrust to account for the mass of fuel burned by the engine. Within a span of 12 minutes, the nominal duration of the powered descent, the AGC incorporates into its logic that the engine is burning off over half the mass in fuel while slowing the vehicle from 5,200 ft/sec (1,600 m/sec) to zero. Guidance targeting, and the design limitations of the engine, also are significant factors in setting the proper engine thrust level. When the AGC finishes computing the necessary throttle setting, the THRUST special register is loaded with a value representing its percentage of rated thrust. The descent engine has a remarkable ability for deep throttling. From 92.5 percent of its rated thrust of 10,500 pounds (46,700 Newtons) down to 10 percent, the useful performance range of the engine is exceptionally wide. Importantly, the AGC must avoid the region between 65 percent and 92.5 percent of thrust its rated which would cause excessive erosion of the engine combustion chamber and nozzle throat.

As the interface between the computer and the engine, the Decent Engine Control

Assembly (DECA) controls the valves that regulate the engine thrust. The DECA reads the contents of the THRUST register and sets the engine thrust accordingly. Using the Thrust/Translational Hand Controller (TTHC), the crew has the ability to augment the computer's commanded thrust. This is strictly a manual override, with the TTHC connected to the DECA directly. The computer does not see any manual changes to the engine settings.

ALTM – analog displays: altimeter and rate meters

When thinking of computer controlled displays in Apollo, it is natural to think of the DSKY and perhaps a few warning and status lights. More surprising is that the computer drives a number of analog displays in the Lunar Module. Despite the precision available in digital displays, humans still prefer an analog representation for observing dynamically changing data. Witness the short-lived fad of digital speedometers in automobiles in the 1990s. Although they accurately described the car's speed, the intuitive understanding of acceleration from watching the continuous motion of a pointer across a dial was lost. Raw digital data, especially data that changes frequently, is ill-suited for human consumption.

In the 1960s, aircraft manufacturers experimented with vertically mounted tapes to represent airspeed, altitude and vertical velocity. One of the legendary aircraft of the era, the F-106 Delta Dart, included such "tapemeters", and pilots were enthusiastic about them. This design carried over to the LM, where tapemeters were used for the altitude and altitude rate displays.[10] Although the tapemeters are a duplication of data displayed on the DSKY during landing and the data is recited verbally by the LMP, the ability to visually assess the landing site and glance at meters conveniently located alongside the window to read the LM's altitude and rates during the descent provides the Commander with important situational awareness. Horizontal velocity during landing is another piece of data that lends itself to graphical representation. At approximately eye level on each man's instrument panel, a square display with two perpendicular bars depicts fore-and-aft and left-and-right motion. This "cross-pointer" is another instrument derived from the astronauts' experience as pilots. Essentially identical to the needles on the aircraft navigation display used for instrument landings, the cross-pointers were a simple and intuitive way of displaying spacecraft motion. Data for the analog displays originates from the IMU and landing radar. Using these sources as inputs, the AGC converts the data into a format suitable for the display, and sends the commands out through the ALTM special register. Like the radar sources that multiplex their data through a single register, analog displays route through bits in Output Channel 14.

---

[10] The Space Shuttle used tapemeters until its conversion to an all-glass cockpit, further confirming their appeal to pilots. What is more, the newer digital displays present the data graphically in the same format at the "old-style" tapemeters.

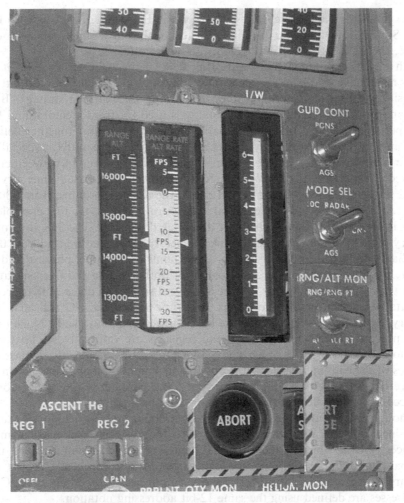

Figure 15: LM tapemeters: altitude and altitude rate

MEMORY ADDRESSING AND BANKING IN THE AGC

Two types of memory are available in the AGC, erasable and fixed. Data can be read from and written to locations in erasable memory, and is analogous to RAM in modern computers. Fixed memory is exactly that – storage whose contents are not changeable. Both memory types use banking to reference their full range of addresses, and three registers are dedicated for this purpose. Two of these, named EBANK and FBANK, perform banking functions for erasable and fixed storage respectively. A third register, BBANK, reflects the contents of both the EBANK and FBANK registers. Finally, to accommodate the ever-growing memory requirements of the AGC beyond the 32K word limit imposed by the FBANK, a departure from the register-based banking scheme introduced the Fixed Extension Bit, also known as the Superbank Bit. Not a banking register at all, but a single bit in an output register, the FEB extended fixed memory to 36K words.

With 12 bits available in the addressing field of the AGC instruction, it might appear reasonable that banking occur on 4K boundaries. Instead, several bits in the addressing field take on multiple roles in defining the banking strategy. Broadly speaking, bits 12 and 11 define how banking is used for fixed memory, and bits 10 and 9 perform the same role with erasable memory. These bits together define the addressing strategy, and directly reference a storage address only in special cases. Storage is implemented as a single address range, with no overlap between erasable and fixed memory. The first 2K area of storage is assigned to erasable memory and is divided into eight 256-word banks that define the addressing range from 00000_8 to 03777_8. As seen in the tour of the special and central registers, the first erasable bank also includes areas for the accumulator and related registers. The 36K area of fixed memory consists of 36 equal-sized banks. Fixed memory starts at address 04000_8, immediately following erasable storage, and extends to the upper physical address of 113777_8. Although banking hardware exists for both erasable and fixed storage, not all storage requires banking for its access. Several situations demand that programs and data be at fixed physical locations. Interrupt processing, I/O registers, restart/recovery, and boot routines all operate at the lowest hardware level. None of these functions can presume that the banking registers are optimally set up for their operation. By defining areas of memory that are addressed directly, bypassing the contents of the banking registers, it is possible to simplify the implementation of the low-level routines. Without the requirement for a banking register, this storage is now accessible regardless of the erasable or fixed banking registers.

Although much of this discussion uses the instruction address field for its examples, addressing is not an issue that is exclusive to the basic instructions. All memory references, whether through the program counter or an address in memory, use the same banking techniques to reference instructions or data in storage. As the 16 bits necessary to access all 38K words of memory cannot be squeezed into an AGC word, all addresses are defined using the same 12-bit addressing notation.

Erasable storage banking
The 11 bits required to directly address the 2,048 words of erasable storage would fit in the addressing field in the AGC instruction. However, it was decided for erasable storage (and later also for fixed storage) to interpret the addressing field not as a memory address but as banking indicator bits and offset into the bank. In creating a banking strategy for erasable storage, only 10 bits of the addressing field in the instruction word are needed. Erasable storage is divided into 8 banks of 256 words, with each bank requiring 8 addressing bits to reference an individual word. Bits 8 through 1 of the instruction word are used to access the word within the bank. Two additional bits are necessary to indicate whether banking is used to access storage, or if the reference is to a common area of erasable storage. Bits 10 and 9 of the instruction word perform this purpose. If these bits contain a binary 00, 01 or 10, the EBANK register is not used and all 10 bits in the addressing field are used to directly

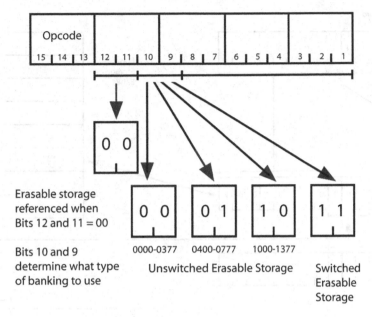

Figure 16: Erasable storage and bits 10 and 9

reference the location in erasable storage. When not using the EBANK register, this area of storage is referred to as "unswitched-erasable".[11] Only when bits 10 and 9 have the value of binary 11 is banking used. In this final case, the erasable storage address is calculated by concatenating the contents of the EBANK register with the 8 address bits from the instruction word, as shown in Figure 16. In all cases where erasable storage is used, bits 12 and 11 of the instruction address field contain binary 00, indicating that the reference is for erasable storage.

Fixed storage banking

A technique similar to that used for erasable storage banking is also the basis for fixed storage banking. Bits 12 and 11 of the instruction word specify how banking is used for fixed memory, and of the four possible values for these two bits, three define the banking for fixed storage. One value has already been mentioned: a binary 00 in bits 12 and 11 indicates that an erasable storage location is referenced. When the binary value 10 or 11 is in these two bits, the entire 12 bit

---

[11]   Confusingly called "Fixed-Erasable" in the literature, we will avoid this use to be consistent with our erasable memory nomenclature. Assigning a descriptive name to the various memory areas often results in a maddening complexity of terms. Here, the first three banks of erasable memory are often referred to as "Fixed-Erasable", and the first two banks of fixed storage are "Fixed-Fixed". Using the prefix "Switched" or "Unswitched" is also historically accurate and defines whether a banking register is used. Our use of these terms is both technically correct and helps avoid confusion.

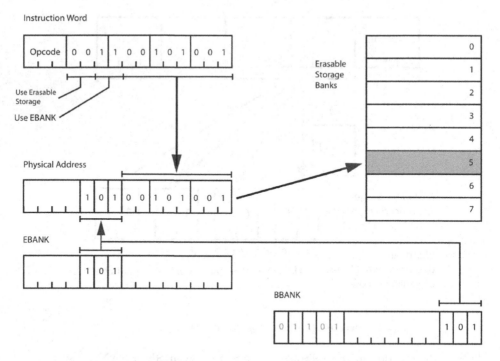

Figure 17: Erasable storage banking

address field is used to reference the "unswitched-fixed" storage, the region of fixed storage where the FBANK register is not used. Only when a binary 01 exists in bits 12 and 11 are the contents of the FBANK register appended to the lower ten bits of the address. As seen with erasable storage, banking in fixed storage allows a limited amount of overlap of the switched- and unswitched-fixed storage. Where bits 12 and 11 are binary 10, indicating that the FBANK register is necessary, and the contents of FBANK are 00 or 01, the address overlaps with the unswitched-fixed storage area.

Common banks

A useful byproduct of the banking implementation is called common or unswitched memory, which defines areas of storage that are accessible without using banking registers. Banking in the AGC allows for only one fixed and one erasable bank at a time, which limits the amount of memory that a program can access. When including the common fixed and erasable storage banks, the addressable storage available at any time increases to three fixed and four erasable banks. Figure 20 presents a view of memory from the perspective of an executable program. Unswitched banks are necessary when a program or data must be accessible to all programs. Operating system routines, especially those that manage jobs, interrupts, and software recovery, are ideally suited for unswitched storage. Locating critical data in a common area of memory allows that data to be read or updated without requiring the program to change its banking environment. For example, flagwords are several

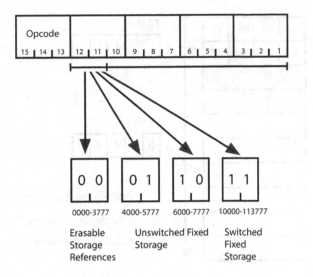

Figure 18: Fixed storage and bits 12 and 11

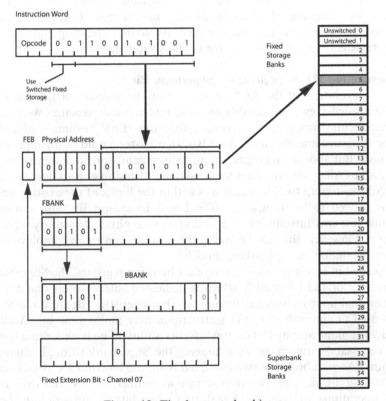

Figure 19: Fixed storage banking

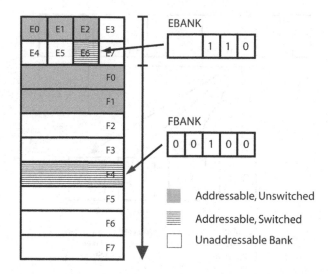

Figure 20: View of both erasable and fixed storage

words in unswitched storage whose bits define various states in the system, and are referenced by programs that reside throughout storage. Locating flagwords in unswitched memory eliminates the overhead of saving, updating and reloading the banking registers every time one of their bits is referenced.

Going beyond the 32K word limit: the Superbank Bit

During the evolution of the AGC, estimates of the amount of fixed storage that would be needed grew considerably as more and more capabilities were added. As the requirements firmed up, it appeared that the "final" estimate of 24K words would be sufficient and that a five bit FBANK register would be an appropriate size for this task. But additional programming requirements forced designers to scramble for more addressing bits even after the fundamental addressing architecture had been frozen. Addressing up to 32K words was within the limits of the existing design, but even this proved to be insufficient. The Fixed Extension Bit, also known as the Superbank Bit, was introduced, and limitations on physical memory were relieved once again. However, this relief came at the cost of even more complexity as it is a significant deviation in the banking model.

Not located in low memory like the other banking registers, the Superbank Bit is found in bit 7 of I/O Channel 7, which is dedicated solely to this one bit. Such an unconventional approach makes for unique characteristics. Because the Superbank Bit uses an I/O channel, only I/O instructions may modify the bit. Additionally, Channel 7 is unique among I/O channels for its ability to be both written to and read from; a necessary requirement for managing the Superbank through interrupts and dispatching. As with both erasable and fixed banking, bits that are already employed for addressing are now pressed into service as indicators to signal that additional banking operations are required. Recall that when bits 12 through 9 in the address field of the instruction are set to binary 0011, the operand resides in erasable storage with its address computed using the EBANK register. A similar decision for fixed storage banking results when bits 12 and 11 of the address are set to 01. Addressing fixed storage using the Superbank Bit occurs when the two uppermost bits of the FBANK register are set to binary 11. The storage address is appended to the Superbank Bit, giving a full 16-bit address. Introducing the Superbank Bit affects eight fixed storage banks, not just the four banks at the high end of memory. When the Superbank Bit is set to zero, the fixed storage banks from 0 to 31 are accessible. The extension to the addressing range occurs when the Superbank Bit is set to one, making fixed storage banks 0 to 27, and 32 through 35 available to the program.

   All of this sounds unnecessarily complicated. Why not simply extend the FBANK register? While this may seem the reasonable solution, at this point the growth in memory had met the fundamental limits in the AGC. With the exception of the accumulator, all of the registers and memory locations in the AGC have a fixed size of 15 bits. The five bits in the FBANK plus the ten bits from the addressing field yields the 15-bit limit for addresses. Quite simply, the architecture cannot support a larger addressing range without extensive redesign. Arguably, the introduction of the Superbank Bit was the best available option for extending the address space. The final result is that three banking "registers" are necessary to address the 38K words of storage; the EBANK, FBANK (and by extension the BBANK) and the Superbank Bit. The cost of managing this architecture was more than a high level of complexity. A non-trivial amount of storage was necessary to manage the banking overhead in the form of both instructions and data, plus the need for additional processing cycles. Given both limited storage and CPU resources, banking is a costly but essential function.

Interbank transfer of control
Notice the order of the FBANK, Z and BB registers in lower memory; the FBANK register precedes the Z register, and the BB register follows the Z register. At first glance, one might wonder why they are placed in this sequence rather than being grouped logically together. This is intentional, and creates a powerful mechanism for transferring control. When switching to another bank, or setting up execution with a new fixed and erasable bank, it is evident that new values should be placed in the FBANK, EBANK or BBANK registers. However, a problem arises. Assume a program is executing at location 01033_8 within fixed bank 07, and needs to transfer to location 02371_8 in fixed bank 13. Changing the Z register first will transfer control to location 02371_8 in the current bank, which now does not have any means to switch to fixed bank 13. However, changing the bank number to 13 first creates a similar problem. Execution would continue at location 01034_8 in the new bank (since Z would have incremented) as soon as the bank register is loaded. This is not a very practical way to move about memory. Clearly, simultaneously changing both the Z register and the banking registers is necessary for any transfer to code in another bank.

   With the Double Exchange (DXCH) instruction, bank switching and interbank transfers become practical. The DXCH instruction takes the accumulator and L register, and exchanges them for the contents of two consecutive locations in storage.

DXCH finds uses throughout AGC programming, but has a special importance for interbank transfers of control. In two cases the accumulator and L register exchange their contents with the Z register and either the FBANK or the BBANK. In both cases the same instruction is used, but the banking registers are different. As a notational convenience to the programmer, two mnemonics assist in coding these instructions: Double Transfer Control Switching Both Banks (DTCB) and Double Transfer Control Switching Fixed Bank (DTCF).

In the first case, DTCB provides the ability to transfer into a new bank while at the same time setting both the erasable and fixed bank registers. Remembering that the BBANK register holds both fixed and erasable bank values, we are able to change both bank registers with a single instruction. When DTCB executes, the accumulator and L register are loaded with the new 12-bit address and a new image of the BBANK register. The address is interpreted in the same manner as the instruction operand address, using bits 12 through 9 to establish whether to use fixed or erasable banking. Similarly, a program may want to transfer to a location within a new fixed bank, all the while keeping the current erasable bank. In this case, the new contents of the FBANK register are placed in the accumulator, and the L register is loaded with the 12-bit address within the new bank. A DTCF replaces the contents of the FBANK with that of the accumulator, and the Z register is updated from the L register.

When the DTCB or DTCF instruction completes its execution, everything is in place to begin execution at a new location using the new set of fixed and erasable banks of storage. The added benefit is that it also provides the means to return to the calling program. As the DXCH instruction exchanges the two locations specified, rather than simply replacing one with another, the original values of the Z and BBANK (or FBANK) registers can be stored in the accumulator and L register! The contents of the accumulator and L register are now invaluable, as they are the only means of knowing where the subroutine should return once it has completed. The called subroutine returns to the calling program by reversing the calling sequence. The accumulator and L register are reloaded with the data it had on entry to the routine, and a DTCB or DTCF is executed to return.

Surprising omissions in the AGC architecture
In this review of the basic and special registers, it is surprising to discover what does not exist. In most computer architectures, there is the expectation of a stack pointer and at least one index register. This is not the case in the AGC, where even the earliest designs of the computer did not include these features. Using stacks for passing parameters and saving the return address requires additional hardware and memory to be implemented effectively. Given the simplicity and constrained resources of the original design of the AGC, with eight instructions and 4K words of storage, stacks were not an optimal use of hardware. In general, managing return addresses without a stack pointer or other linkage mechanism is not particularly difficult in a small and resource constrained system. However, as system requirements grow, especially in the environment of a real-time processor like the AGC, the lack of a stack pointer seriously complicates the software design problem.

Through careful and creative coding, programmers limited the number of calls to subroutines and used global variables for parameters. Effectively, this bypassed the need for stack-based data management.

Computer programmers use index registers to ease the task of referencing data organized tables, lists or more complex datatypes such as matrices. The most common example is that of a table of numbers in a spreadsheet consisting of several rows and columns. All data is stored in the computer's memory in a linear fashion, creating a simple mechanism to reference a specific row in the spreadsheet. By knowing how much storage is used for each cell in the spreadsheet, and the number of columns, a straightforward calculation can direct the program to the exact memory location of a specific cell. Placed in an index register, the spreadsheet program now has a simple and effective way to access any cell in the table. Although an index register is not part of the AGC's basic architecture, tables and other complex datatypes abound in the software and the AGC must process them without unnecessary overhead. In the instruction set discussion, the elegance and versatility of the INDEX instruction effectively eliminates the need for an index register.

One of the most notable capabilities in the AGC is that it uses a priority scheduled multiprogramming operating system called the Executive. While a thorough discussion of the Executive is deferred to a later section, one function is decidedly missing from the hardware. Multiprogramming operating systems are successful because of their ability to share the various resources in a computer with several programs. If two programs are trying to update the same piece of data, chaos results if each program assumes (reasonably) that its update is valid, whereas in reality, another program has changed it. For example, imagine that two clerks are unknowingly trying to update a personnel record of an employee who has just married and has moved to a new house. Jane Smith, originally living at 42 Main Street, is now living in marital bliss as Jane Brown at 39 Center Street. Without a means to prevent one clerk from modifying the personnel record while the other is already in the process of a update, Jane Smith could be recorded as living at 39 Center Street. A technique called locking or serialization is absolutely essential to assure the integrity of the data in multiprogramming computer systems. In this example, the program would first "lock" the employee record, then allow the first clerk to make the update, and finally unlock the record. The second clerk is not able to make any changes until the first clerk is finished, and then would notice that Jane Brown's record is already changed.

Instructions to perform locking are complex. The problem they address is the case where a program is in the middle of reading or updating a section of data,[12] a process that requires many steps. During this time, an interrupt can occur and the Executive can pass control to a higher priority program. This program could then update the same area in memory and finish processing, allowing the first program to resume execution. Unaware that the data has been changed, the first program continues

---

[12] Although memory is used as the example of a resource, many all other computer components must also be serialized, for example devices such as the DSKY or the IMU.

blithely on, producing unpredictable results. A solution might be to set aside a word in storage, often called a lockword, that is set to a particular value if a resource is in use. If it is in use, the program must wait, and check again later. Upon finishing the update, the program must clear the lock, allowing the next program to access the resource. Unfortunately, the process of testing, updating and clearing the lockword requires several instructions, any one of which can be interrupted, returning us to the original problem. The proper solution is to implement the entire locking process as a single instruction that is not interruptible. A single instruction will either recognize that the resource is available, and lock it for its own use, or that it is busy and the program should wait. Without such an instruction in the AGC, it is necessary to write the various programs to ensure that they do not attempt to use the same resource simultaneously. Bypassing the problem was possible only by very carefully testing the software, and establishing strict rules on the sequence in which programs could execute.

While index registers and stacks are not part of the basic AGC architecture, they are included in a powerful software component called the Interpreter. By creating a radically different and independent system architecture "on top of" the AGC, the Interpreter overcame a number of the limitations found in its host's design.

INTERRUPT PROCESSING

As a program is executing, it is blissfully unaware of events that may demand the immediate attention of the system. Timers and other I/O devices operate independently of whatever program is executing, and there is no facility to coordinate their execution with other software in the computer. Raising an interrupt abruptly halts the execution of the currently executing program and transfers control to the address assigned by the hardware. Since an interrupt from a timer requires different processing than that from a keyboard interrupt, a means to differentiate between them is also necessary. By assigning a unique address in storage to each device or timer that can generate an interrupt, the resource that requires attention is identified. For example, powering up the AGC or performing a fresh start causes an immediate transfer of control to location 04000_8, while pressing a key on the DSKY routes control to location 04024_8. Note that these interrupts are very different from counter interrupts produced by the IMU, radar and optics. Counters do not perform a context switch (which suspends a program and permits executing an interrupt routine) and do not require saving and restoring registers. The types of interrupts discussed here require logic that is more complex than counter increments or decrements, and frequently involve scheduling new work in the system. During the processing of an interrupt, the hardware prevents any other interrupts taking control of the system, and devices or timers that need attention must wait until the current interrupt has completed.

A program running at the time of the interrupt can never know when, and for how long its suspension will last. When an interrupt routine completes, no trace of its work can be visible to the resumed program. From the perspective of the interrupted program, absolutely nothing has occurred. To create this magic, the interrupt routine must save all of the registers it will modify, and restore them before returning control to the interrupted program. Recalling from the tour of lower memory, several locations are set aside for storing registers that may change in the course of interrupt processing. Immediately upon receiving the interrupt, the hardware saves the program counter of the interrupted process and the next instruction to execute in the ZRUPT and BRUPT locations. As the interrupt routines are located in storage areas outside of the common fixed banks (where the interrupt vectors are located), the first task in interrupt processing is to save the BBANK register. Then an interbank transfer is used to call the program performing the actual work of the interrupt. To limit the amount of time spent with interrupts inhibited, there is usually very little code in the interrupt routine. More extensive processing, if necessary, requires scheduling a job to continue processing outside of the interrupt.

Often there are times when a program cannot tolerate an interrupt. For example, fetching data from the IMU often requires reading all three accelerometers in quick succession, otherwise navigation calculations may not be valid. Tasks requiring critical timing of events or updating sensitive data areas must inhibit interrupts using the INHINT instruction. Overflow in the accumulator is another case where an interrupt is not acceptable. Without the ability to save all 16 bits of the accumulator, an interrupt could clear the overflow condition and upon returning to the interrupted task the value in the accumulator would no longer be the same, violating the fundamental assertion that interrupts are invisible to other programs. Inhibiting interrupt processing when an overflow exists in the accumulator guarantees that the accumulator's state is preserved.

Although the AGC's non-addressable 'internal' registers were said to be beyond the scope of this book, the fact that the B register is involved in interrupt processing requires that we make an exception. When following the internal execution of an instruction, the first step is to load the instruction into the CPU where it is ready for decoding. This decoding process takes place in an internal register called the B register, which is inaccessible to the programmer and whose particular attributes and operations are normally unimportant. During the execution of any instruction, the AGC is fetching the next instruction pointed to by the program counter (the Z register). This instruction is stored in the B register, and is ready for execution as soon as the currently executing instruction finishes. If an interrupt were to occur at this point, the B register is replaced with the instruction located at the beginning of the interrupt processing code, and execution continues from that point. When interrupt processing is finished, the AGC needs to return to the interrupted program and resume execution as if nothing has happened. While losing the contents of the B register does not affect the execution of the interrupt code, its previous contents are necessary when resuming a program after interrupt processing has completed. When the RESUME instruction is executed to end the interrupt processing, the Z register is reloaded from the ZRUPT storage location. The Z register will reference the next instruction fetched from memory, and the first instruction to be executed after resuming from the interrupt is taken from the BRUPT location.

Interrupt vectors

Appendix B lists the memory addresses reserved for interrupts in the AGC. Most are associated with timers or data transfer, but what is particularly interesting are the interrupts that do *not* exist. A quick inspection reveals that interrupt vectors for common errors are missing. Invalid operation codes or invalid address are common programming errors when writing software in assembly language, but no such interrupt vectors exist to trap them. The reason for this omission is as simple as it is surprising. The instruction set of the AGC is so dense that each opcode and quarter code exists as a legitimate instruction; no illegal operations are possible. Addressing errors are equally impossible, as all combinations of address fields and bank register values are also valid. One condition that does demand an interrupt (yet does not exist) is an attempt to divide by zero. We can escape the need for such a trap by requiring that the divisor be larger than the dividend, as otherwise the result is undefined. With no indication that a division result might be amiss, an assertion that an instruction leaves an undefined mathematical result is particularly unsatisfying. Regardless, the task of ensuring that division receives only valid inputs is solely the programmer's responsibility.

THE INSTRUCTION SET

Instructions as tools in the hands of an artist

Painters, regardless of their style or skill, begin with basic tools. Canvas, paint and brushes, elements common to them all, are readily available at art supply stores. Purists may squabble about the comparative merits of one brand over the other but, in the end, the tools used to create the Mona Lisa would not be out of place in the studio of a 21st century modernist. It is now time move to a studio of a different sort. Like artists, computer programmers use essentially identical tools to create vastly different works. Whether the finished product is for payroll processing, a cardiac pacemaker or a weapon system, the constituent tools of software developers are interchangeable. So, what is the point of this philosophical perspective?

Much ink is spilt by purists in debating the relative merits of one machine architecture over another, especially in the instruction set. In this context, the AGC can arguably be criticized as having an inefficient and overly complex instruction set; the result of too many attempts at extending a limited architecture. This may be true. It is also irrelevant.

All computers that meet a basic design criterion are functionally equivalent to each other.[13] Memory and speed requirements of different designs may vary considerably, but no design holds an exclusive advantage over another. Yes, a "clean sheet of paper" would create a more workable and elegant AGC, but at the expense of discarding huge amounts of existing work in designing, building and testing hardware and software. Computer development was becoming a pacing issue in

---

[13] Known as a "von Neumann" architecture, it now dominates computer design.

Apollo, and there was not the luxury of time to create a new design that was optimized to each new requirement. In the end, Apollo engineers and programmers adapted to new iterations of the AGC architecture, preserving their existing work as much as possible and then exploiting the new features to expand on it.

Understanding the complexity of the AGC begins with the instruction set, where most of the compromises are apparent. In the vernacular of the day, the first three bits were known as the "order code" (today, this field is universally referred to as the "operation code", or truncated to the "opcode"). Only eight codes are possible, numbered from 0_8 to 7_8. Such a small number of instructions make it difficult for programmers to develop efficient software. In such a constrained machine, several simple instructions are needed to perform the task of a single complex instruction, and they impose a cost of memory and execution time. Increasing the size of the opcode field is necessary to grow the number of instructions, and the effect of just one additional bit is profound: each bit added to the opcode doubles the number of possible instructions. In fact, the AGC designers managed to devise ways of extending the instruction set whilst living within the limitations of the existing hardware.

Creating additional instructions: quarter codes, special cases and EXTEND
Keenly observing the type of memory referenced by a given instruction reveals a novel method to "find" a few extra bits, thus extending the opcode. Specifically, consider the Transfer to Storage (TS) instruction, which saves the contents of the accumulator in the memory location referenced by the address field. So far, there is nothing exceptional with this instruction, and the same observation is possible with the Add to Storage (ADS), Increment (INCR) and Exchange (XCH) instructions. Yet, a pattern begins to emerge. Unsurprisingly, all of these instructions modify a location in erasable memory. Recall from the section on memory management that a reference to erasable memory requires zeros in bits 12 and 11 of the address field. From that earlier discussion, the expectation is that the contents of bits 12 and 11 singularly determine whether the instruction will reference fixed or erasable memory. Exploiting this important fact permits the introduction of a new instruction format that anticipates the type of memory it will access. By reinterpreting bits 12 and 11 as an extension to the opcode, a new field known as the "quarter code" is introduced. When the two bits of the quarter code are appended to the three opcode bits, the number of machine instructions increases considerably.

Quarter codes do not exist for every AGC instruction. An opcode of 000 corresponds to a large number of special operations, and the binary opcodes 011, 100 and 111 each implement a single instruction. The four remaining binary opcodes (001, 010, 101 and 110) use quarter codes to varying degrees. In most cases, the instructions created by concatenating the opcode and quarter code reference erasable, not fixed memory. With an addressing field reduced by two bits, only 10 bits remain, limiting references to only erasable memory. As quarter codes exploit the knowledge that the instruction only references erasable storage, the requirement that bits 12 and 11 equal zero is now redundant. Not all instructions that use quarter codes reference erasable storage exclusively. For example, instructions that use an

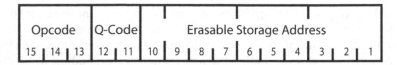

Figure 21: Instruction format including the quarter code

opcode of 001 harken back to the traditional memory addressing scheme. Here, an instruction that has an opcode 001 and a quarter code of 00 (the CCS instruction) uses an erasable storage location as its operand. The Transfer Control to Fixed (TCF) instruction shares the same opcode, but uses 01, 10 and 11 for the quarter code. As expected, the TCF instruction operates only with fixed addresses. Inspecting the quarter code for TCF, it is apparent that the bits correspond to the expected values for referencing fixed storage.

Clearly, the need to expand the instruction set has created an elegant solution that exploits prior knowledge of where an instruction will reference its data. But what about instructions that must reference erasable and fixed memory? Is it reasonable to assign quarter codes to them? Consider the instruction Clear and Add (CA). It does not modify erasable memory, and its operand can be anywhere in fixed or erasable storage. The broad range of addresses needed for the instruction makes it necessary to use all 12 bits, prohibiting the use of quarter codes.

In some computer architectures, the number of possible operation codes exceeds the number that are actually implemented. This usually arises because system architects wish to limit the cost of complexity, or to reserve capabilities for future version of the hardware. Attempting to execute an unimplemented instruction code invariably is an unrecoverable error and causes the program to terminate. Properly handling an unimplemented operation introduces additional complexity as hardware and software are necessary to trap and recover from the error. The AGC does not have this problem, as the engineers implemented every possible opcode and quarter code. However, specific cases still exist in which an opcode plus certain operands create completely undefined results. The principal example is in the Transfer Control (TC) instruction. Used to set up the return address and branch to a subroutine, TC reasonably expects the target of the branch to be a valid instruction. At the same time, TC can branch anywhere in the central and special register area of lower memory, including locations that make no sense as a branch destination. Location 00003_8, for instance, contains the Erasable Bank Register. As the EBANK register uses only three bits, it cannot represent any valid instruction or data. To transfer control of the program to any of the other banking registers would also give rise to unpredictable results. In a classic example of "making lemonade when all you have are lemons", the AGC turns instructions such as a TC to the EBANK register into useful operations. Attempts to transfer to locations 00003_8, 00004_8, and 00006_8 (the EBANK, FBANK, and BBANK special registers) are reimplemented as special cases of the TC instruction which create completely new operations. For example, a TC to the EBANK now becomes an "Enable Interrupts" instruction. Details on these and other "implied address" instructions are discussed in a later section.

While quarter codes are a clever means to implement additional instructions, several essential capabilities are still missing. It is possible to use software routines in place of new instructions,[14] but such routines are quite slow in comparison to the hardware. With no remaining places available from which to steal bits, the AGC designers had to resort to a brute-force technique to extend the instruction set, using a new instruction, appropriately named EXTEND. As a TC to location 00006_8, it does not perform any operation per se; rather, it directs the AGC to execute the next instruction in storage using an entirely different set of opcodes, effectively creating a second instruction set. Like "normal" instructions, "extended" instructions provide for quarter codes and special cases of a single opcode. As expected, creating a second instruction set is not without additional cost. Executing an extended instruction involves additional processing overhead because two instructions, the EXTEND and the "extended" instruction itself, are executed. To limit the impact of the overhead, opcodes are distributed between the two types of instructions according to their frequency of use. The most common instructions occupy the "normal" instruction pool, and the remainder of the instructions are "extended". Thus, the occurrence of extended instructions is already small, and so the additional processor time and memory resources used are limited as much as possible.

The basic AGC instructions
Fundamentally, the instruction set defines the architecture of a computer. The AGC implements a collection of 41 distinct instructions, using a variety of techniques to stretch the limited instruction format. Quarter codes, extended instruction sets and special cases of existing instructions all are used to create the basic, yet fully capable repertoire of the AGC. Many instructions have similar characteristics, so it is best to break them into groups and discuss the operations within that group. Of course, the AGC has several unique instructions that are worthy of a comprehensive discussion, and they will be explored in more detail. The six groups of instructions are:

Arithmetic and logic
Sequence changing
Moving data
Instruction modification
I/O instructions
Miscellaneous

Arithmetic and logic
AD   Add
AUG   Augment
ADS   Add to Storage

---

[14] The need to extend both the instruction set and the entire hardware architecture resulted in the Interpreter. Discussed with the Executive, the Interpreter presented a powerful "virtual machine" with an architecture that was radically different from the native AGC.

DAS	Double Add to Storage
DIM	Diminish
DV	Divide
INCR	Increment
MASK	Mask A by K
MP	Multiply
MSU	Modular Subtract
SU	Subtract

Increment, Augment and Diminish

Increment (INCR), Augment (AUG) and Diminish (DIM) are related instructions that are useful for counting, and are the closest the AGC has to an immediate instruction. As the accumulator and other special registers are mapped to lower memory, INCR, AUG and DIM can operate on these registers as they do with other erasable storage locations. INCR simply adds 1 to the value referenced in storage by the address field. AUG and DIM are similar to INCR, but operate on the absolute value of the operand. Executing an AUG against a positive number such as 4 will result in 5, but if it is -4 then AUG will change its value to -5. Likewise, DIM takes a value such as 4 and reduces its value to 3, and starting with -4 in storage, it changes the number to -3. Accumulator overflows are possible with these three instructions, and interrupts are inhibited until the overflow condition is cleared.

Add and Subtract

Add (AD) and Subtract (SU) operate on the contents of the accumulator using the rules for modified one's complement. Overflows and underflows are possible with these arithmetic operations, so the programmer must be especially mindful of the scaling used in the data.

Add to Storage, Double Add to Storage

Only a few instructions in the AGC operate directly on data in memory. Although the accumulator can perform all necessary arithmetic operations, it is frequently more efficient to perform these operations directly on data in storage, as often an arithmetic operation is followed by another instruction to save the result in storage. Combining this common sequence into a single instruction produces benefits in processing time and storage requirements. Add to Storage (ADS) and Double Add to Storage (DAS) eliminate the need to code an additional instruction to save the contents of the accumulator (and the L register, when executing a DAS instruction).

While it might appear that the two instructions would operate in a similar manner, this is not the case. Both instructions begin with taking the contents of the storage referenced by the operand and adding it to the accumulator, and in the case of DAS the L register is added to the word following the operand specified in the instruction. ADS copies the result back to storage, replacing the original value. After DAS adds the contents of the two words referenced by the instruction to the accumulator and L register, it also places the result back into storage. However, their overflow processing differs. ADS retains the result in the accumulator, so normal overflow processing occurs, including the storing of an overflow corrected result back into memory. DAS sets the accumulator to 0 if there is no overflow, and to $+1$ or -1 for positive or negative overflow respectively; and clears the L register in all cases.

Multiply and Divide

Moving past simple addition and subtraction, it is now time to look at the multiplication (MP) and division (DV) instructions. Special challenges exist for multiplication operations, as the range of possible data values can be quite large and proper scaling is especially important. Unlike the instructions that perform addition and subtraction, multiplication of one fraction by another creates a number smaller in magnitude. To help limit the loss of accuracy during multiplication, the result is always double precision, with the most significant word in the accumulator and the least significant bits in the L register.[15] As both of the input values were single precision and the result is likely to be single precision as well, there is a loss of precision when saving only the contents of the accumulator. Such loss of precision may not be acceptable, and rescaling the result might be necessary to recover some of the bits now stored in the L register. The original value of the multiplicand is lost after the instruction executes. There is one small consolation for the extra steps that might be involved in recovering some of the precision after a multiply: because the result is always a smaller fractional number, there is no possibility of an overflow.

In some respects, division operates in the opposite manner as multiplication. Because division requires two words to hold the quotient and remainder, DV starts with a double precision number as the dividend. The divisor is referenced by the DV instruction and is always a single precision value. Like multiplication, the values in the accumulator and L register are lost after the operation, with the quotient replacing the most significant bits of the dividend in the accumulator. The remainder overwrites lower order bits of the dividend in the L register. Overflows can occur in any case where a dividend is smaller than the divisor, such as in the case of 7 divided by 2. There are few other places in the AGC where the need to carefully manage data scaling is more apparent than in division, as an overflow error from division is undetectable. Preventing an overflow is assured only when the dividend is a larger number than the divisor contained in the accumulator and L register. But the hardware does not enforce this requirement, leaving the programmer to assume the burden of assuring the data is valid for this operation.

---

[15] Storing a doubleword with the most significant bits in the lower storage address, followed by the least significant bits in the next higher storage location is called "big endian" ordering.

Modular Subtract

Up to this point, all of the discussions of data representations said that the one's complement number system is used throughout the AGC. The major drawback of one's complement is distinct "positive" and "negative" zeros, which is inelegant and requires extra programming to accommodate the two values. While this problem may be manageable from a software developer's perspective, the fact remains that there is no mechanical analogue for multiple encodings of zero.[16] Counting using one's complement uncovers a special problem. Specifically, the gimbals in the IMU contain sensors called angular resolvers that send counter-type interrupts to the AGC. Each interrupt corresponds to a positive or negative rotational movement of the gimbal. Using the one's complement system for tracking the movement of the gimbal fails in one critical case: defining when the gimbals have reached exactly zero degrees since the next interrupt will merely flip the sign of the zero. Remember that the counters do not receive data containing absolute position or velocity data, rather they simply count up or down with each increment corresponding to a small relative movement. But with two values between +1 and -1, how can zero be uniquely defined? Each counter pulse from the gimbals is now defined to represent 1/32,768 of a complete revolution of the gimbal, with the values of the counter ranging from zero to 32,767. This is easily implemented as an unsigned 15-bit two's complement number. A dilemma occurs when we want to calculate the change in angular rotation of a gimbal. A simple subtraction of the past and current counter values appears straightforward, but a conversion from two's complement to one's complement is necessary for computations in the AGC. Simplifying this process is the Modular Subtract (MSU) instruction. It takes an unsigned two's complement number from storage and subtracts it from another unsigned two's complement value (presumably a gimbal counter) in the accumulator. MSU converts the resulting value to one's complement, making it usable for computations throughout the AGC.

MASK

As the AGC word is not divisible into smaller entities such as bytes, there is the need for instructions to manipulate individual bits. The logical operations AND, OR and Exclusive OR begin with a word in storage or the accumulator, and combine it with another word according to the rules of binary logic. Figure 22 shows the operations that are possible with these functions. A fourth operation is a logical NOT. Most commonly, logical ORs set bits, ANDs reset or mask out bits, and Exclusive OR (or XOR) and logical NOT flip bit values. NOT is unique among logical operations by being a unary operator that does not need a second operand to negate a bit.

---

[16] Engineers have long been acquainted with values that have no physical counterpart. For example, the imaginary number i, defined as the square root of -1, is used extensively in electrical engineering.

AND	0	1		OR	0	1		XOR	0	1
0	0	0		0	0	1		0	0	1
1	0	1		1	1	1		1	1	0

Figure 22: AND, OR and XOR logical operations

The only logical operator available in the basic AGC instruction set is the logical AND operation, called the Mask (MASK) instruction. "Masking" of bits within a word is a useful way of visualizing how a logical AND works, and is a very appropriate name for this instruction. The AND truth table shows the result of ANDing a bit in the accumulator with a corresponding bit in the operand. Only in the case where both that bit in the accumulator and its counterpart in storage are 1 will the bit remain set in the accumulator. In systems with limited memory, such as the AGC, it is wasteful to dedicate an entire word to indicating a particular state of the system. It is a far more efficient use of storage to dedicate only a single bit within a word to indicate such status information. The remaining bits in the word can then hold other information on system states or conditions. This is known as a flagword.

Because the AGC lacks the capability to address bits directly, the desired bits must be extracted from the word and isolated so they can be tested or manipulated. For this, the MASK instruction is used to isolate the bits within the accumulator. Consider a flagword in the accumulator, and that bits 7, 6 and 5 are of interest. A masking word is defined in storage containing ones in bits 7, 6 and 5, and zeros in all other bits. MASK performs a logical AND of the accumulator and mask word in storage, and places the result in the accumulator.

Accumulator	010 110 110 110 000
Mask word (in storage)	000 000 001 110 000
Result	000 000 000 110 000

All bits in the word are cleared to zero, and only the values of bits 7, 6 and 5 remain unchanged in the accumulator. These bits are now available for testing by the programmer. Given the limited instruction set, the choice of AND as the sole logical operator is deliberate, because the other logical operators, OR, NOT and XOR, are possible using other AGC instructions and are described later.

Moving data
CA	Clear and Add
CS	Clear and Subtract
DCA	Double Clear and Add
DCS	Double Clear and Subtract
DXCH	Double Exchange A and K
LXCH	Exchange L and K
QXCH	Exchange Q and K
TS	Transfer to Storage
XCH	Exchange A and K

Clear and Add, and Clear and Subtract and their Double counterparts
While Clear and Add (CA) and Clear and Subtract (CS) are more closely associated with arithmetic operations, the names are somewhat misleading. Both instructions clear the accumulator (including the S_2 bit) before any addition or subtraction occurs, effectively removing any overflow indication. A more proper description of these two instructions could be Load and Load Complement. Double Clear and Add (DCA) and Double Clear and Subtract (DCS) are very similar to the CA and CS instructions, but operate on two adjacent words at once. Data pointed to by the instruction address defined as K is loaded into the accumulator, and the L register is loaded with the data at K + 1. Like the CA and CS instructions, a more descriptive name for these two instructions might be Load Double and Load Complement Double. Because of the clearing of the accumulator (and the L register in the case of DCA and DCS), these instructions are *not* double precision addition or subtraction operations.

By their names, Clear and Subtract and Double Clear and Subtract operate as data movement or arithmetic instructions. However, subtracting a one's complement number from zero is the same as taking the logical NOT of the word. When combined with the MASK instruction (a logical AND), it is possible to implement the OR and XOR logical operations.

Transfer to Storage
With so much processing occurring in the special registers, eventually the data must be moved to storage. The simplest of these, Transfer to Storage (TS) copies the contents of the accumulator to erasable memory. In the absence of an overflow in the accumulator, there is nothing else for the TS instruction to do. However, what if an overflow exists? An overflowed accumulator not only contains invalid data, it also inhibits interrupts in the computer. The TS instruction provides two different approaches to handling accumulator overflows. In the first case, the TS instruction saves the accumulator in erasable storage. An "overflow corrected" result consisting of the S_2 bit and bits 14 through 1 of the accumulator is saved in storage, thereby preserving the data's original (pre-overflow) sign. After saving the data, the accumulator contents are replaced with a +1 or -1, depending on whether a positive or negative overflow was present. Replacing the contents of the accumulator with ±1 also resets the overflow indication, thereby allowing the AGC to process interrupts again. Finally, the instruction immediately following the TS instruction is skipped, and control transfers to the location immediately following the skipped instruction. Skipping the next instruction is exploited further in the second case of TS overflow processing. When the operand of the TS instruction references the accumulator rather than erasable storage, the contents of the accumulator are preserved. The overflow condition remains, and interrupts remain inhibited. Skipping the next instruction is still performed, and transforms the TS instruction into an overflow testing instruction. Given the name of Overflow Skip (OVSK), this special case of TS creates important functionality without the expense of dedicating a new opcode.

```
              TS    TEMP                      SAVE ACCUMULATOR INTO STORAGE
    NOOVF     B     NOOVRFLO                  NO SKIPPING IF NO OVERFLOW
    OVF       B     OVRFLO                    SKIP HERE IF THERE IS AN OVERFLOW
    NOOVRFLO  ...
```

Figure 23: The TS instruction and overflow processing

The exchanges:

There are often occasions where we must exchange two data objects. This is seen in the simple case of a bubble sort, where swapping table elements is necessary if they are not in the correct collating sequence. Four instructions are available for exchanging data between the special registers and storage. Exchange Accumulator and Storage (XCH) writes the contents of the accumulator to erasable storage, and the contents of the erasable storage location are placed in the accumulator in a single instruction cycle. Exchange L and Storage (LXCH) is essentially identical to XCH, but the L register rather than the accumulator is exchanged with storage. Likewise, Exchange Q and Storage (QXCH) exchanges the Q register and erasable storage. Double Exchange (DXCH), exchanges both the accumulator and the L register with two consecutive words in erasable storage. Introduced previously in the discussion on managing fixed and erasable storage banks, DXCH plays an important role in transferring control in interrupt processing and job scheduling.

At first glance, exchanging data between the special registers and memory appears trivial, and hardly worth dedicating four precious opcodes to the task. True, an exchange is conceptually simple, but it actually replaces several more basic steps and not only makes the program shorter it will also execute faster. Consider the situation where you pick up a pen and a pad of paper in each hand. On realizing that you have placed the pen in the wrong hand, you must swap it to the other hand to begin writing. However, exchanging the pen and pad is a juggling act, and cannot be done in one simple step. We require a temporary holding area for the pad while placing the pen in the other hand. This temporary "storage" can be anything – the pad can be placed on a table, under your arm, held in your mouth or tossed in the air – any of these are acceptable, so long as one hand is freed to take the pen. So, our sequence is, 1) put the pad of paper on the desk, 2) put the pen in the other hand, and 3) pick up the pad with the, now empty, other hand. Performing these individual steps in software requires several instructions, consuming processing time and memory. In a single operand system such as the AGC, swapping the accumulator with storage can take up to six instructions. A far more important issue than the overhead is that of interrupts.

System integrity is easily lost if exchanges in Executive level operations, such as manipulating the job queue or banking registers, are interrupted. This issue becomes even more important because of the AGC's lack of any capability to serialize on, or lock, resources such as memory. Consider the case of manipulating a double precision quantity as part of a multiplication or division operation. Without a single instruction to move two words of data, there is a window of opportunity for an interrupt to occur. During this small but sensitive period, one word could be moved, an interrupt could occur, and another job might attempt to use or update the data in the expectation that it is valid. This data corruption will be extremely difficult to detect. Adding instructions to inhibit interrupts during the exchange is certainly possible, but at the cost of even more processing time and memory overhead. By combining all of the steps in a single exchange instruction, any possibility of data inconsistencies across interrupts is eliminated.

Sequence changing

During instruction processing, the AGC increments the contents of the Z register in expectation of program execution continuing with the next instruction in storage. However, program execution rarely progresses in a linear manner for any great length of time. In many cases, data is tested against another value, with the sequence of program execution changing based on that test. Or, a subroutine to handle a common function must be entered and returned from. In these and similar cases, the sequence of instruction execution is altered. We can classify these scenarios into three broad classes of branching instructions:

- Conditional branches
- Unconditional branches
- Unconditional branches with return.

Branch Zero to Fixed and Branch Zero or Minus to Fixed

Four branching instructions are available in the AGC. Three instructions, all with the suffix "fixed", are limited to transfers of control to fixed storage. The actual "transfer of control" in any of the branch instructions occurs by replacing the contents of the Z register with the instruction's operand address. Execution of the program continues at the location specified by the Z register. Two of the branch instructions are called "conditional branches" because they test the contents of the accumulator and branch depending on the results of that test. Branch on Zero to Fixed (BZF) compares the accumulator against both positive and negative zero. If the test is successful (that is, if the accumulator does indeed contain a positive or negative zero) control transfers to the location referenced by the BZF operand. Testing for negative numbers as well as ± 0 is possible through the Branch on Zero or Minus to Fixed (BZMF) instruction. Again, the program will perform the branch if this test is successful. With the exception of ± 0, the AGC cannot do a direct comparison against a specific value, such as a check to see if the accumulator contains 0.625. To perform such a comparison, 0.625 must be subtracted from the accumulator and the result tested for zero.

Transfer Control to Fixed

Unconditional branching alters a program's execution sequence by changing the location where the next instruction is fetched. For example, consider the case where a program has to make a decision whether to execute a piece of logic, such as to determine whether a leap year calculation is necessary. If it is not a leap year, the leap year calculations are bypassed and the program continues. An unconditional branch is useful to transfer control around software logic to avoid executing it. The

Transfer Control to Fixed (TCF) instruction simply replaces the contents of the Z register with the instruction operand, forcing the fetching of the next instruction from that address. Like the BZF and BZMF instructions, TCF only transfers control to locations in fixed storage.

Transfer Control

Underlying the basis of well structured programming is the notion that programmers should write a specific routine only once, and all programs use this common piece of code by transferring control to the routine. The subroutine must have a means of knowing which program called it, in order that it can return control. Transfer Control (TC, or the alternative notation, Transfer Control with Return – TCR) is an unconditional branch that also saves information on where to return in the calling program. At this point, the Z register has already been incremented, and points to the instruction immediately following the TC instruction. First, the contents of the Z register are copied to the Q register. With the return address now safely preserved, TC unconditionally branches to the subroutine by replacing the contents of the Z register with the TC operand. Unlike the other three branching instructions, TC may transfer control to either an erasable or fixed storage location.

Complicating the use of the TC instruction is the problem of subroutines calling other subroutines. This "nesting" of subroutines requires each routine to save the Q register before executing another TC instruction. If the programmer fails to perform this task, the ability to return to the calling program is lost. On a practical level, the lack of a hardware stack, plus the constraint of limited erasable storage, places a practical limit on the depth of subroutine nesting.

Returning to the calling program requires placing the return address into the Z register. When a TC instruction is executed using the Q register as its operand, the contents of the Q register are copied to the Z register. Processing continues at the instruction immediately following the TC of the calling program. With these steps as background, an important subtlety in the TC instruction comes to light. Returning to the calling program requires that the address of the resumption point be placed into the Z register. On inspecting the Q register before executing the TC Q instruction, a value like 01234_8 might be present. The assumption is that 01234_8 is a valid address that is moved into the Z register, and voilà! Control passes back to the caller. The details of this return, however, are not quite as simple. Rather than copying the contents of Q to the Z register, the contents of the Q register are interpreted as an AGC instruction. If interpreted as an instruction, the return address of 01234_8 has a 12 bit operand, and the first three bits of the word are zeros. It is no coincidence that this is decodes to the instruction TC 01234_8. We now see that the mechanism for returning from a subroutine is exactly the same as calling the subroutine itself. In "executing" the return address in the Q register, we are indirectly executing another TC. Using the mechanism already defined by TC, we eliminate the need to create a new "Return From Subroutine" instruction, or a case where TC Q is designed as a special case of TC. This property of indirectly executing an instruction is exploited to create new and valuable capabilities.

TC also offers another powerful capability in its design. Subroutines are usually designed to perform a very specific task, such as calculating taxes on an income. For such a subroutine to operate effectively, there must be a means to supply, or "pass", data to that routine. With the Q register pointing back to the storage location immediately following the TC instruction, an elegant means of passing a parameter list is created. A convention in AGC programming is to have parameter words immediately following the TC instruction, where the called routine can reference them directly through the Q register. Any number of parameter words may be passed, but it is the responsibility of the called program to increment the Q register to point to the code immediately following the parameter list. Otherwise, the Q register will point to the parameter list and cause the resumed program to fail as it tries to execute data as a legitimate AGC instruction.

Count, Compare and Skip

Repetitive execution (looping) of a section of code is a common programming task. Whether traversing the rows of a table or limiting the number of iterations in a mathematical function, all require a common set of capabilities:

- A counter or index variable used to control the loop
- The capability to alter the index variable each time through the loop
- Testing the counter against another value
- Branching to a new location based on testing the index variable.

All high-level languages combine these tasks into a single construct, usually called a "FOR" or "DO" loop. Intuitively, these kinds of loops come to mind when we do everyday tasks, such as fastening the buttons on a shirt. Let us start at the top of a shirt that has six buttons. Fasten the button at the top. We now have one button completed, and being rather pleased with our performance so far, ask ourselves, "I have six buttons – am I done yet?" Having just started, we continue to the second and repeat this process with the other buttons. With each iteration, we continue to ask if all six buttons have been fastened or if we should continue to the next one.

FOR and DO loops enclose a sequence of instructions that are repeated while changing the value of one variable, called an index, which is used to count the number of times through the body of the loop. Syntactically, they all begin with creating and initializing the index variable before beginning to execute the code in the body of the loop. An ending condition also is necessary for exiting the loop once the index variable reaches a particular value. After each execution of the code within the body of the loop, the index is updated by incrementing or decrementing. Testing the index is necessary to determine if the desired number of loops through the code has been completed. Upon reaching the exit condition, processing continues with the instruction following the loop.

Count, Compare and Skip (CCS) combines many of the requirements for controlling program loops into a single instruction. Programming code using the CCS instruction begins with the assignment and initialization of an integer counter to some value. The code in the body of the loop follows, and may or may not use the counter when it executes. The body of the loop ends with the CCS instruction. CCS loads the counter from erasable storage into the accumulator, takes its absolute value to ensure a positive value, and decrements the accumulator. The copy of the counter used by CCS that still resides in storage remains unchanged by CCS. Next, the value in the accumulator is tested against four possible cases, and control is passed to one of four storage locations immediately following the CCS. Depending on the value of the counter in storage, a different number of words are "skipped" after the CCS instruction.

Number of skips	Value of CCS operand
0	> 0
1	= +0
2	< 0
3	= -0

After the CCS instruction updates the accumulator, it is common programming practice to save the counter back into storage, freeing the accumulator for other operations. Although complex, CCS is a powerful looping instruction. Sections of code can now be executed a variable number of times, and several termination conditions are available.

Indexing

A common data structure in computing is the table, which is very similar to the cells in a spreadsheet. Tables can exist as several "rows" of data, each with one or more "columns". Each row contains related information, such as in an invoice. Here, each column contains a single type of data, such as an invoice number, with additional columns holding quantities and unit costs. Implemented in the AGC, each table entry may require one or more words to hold the data. Referencing each row in the table is difficult, since the address in an instruction is not changeable, yet the table reference is not known until the program has begun executing. Most computers provide a special register known as an index register to assist in these storage references. Index registers usually hold an offset into the table, which is computed as part of the program's execution. This value is added to the instruction's operand address, yielding a generalized means to access any data in the table. But an index register is notably absent from the AGC architecture. This omission is due to the need to allocate additional opcodes to reference the index register, or bits in the instruction word that specify whether the index register will be used. However, there is a simple, easily implemented solution available.

First, carefully restating the original problem is necessary, so as not to lose sight of the actual objective: There is a need to access individual elements within a table that is stored in memory. The address in storage of each element is obtained by adding the starting address of the table and the offset of the element within the table. This offset is maintained as a separate variable, called an index, and is independent of both the type of operation being performed (addition, storing, comparison), and the type of data in the table. Since indexing functions are still an essential part of software, a novel solution is to introduce a new instruction, called INDEX, which, like EXTEND, operates on the next instruction in storage. During processing, the INDEX instruction dynamically modifies the target instruction as it is being executed, despite the fact that the instruction resides in fixed memory. During the execution process, the instruction is fetched into the AGC's inaccessible internal registers, where the index value is added. This summed value now points to the desired table element, and can be executed as any other instruction. The INDEX instruction, while complicated, is tremendously powerful, and makes available an indexing capability to virtually every instruction in the AGC. More importantly, this capability is available without altering the instruction format or adding hardware for an additional register, and every erasable storage word is now a potential "index" register.

When the powerful combination of the INDEX instruction is combined with the Count, Compare and Skip (CCS) instruction, this provides an efficient mechanism to loop through tables and arrays. Three steps are necessary to index through a table. The first task is to define and initialize an index variable, and the counter in the CCS is an ideal choice for an index variable. Once the index variable is prepared, it is followed by the INDEX instruction, specifying the index variable as its operand. Finally, the instruction to be indexed is coded, with its operand pointing to the start of the table. Figure 24 shows what such a sequence of instructions might look like. Note that because the CCS instruction continuously decrements the index we will work our way from the "bottom" (highest storage address, with offset 4) to the "top" (the starting address of the table, with offset 0) of the table.

All "normal" and "extended" instructions can be indexed, as INDEX is unique in that it does not clear the EXTEND indication in the processor and interrupts continue to be inhibited. As a result, the effect of EXTEND "passes through" the INDEX instruction, altering the target of the INDEX instruction. Because an EXTEND indication might be present, the "normal" and "extended" versions of INDEX, called INDEXE, perform identically.

Dynamically changing an instruction during its execution is called "self-modifying code". As with other aspects of the AGC, 21st century programmers are excused if they fall back in horror at the thought of altering an instruction while it is executing.[17] Self-modifying code has a tradition of being difficult if not impossible to maintain, but the use of INDEX is generally limited to table accesses, and when used for this singular purpose most of the objections disappear.

---

[17] We often forget that the concept of "self-modifying code" still exists, but has been transferred from machine instructions to much higher level constructs. In fairness, it can be argued that some features in object-oriented languages, such as operator overloading, are cases of self-modifying code.

```
          CA      FOUR            PUT NUMBER OF TABLE ENTRIES (MINUS 1)
                                  INTO ACCUMULATOR
          TS      TBLINDEX        SAVE # OF TABLE ENTRIES IN THE INDEX VARIABLE
LOOP      CA      TBLSUM          GET THE CURRENT SUM OF THE TABLE ENTRIES
          INDEX   TBLINDEX        INDEX THE NEXT INSTRUCTION, ADDING THE
                                  CONTENTS OF TBLINDEX TO THE NEXT INSTRUCTION
          AD      TABLE           ADD THE CONTENTS OF THE TABLES (INDEXED BY
                                  TBLINDEX) TO THE SUM IN THE ACCUMULATOR
          TS      TBLSUM          AND SAVE THE RESULT
          CCS     TBLINDEX        TEST TO SEE IF WE ARE DONE LOOPING, AND
                                  DECREMENT THE INDEX
          TCF     LOOP            TBLINDEX > 0 BEFORE CCS INSTRUCTION,
                                  SO WE CONTINUE LOOPING
          TCF     EXIT            TBLINDEX = 0 BEFORE CCS INSTRUCTION,
                                  SO IT'S TIME TO GO HOME
FOUR      DEC     4               NUMBER OF TABLE ENTRIES
TBLINDEX  DEC     0               TABLE INDEX VARIABLE
TBLSUM    DEC     0               SUM OF ALL ELEMENTS IN TABLE
TABLE     DEC     8               0TH ENTRY ON THE TABLE
          DEC     1               1ST ENTRY IN THE TABLE
          DEC     3               2ND ENTRY IN THE TABLE
          DEC     14              3RD ENTRY IN THE TABLE
          DEC     2               4TH ENTRY ON THE TABLE
EXIT      .....
```

Figure 24: Summing five entries in a table

Special case and implied address instructions

As a consequence of having special registers mapped to addressable storage, not all storage locations are valid for instructions. The earlier discussion of the Transfer Control (TC) instruction highlighted the cases where a TC to a banking register makes little sense. For example, assume that a TC instruction executes using the erasable memory bank register (location 00003_8) as the storage address of the called "program". Having only three bits in the middle of the word, the erasable bank cannot be valid as an instruction. To specify other banking registers is similarly nonsensical. As the AGC does not have any means to trap illegal instructions, some assurance is necessary to prevent using these special registers as the operand address of a TC instruction. The solution is to exploit an otherwise invalid TC to create an entirely new instruction.

This elegant solution solves two pressing issues at once. First, there is no longer a concern about invalid operands when executing a TC. Secondly, and even more significant, a new mechanism becomes apparent to create additional instructions. In Figure 25, four new instructions are created from special cases of existing operations. Not only are all the instructions defined as part of the "normal" instruction set, the EXTEND instruction itself is one of the new operations.

Of the four special case instructions, three are dedicated to interrupt handling. Inhibiting and releasing (i.e. allowing) interrupts is common during the normal operation of the AGC and a necessary part of maintaining system integrity. For

Assembled Instruction	Special code	Description
TC 3	RELINT	Release Interrupts
TC 4	INHINT	Inhibit Interrupts
TC 6	EXTEND	Set Extracode Flag
INDEX 17	RESUME	Resume Interrupted Program

Figure 25: Instructions created from special cases of existing instructions

example, an overflow condition in the accumulator will automatically inhibit interrupts, and execution of the interrupt handlers (from the DSKY, timers, etc) is with interrupts inhibited. Frequently, critical system code must execute without the distraction of other processes competing for the AGC's attention. Preparing the system environment for a new job to run in the AGC is an excellent case of a process that cannot be interrupted. As we will see in more detail with the AGC Executive, memory allocation and other housekeeping tasks are necessary before the new job is ready to run. An interrupt during this activity may cause data corruption in system tables and variables, and may well force an AGC restart. To illustrate, assume that a job is attempting to start, and the Executive is performing the steps to prepare the program for execution. The allocation of an area of memory called a Core Set has just been completed, but the banking registers are not yet set up. Next, assume an interrupt occurs, with another process trying to start a job during this interval. Much of the information necessary to start the first job is corrupted and lost, since the original work was not preserved. Later, the AGC returns to its original effort to schedule a job, which will certainly fail because of the loss of critical information. Manually inhibiting interrupts (with the INHINT instruction) at the start of a job's setup ensures that the data is not corrupted by another task. After the critical code has completed executing, the software can allow interrupts again using the RELINT instruction. Of course, it is important to limit the time spent with interrupts inhibited. Several routines in the computer, notably the Digital Autopilot, expect to be serviced frequently. While occasionally postponing their regular execution is tolerable, overall system performance degrades if interrupts are inhibited for extended periods.

The process for returning from an interrupt routine is straightforward, but requires that several steps be completed within one instruction cycle. From the previous description of interrupt processing, the Z register (the program counter) is saved in the ZRUPT special register and the BRUPT special register holds the interrupted instruction. When returning from an interrupt routine, the Resume Interrupted Instruction (RESUME) instruction moves the contents of ZRUPT back to the Z register, releases the hold on interrupts,[18] and executes the instruction saved in the BRUPT register.

The group of instructions called "implied address" operations do not actually

---

[18] Unless interrupts are being held manually by a previous INTINH.

create a new or different type of instruction. Here, previously familiar instructions are written using an alternative form, or alias, for an existing operation. Using instructions in this manner does not create new capabilities in the AGC; rather, their advantage is to make common uses for existing instructions more intuitive for the programmer to read and debug. By using aliases to define a frequently executed instruction, it is possible to eliminate the requirement of specifying the operand address since it is redundant. For example, doubling the value in the accumulator is possible by multiplying it by two, or alternatively by storing it in the Cycle Left special register and reloading it back to the accumulator. Both techniques will work, but they take more processing cycles than desired. A faster and easier way is to add the contents of the accumulator to itself:

AD   A   ADD CONTENTS OF STORAGE AT "A"
                   TO THE ACCUMULATOR ("A" IS
                   THE ACCUMULATOR ITSELF)

Doubling is a common operation, and the statement shown above is the preferred method to accomplish the task. Enforcing a consistent use of this method is possible by creating a simple shorthand for AD A. The mnemonic, DOUBLE, exists only in the assembler language, which is used to assemble the source code into AGC machine language. When converted to binary, DOUBLE is saved in storage as 60000_8, which is the same as AD A.

Aliases for 16 different instructions and operands are assigned to implied address instructions, and all but one reference the special registers in the lowest range of storage. Although a number of these instructions have been discussed earlier, Appendix A provides the complete list. Not all of the implied address instructions are commonly used. A few, such as XXALQ (Execute Extracode using A, L and Q) are sufficiently esoteric that it is difficult to imagine a problem that would require such an operation.

COMMUNICATING WITH THE OUTSIDE WORLD: THE I/O SYSTEM

Ultimately, a computer exists to process data, usually from a wide variety of sources. Elaborate hardware configurations are designed to maximize the amount of data flowing into and out of the processor, ensuring each component is used to its fullest. The amount of data processed for even the most common of consumer items is tremendous, as analysts look for subtle trends that might help improve market share. A 21st century data warehouse system comfortably process databases of several terabytes in size, analyzing the buying habits for items ranging from toothpaste to diet colas. At the other end of this computational spectrum is the AGC. As a control computer, the AGC is not burdened with the need for the high I/O throughput necessary in commercial data processing. Virtually all of its work is in the form of numerical calculations, not the manipulation of character-based data commonly found in large databases. Further, the AGC does not use external devices such as disks or tapes, reducing its I/O requirements to a problem of setting or resetting bits, or moving single words in and out of memory. The instructions used to manage the AGC's I/O channels are a reflection of this reality.

I/O devices

I/O devices in the AGC are not the expected collection of disks, scanners, printers and network connections found in home computers. Rather, they take the form of hardware one would expect to see in a spacecraft: the inertial platform, rocket engines, radars, the DSKY and panel switches. Unsurprisingly, only hardware related to guidance, navigation and control are within the scope of the AGC's duties. The computer does not manage other spacecraft systems, such as environmental control and the electrical power system at all. These systems are operated manually through switches on the control panel, with no provision for automation or remote control from the ground. The technology for mass storage devices, such as disks, was too immature for use in a spaceborne computer.[19] Magnetic tape storage, used in the Gemini spacecraft to hold programs that otherwise would not fit in memory, was not carried over to Apollo.[20]

Control computers like the AGC differ significantly from systems used to process large databases. Consider a modern server in a major company. In large database applications, the processor can transfer a huge amount of data between disk and memory. The speed of these data transfers is one of the key variables defining overall system performance. Processors, disks, and their related hardware, are designed to complement each other, to squeeze the greatest amount of performance out of the hardware. Usually, several I/O channels exist to distribute the I/O load across several paths, much as a multi-lane highway has a far greater capacity for traffic than a single lane. All of this requires large, complex and expensive hardware, but the resulting performance justifies the investment.

Contrast this architecture with the AGC. Rather than optimizing the I/O system for a large volume of I/O traffic, the AGC design focuses on servicing a large number of discrete "devices". Three types of I/O interface are found in the AGC: counter inputs, individual control bits, and 15-bit data words. A comparatively simple architecture supports all of these types of I/O. Data is not transferred in large quantities in and out of the processor, and data transfer speed is not a significant factor in AGC performance. Indeed, much of the I/O performed is only reading or writing a single bit in a channel. Because single-bit I/O is so prevalent, a set of logical, or Boolean operations are included in the I/O instructions.

---

[19] Even today, disks are not used in spacecraft control computers for a variety of reasons. A notable exception is the Autonetics D-17B used in the Minuteman I missile, where a rotating disk doubled as data storage and main memory.

[20] A magnetic tape storage system was considered for the next generation AGC, but it never evolved past the prototype stage.

I/O channels and counters

Control data moves in and out of the AGC through the I/O channels. Channels are superficially similar to erasable storage words, being 15 bits wide and referenced with the "address field" of the I/O instruction. The first observation is that the I/O channels appear to overlap the lower range of erasable storage, especially the accumulator, banking and other control registers. In fact, only the accumulator and L register overlap the I/O channel addressing range, and do not serve as I/O channels themselves. However, this overlap produces useful new capabilities that extend the AGC's instruction set. In all I/O instructions, data moves either from the channel to the accumulator, or from the accumulator to the channel. Data in the channels is available for the AGC to reference at any time. With one special exception, I/O channels are unidirectional. Software in the AGC cannot change the data in an input channel; only a change in the state of the input device can alter these bits. Similarly, data placed in an output channel maintains its value until the software modifies it. The output device has no control over the value of the data it receives from the output channel, and cannot modify it in any way.

Perhaps notably lacking in this discussion of I/O interfaces are the counter registers that are used by the IMU, optics and LM radar. Strictly speaking, both counters and channels are I/O interfaces, and are sufficiently similar that they should be discussed together. However, the differences in how their data is referenced and processed are sufficient to consider them separately. Both channels and counter registers are defined as input- or output-only, and software reads and writes data to/ from devices elsewhere in the spacecraft. Perhaps the most important differences between channels and counters are the types of data they represent and the amount of processing they require in the AGC. Counters operate exactly as their name implies, counting up or down along a continuous range of values such as velocity or angular position. Input from the device is only a pulse commanding the AGC hardware to increment or decrement the current value in the counter register. As seen earlier, this steals a cycle to suspend (but not interrupt) AGC processing, and an "unprogrammed sequence" executes to adjust the counter. Contrast this with I/O through the channels. Almost all channel data is represented as a single bit, and only a discrete "on/off" state is sensed or commanded. No additional processing is necessary for the AGC to interface with the device. Figure 26 summarizes the differences between the two.

Strictly speaking, counters are input-only devices to the AGC. Using the term "output counter" is somewhat misleading; they neither count nor reflect the status of a device. Output counters are used only to command *movement* of a device such as a gimbal. Once the gimbal is commanded to move, its input counter continuously reflects its current position. No testing of the output counters is necessary, since they do not change while the device is moving.

Channels	Counters
Synchronous I/O	Asynchronous I/O
Data represented as a discrete state	Data is one point over a continuous range of values
Single bit assigned for most devices, requires Boolean functions to extract a device's bit	Full word dedicated to each device
Data is accessible only through special I/O instructions	Data is accessible using any normal or extended instruction
Multiple devices managed with one channel	Single devices represented by each counter
Channels connect "directly" with a device; no intermediate processing is done in the AGC	Counter pulses cause "unprogrammed sequences" (AGC internal instructions) before data is presented in the counter register

Figure 26: Characteristics of counter registers and I/O channels

Performing I/O operations

For a computer such as the AGC to operate a device, an interface or channel to the outside world is necessary. I/O devices, or more properly, the bits that represent the devices, are visible through the channels, which in turn are manipulated by the I/O instructions. Setting or clearing a bit is a signal to the device to *change its state*, such as by firing or shutting down a thruster. Similarly, hardware external to the AGC can report on its status, such as the liftoff 'discrete' in the Command Module. These are examples of binary operations – either the firing command is sent to the thruster or it is shut off; either the booster has lifted off the pad, or it is still firmly on the ground.

Performing I/O in the AGC is a trivial task, requiring a single instruction to move data between the accumulator and the I/O channel. Unfortunately, understanding the simplicity of an I/O device in the AGC is clouded by experience with today's hardware. A disk or video card in a modern computer is a complex device, often requiring the processor to use sophisticated protocols to send commands and data. With few exceptions, devices connected to the AGC are simplistic in the extreme, with a single bit able to control a piece of hardware. It might seem incredible that a single bit, when set to zero or one, can stop or start a large rocket engine. But why not? After all, that bit operates as a switch. This is best illustrated using a hypergolic rocket engine as an example. By computer standards, these engines are truly "dumb" devices. Igniting the engine requires only that a "start" signal is sent to the engine controller, which in turn commands the fuel and oxidizer valves to open. The two propellants do not require elaborate hardware to initiate combustion, the mere act of the two mixing together is sufficient for them to ignite. As a result, the start signal does not have to manage a sophisticated sequence of events; only a single wire carrying a binary '1' to the controller is necessary. Shutting the engine down is equally straightforward. Resetting the signal back to '0' is sufficient to direct that the controller close the propellant valves, causing the engine to cease firing.

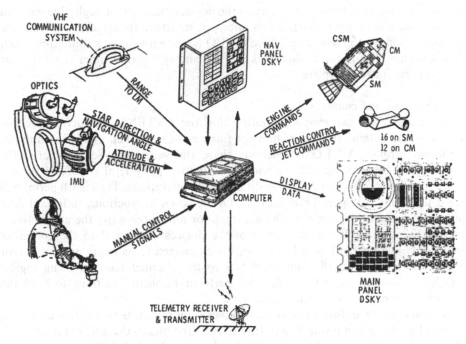

Figure 27: The AGC and its I/O devices and interfaces

A basic assumption of the I/O system is that the time required to complete an I/O operation is very short. This assumption is possible because the electromechanical devices connect to controlling electronics, which in turn connect to the AGC. Latency to these controllers, which the AGC sees as the actual "I/O device" is very short. With the controllers "hiding" the physical characteristics of the device, the software does not need to concern itself with the complexities of the external hardware.

Perhaps surprisingly, no mechanism exists for testing whether the data sent is actually received and the device successfully performs its operation.[21] For example, if a bit to fire a thruster is set, and a mechanical failure prevents the jet from firing, the AGC will be unaware of this as there is no direct feedback or status information. The only notification the crew receives is through the Caution and Warning system (which usually gets its information through device controllers) or by panel displays, but the computer itself remains oblivious to the problem.[22] With

---

[21] There is one exception. The Lunar Module landing radar indicates which of the two possible positions it is in. When the radar is commanded to rotate from the descent to hover position, the AGC can test later to verify that the radar is in the correct position.

[22] More precisely, in the event of a thruster failure, the Digital Autopilot (if it is enabled) will eventually sense the failure, but only indirectly. It takes several failed firing commands for the DAP to recognize that a given thruster is not firing correctly.

no status indicators returned from any of the devices, and with a negligible delay in execution there is no need to design an interrupt structure to signal the completion of an operation. In the few cases where the I/O is known to take a finite time, such as repositioning the radar or downlinking a telemetry word, a timer is set to wait for the operation to complete.

I/O instructions and channels

I/O instructions are extracodes, requiring that the EXTEND instruction precede them. To help extend the instruction set further, I/O instructions define another instruction format. All I/O instructions share the same opcode, a binary 000. Following the opcode, bits 12, 11 and 10 form the "peripheral code", to further qualify the instruction as one of seven different I/O operations. The eighth peripheral code is used only for special ground tests. Of the seven instructions, four read data from the channel and place it in the accumulator and three write the accumulator data to the channel. Recall that most of the devices connected to the computer require only one bit, and each I/O channel is connected to as many as 15 different devices. To operate on the individual bits in the channel requires using logical operations. Most of the I/O instructions perform Boolean functions to limit the scope of their reading or writing operations.

Using six bits to define an operation leaves only nine bits to address a specific channel, but this is not a significant drawback. Nine bits in the address field yields 512 possible channels, far more than the 16 channels that are actually implemented in the AGC. I/O channels are not accessible in the same manner as the special and central registers. The central registers occupy a range of erasable memory, and are accessed in the same manner as any other erasable memory location. Channels are different from registers in that they are not addressable as a memory location and are accessible only through I/O instructions. Each channel is 15 bits wide, reflecting the size of the accumulator, and operates either as an input or output device. Bi-directional data channels, where data can be both read from and written to, are not implemented in the AGC architecture.[23] Channels are not arranged contiguously, but are dispersed irregularly in the range 00003_8 through 00035_8.

Figure 28: I/O instruction format

---

[23] The Fixed Extension Bit, or Superbank Bit in channel 7 is an exception, but this discussion considers it as part of the memory banking environment, not the I/O system.

Address	Input or Output	Primary Function
00000_8	Bidirectional	Accumulator
00001_8	Bidirectional	L Register
00002_8		
00003_8	Input	High Scaler
00004_8	Input	Low Scaler
00005_8	Output	RCS Jets
00006_8	Output	RCS Jets
00007_8	Bidirectional	Fixed Extension Bit
00010_8		
00011_8	Output	
00012_8	Output	
00013_8	Output	
00014_8	Output	
00015_8	Input	DSKY Input
00016_8	Input	
00017_8		
00020_8		
00021_8		
00022_8		
00023_8		
00024_8		
00025_8		
00026_8		
00027_8		
00030_8	Input	
00031_8	Input	
00032_8	Input	
00033_8	Input	
00034_8	Output	Downlink 1
00035_8	Output	Downlink 2

Figure 29: I/O channel usage

Boolean operations in I/O instructions

Input and output instructions come in two basic forms. In the first, all 15 bits of the channel are read from or written to, and are not modified by a logical operation. The READ instruction moves data from the channel to the accumulator, while the WRITE instruction does the opposite; all of the data in the accumulator is placed in the channel. The second type of I/O instruction performs a logical operation on the data after it is read from the channel, or before it is written to the channel.

Because most channels control several I/O devices, channels consist of a set of single bits or small fragments of the entire word. The first step in determining the state of a device is to isolate the individual bits that represent the I/O device so that it can be tested. Setting or extracting these values requires using the logical functions AND, OR and Exclusive OR as the data is moved between the channel and the accumulator. Although a logical AND is available through the MASK instruction, many I/O instructions can combine reading or writing and a logical operation in one instruction. For example, assume there is the need to set bit number seven in an output channel without disturbing the remaining bits. Here, the binary value 000 000 001 000 000 is placed in the accumulator and the WOR (Write with OR) instruction is executed to logically OR the contents of the accumulator with the channel. Conversely, resetting bit seven is performed by placing the value 111 111 110 111 111 in the accumulator and executing the WAND (Write with AND) instruction. With the exception of bit seven, which is set to zero, the other bits in the output channel word retain their status. Reading the channel and performing a logical OR on the data is also possible using the ROR (Read and OR) instruction. Bits in the accumulator are OR'ed with the channel data, with the accumulator containing the result.

A more interesting logical operation is available with the RXOR instruction, which performs an Exclusive OR between a bit string in the accumulator and the channel data. If a bit in both the accumulator and channel are one, a zero is placed in the accumulator. If either the bit in the channel or the accumulator is one, but not both, then the bit in the accumulator is set to one. In this manner, Exclusive OR toggles bits from zero to one, or one to zero irrespective of their initial state. The most familiar example is where an indicator light must flash on and off. In this case, it is not important that the light is on or off; the only thing that matters is flipping its state. Exclusively ORing is available only with data being read from the channel into the accumulator, as there is not a similar instruction to perform this operation when writing to the channel.

I/O instructions must reference the accumulator as a source or destination for the data being moved between the AGC and external devices. By mapping the accumulator to the I/O address space, moving data between it and the channel becomes automatic. Since data transfers are only between the accumulator and the channel, the utility of mapping the L (low-order accumulator) central register into the I/O space comes into question. When an I/O instruction references the L register as the "channel" it provides an elegant way of exploiting the logical operations (AND, OR and XOR) that are only available with the I/O instructions. Recall that only the Logical AND (MASK) instruction is available in the normal and extended instruction set. While it is possible to perform Logical ORs or XORs by using additional instructions, this solution is unnecessarily cumbersome to implement. An alternative operation is to load data into the L register and the bit string for the Boolean operation into the accumulator. Executing the ROR, RAND or RXOR instructions with the L register as the channel being read places the result of the operation in the accumulator. Conversely, we can use the WAND and WOR instructions to perform logical operations, leaving the result in both the accumulator and L register. Simply by extending the accumulator and L register into the I/O channel range, a full set of Boolean functions are made available to the entire system, not just to I/O channels. Very nifty!

The EDRUPT instruction

Earlier, interrupts were introduced as events generated by expiring timers or by external devices requiring attention. By their very nature, interrupts are beyond program control, and the timing of their occurrences is generally unpredictable. As is frequently the case, an exception to this rule exists. EDRUPT, or "Ed Smally's Interrupt", was originally created to assist in hardware checkout before flight.[24] When executed, EDRUPT does not actually generate an interrupt, but performs two actions that are common to interrupt processing. First, further interrupts are inhibited, and a RESUME instruction is necessary to enable them again. Second, the ZRUPT register is loaded with the current value of Z (the program counter). EDRUPT is used only in the Lunar Module Digital Autopilot code, to set up the DAP cycle termination sequence.

Nonprogrammed sequences

Up to this point, the description of the AGC has been about the environment that is visible to the programmer, especially instructions that reside in storage. An entirely separate group of instructions are not accessible to the programmer. These "nonprogrammed sequences" are functions that are responsible for incrementing, decrementing and shifting the counter registers that the AGC manages internally. Nonprogrammed sequences are casually referred to as instructions because from the AGC's perspective they execute just like any other machine instruction. Non-programmed sequences cannot run while interrupts are disabled, since an interrupt routine might be accessing the counter in question during that time. Only one or two nonprogrammed sequences are possible, or even reasonable, for each counter. Counters are found in storage locations 00024_8 through 00060_8, and include the timers, input counters and output registers. It might seem odd that TIME1 through TIME6 are included here, but timers share important characteristics with other counters. A timer receives a pulse and its value is incremented. Only a positive increment is required for timers because in the AGC, as in reality, clocks cannot run backwards. Rather than using an external device as the source of a pulse, timers use a signal derived from the AGC's main timing oscillator.

PINC (Positive Increment) and MINC (Minus Increment) modify a counter by incrementing or decrementing its value. These instructions operate on counters that track linear quantities such as velocity, rather than the rotational motion of gimbals, optics and radars. Servicing routines integrate the changes in the counter's value over time, to determine the overall state of the vehicle or device. As each velocity increment is read, its value is added to the total velocity recorded in memory, and the counter is zeroed. Summing these small velocity increments over time creates an accurate value of the accumulated velocity, which is then fed into the state vector calculations. All arithmetic is performed using the one's complement system, where the problem of positive and negative zero is negligible. Accelerated flight usually

---

[24] Ed Smally, at that time an engineer at the MIT Instrumentation Laboratory, is noted for his contributions to the AGC's self-checking and diagnostic routines.

does not create a situation where the counter oscillates between positive and negative values, so the issue of an extra value for zero is of no concern. But counters for rotating devices, such as gimbals, radars and optics, have the property of oscillating between values on either side of zero during their normal operation. As the crew moves about inside the cabin, or fuel lazily sloshes back and forth in its tank, the spacecraft can experience a barely perceptible change in attitude. Gimbals, capable of sensing extremely fine changes in position, report these changes by updating the counters in the AGC. Most of the time, these small motions tend to cancel themselves out, resulting in a net attitude change of zero. However, devices whose motion oscillates about zero will "lose" a counter pulse if one's complement is used. Pointing accuracy in this hardware is especially critical, and the error induced by the mathematical mess of positive and negative zero is unacceptable. For these counters, the PCDU (Plus Increment the CDU) and MCDU (Minus Increment the CDU) adjust the counters using the two's complement number system, which has a single value for zero. After these counters are read, the MSU (Modular Subtract) instruction converts the value to one's complement, making it compatible with the rest of the AGC arithmetic.

Several devices present a continuous range of data through their counters. The most important of these devices, such as the IMU's gimbals, the optics (in the CM) and the radars (in the LM) can also have their movement commanded by the AGC. All such devices use an analog to digital converter, the CDU, as an interface with the computer, and their output is presented to the computer as a counter. Although the CDU counters are not bi-directional, the devices they represent do have the capability to be commanded to new orientations. A set of special registers for commanding the devices reside in low memory, at addresses 00050_8 through 00054_8. By placing a non-zero value in the output register, a device connected to a CDU can be commanded to rotate. A positive value, for example, will command the gimbal to move in the positive direction, while placing a negative value in the register will move the gimbal in the opposite direction. The number placed in the register determines the amount of motion requested, and is scaled the same as the counter registers. As the rotation progresses, the output counter tracks its progress by counting down to zero from the starting value. DINC (Diminishing Increment) decrements a positive value in the counter, or increments a value that is negative. Once the counter reaches zero, DINC leaves the counter at zero and no further processing is done to the counter.

The ability to command a device to a new orientation has many important applications. Tracking the CSM from the LM, for example, requires knowledge of the position and movement of both vehicles through space at all times. Using this positional knowledge and the attitude of the LM, the computer calculates the radar antenna's orientation and loads the appropriate values to its output counters. When the data is loaded into the output registers, the CDUs command the radar to point towards the CSM. Once oriented in the general direction of the CSM, electronics in the radar "lock on" and track the CSM, all the while reporting the radar's position through the counters. Gimbals are also perceived as output-only devices, reporting on their rotation angles through the counters. But the inertial platform has the intractable problem of drifting away from its known orientation in space. After using the optics to determine the platform's correct orientation, the AGC must rotate the gimbals to their corrected positions. Like the LM radar, gimbal movements can be in either a positive or negative direction, requiring the placement of a positive or negative quantity in the output counter for each axis.

The last two nonprogrammed sequences are used for an entirely different purpose: to convert between serial bit streams and parallel data words. Serial data is presented as a stream of bits over a single wire, which sacrifices speed over simplicity in its hardware design. Parallel data transfers, while very fast compared to serial data transfers, require a wire for each bit. In a design such as the AGC, which uses 15-bit words, the hardware and cabling requirements would be unacceptably large. Only two sources of serial input exist, data uplink to the computer, and rendezvous and landing radar data in the Lunar Module. In both cases, a serial data interface is quite reasonable because the delay involved in assembling the data word does not affect its processing routine significantly.

Data transmitted in serial bit streams requires reassembly into the familiar 15-bit word used by the AGC. At the beginning, the counter register is all zeros, with the exception of the lowest bit, which is set to one. This serial to parallel conversion is performed by repeatedly shifting the contents of the register to the left, while inserting the incoming bit into the rightmost location in the word. Upon receiving a data bit, the AGC executes one of two shifting operations. A zero in the bit stream triggers the SHINC (Shift Increment) nonprogrammed sequence, causing the contents of the counter to shift left one bit, while filling the rightmost bit with a zero. A one bit in the data stream causes SHANC (Shift and Add Increment) to execute. SHANC also shifts the counter contents one bit to the left, but places a one in the rightmost bit position. When the last of the bits is placed in the counter, the counter overflows, raising an interrupt to signal that the data is ready.

Special cases in I/O

Up to this point, the descriptions of I/O have been carefully qualified, defining operations as single-bit reads or writes from the channel. The next step is to move beyond this basic scenario to I/Os that involve more than one bit, or are serviced only when an interrupt is raised. The first special case is with DSKY inputs.[25] Key codes from the DSKY are 5-bit quantities, this being the minimum number of bits necessary to represent 18 of the keyboard's 19 keys.[26] Pressing a key on the DSKY generates a KEYRUPT interrupt, and places the key code in the input channel. During the KEYRUPT processing routine, the software passes the key code to the keyboard processing routine from the input channel.

Selection of the data source for the Lunar Module radar input counters also requires multiple bits. The Lunar Module, with its complex rendezvous and landing

---

[25] Telemetry updates from the ground are handled like DSKY inputs, and for the purposes of this section are equivalent to manually entered DSKY keystrokes.

[26] Unlike the other DSKY keys, the PROCEED key does not have a key code associated with it.

radars, presents a large number of data sources for the AGC to process. This creates an important design issue: A primary requirement for the AGC is that the hardware used in the Command Module and Lunar Module be identical. With the number of counters in the LM far outnumbering those in the CSM, multiplexing a single counter for all radar data becomes necessary. This is not a large problem, as data from the counters are sampled over a time scale of seconds, which is a leisurely rate when compared to other programs. Processing each component of radar data and integrating it into the targeting routine is a lengthy process, and the overhead of selecting the data source is trivial. Such a disparity in processing time allows one counter to be used to service multiple data sources.

The second case where the AGC transfers multiple bits in and out of the computer is in support of data uplink and downlink. To keep flight controllers in Mission Control aware of the internal state of the computer, several words of diagnostic data are telemetered every second. Two output channels in the AGC are dedicated to the task of "downlinking" data. Essential process information, flagwords and DSKY entries and displays are formatted into several words of data. Once assembled, the data words are written to the downlink channels, and these are multiplexed by the instrumentation system with other engineering data and transmitted to Earth at either 51,000 or 1,900 bits per second, depending on the data rate selected by the crew.

2

The Executive and Interpreter

INTRODUCTION TO THE EXECUTIVE

In the most elementary computers, executing a program is a relatively easy task. Without having to accommodate multiple programs with potentially conflicting requirements, the user only has to load the instructions into memory, set the starting address, and watch it run. Indeed, for early computers (and specialized controller chips today), this "load and go" methodology works acceptably well. For the first decade of the computer age, the modern concept of an operating system simply did not exist. Programmers were required to write software to operate tape drives, printers and other peripherals in addition to the logic required to solve the business or scientific problem at hand. Over time, primitive "executive" programs were introduced to reduce the burden of managing hardware. Rather than directly commanding a device such as a card reader, a program would invoke a pre-written routine to operate the external hardware on its behalf. Delegating such "low level" operations to standardized and (hopefully) reliable interfaces improved productivity tremendously by eliminating the tiresome chores of system management.

Simultaneously, software products called "compilers" created the ability to express a problem in a language that better reflected the problem to be solved. The introduction of high level languages such as FORTRAN and COBOL provided a huge improvement in programmer productivity, and had the additional advantage of insulating the programmer from the idiosyncrasies of the underlying computer architecture. Of course, prepackaged system routines and high level languages come with a price: Additional memory is needed to store the executive software, and the code generated by compilers is not as efficient as hand-coded machine instructions. Nonetheless, the huge boost in productivity and quality of code greatly outweighed the additional hardware cost.

Running a single program at a time was perfectly acceptable in the early 1960s, where hour-long turnaround times were the norm. In a real-time computer such as the AGC, users do not have the luxury of feeding cards into a hopper, mounting and unmounting tapes, and performing multiple software loads to run a single program.

By its very nature, a real-time system must have a short response time while it processes the constant stream of data coming into the computer. Additionally, the variety of tasks under the computer's control demands that a number of programs run in parallel. These requirements help establish two notable characteristics of control computers: memory resident software and multiprogramming. Given the limited amount of memory in the AGC, it would be reasonable to assume that much of the software would be stored offline on magnetic tape, with programs being read from tape as needed. This is a practical solution, and was demonstrated in the latter part of the Gemini program.[1] However, tape is bulky, slow and not always reliable, making it undesirable for the rapid response needs of a real-time system. Eventually, memory technology progressed to where the mission software could (just barely) be squeezed into a set of read-only magnetic cores. With all of the programming now resident in memory and accessible at all times, there was no need for a tape system and the complex hardware and software required to manage it.

Multiprogramming, the ability to run multiple programs in parallel (all of which are demanding attention), is key to efficiently managing the diverse problems of spaceflight.[2] It is easy to visualize the different types of problem that a computer must solve simultaneously. Perhaps the best example is during the lunar landing, where the processing demands are the highest. First, the spacecraft must know where it is and how it is moving, so a set of variables called the state vector must be continuously updated. Next, the Digital Autopilot must ensure that the spacecraft maintains the desired attitude commanded by the crew or another program. Providing this attitude data is the job of the landing guidance routines, which calculate targeting solutions based on position, velocity, attitude and engine performance data. At the same time, continuously recalculating the optimal abort trajectory might also be desirable. All of these tasks and many others must execute concurrently. While it is possible to write a large, monolithic program to do all of this, the development and testing of such a monster is difficult at best, and unmanageable at worst. A basic principle of programming is to divide the overall task down into smaller, more manageable modules that are far more reliable, verifiable and maintainable.

Although breaking software down into smaller modules greatly reduces the design and development effort, many of these programs must still run at more or less the same time. With the requirement for so many simultaneous operations, how does a single-processor computer perform them? The magic of running several programs at once, or multiprogramming, exploits a combination of processor speed, the relative slowness of human perceptions, and the reality that not all processing is so time-

---

[1]   Storage of AGC software on tape was seriously considered, and was close to becoming a reality. Eventually, the project was cancelled, and development did not progress past the prototype stage.

[2]   Note that multiprogramming (also known as multitasking or time-sharing) is very different from multiprocessing, which uses multiple CPUs within a single computer. Multiprogramming, multitasking and time-sharing are all terms describing the "simultaneous" execution of software on one processor. We will use these terms interchangeably throughout this chapter.

critical that it cannot be interrupted. Insisting that multiple programs execute simultaneously is rarely necessary, and in the case of single-processor computers, physically impossible. Fundamentally, the most important requirement in a real-time computer is being able to handle the worst case processing demands, which in turn drives all the other basic system requirements. With sufficient processing power, the most critical application can finish well before its deadline, with the "leftover" CPU cycles available for other work. Assuming that the hardware is powerful enough to run all of the work within prearranged deadlines, it is possible to ask basic questions about how work is processed. Does it really matter which process runs first? What is the impact if one process hasn't finished executing, and we turn the CPU over to another process? What mechanisms are available for dispatching work to the CPU?

SCHEDULING: PREEMPTIVE AND COOPERATIVE MULTIPROGRAMMING

Scheduling work: Christmas at Macy's

In life, as in a computer program, no-one likes to wait. The image of a harried sales clerk trying to accommodate countless impatient customers during the holiday season is almost too painful to witness. How is it possible to manage this situation with some degree of "fairness", however *that* might be defined? The frazzled clerk could take the customers one at a time and in sequence, with no regard to how much attention each person might require. This "first come, first served" model is the most basic form of scheduling "fairness", yet it ignores the fact that someone might have only a quick question, or is likely to spend a large amount of money. At this point, it is time to move past the "first come, first served" philosophy, with the concept of fairness evolving to include the notion of priority. How much should priority play when selecting customers for service? Next, what should the salesperson do when the customer's attention is directed elsewhere, such as when they leave to try on their new clothes? Should the clerk wait dutifully by the fitting room, or take the opportunity to help another customer? When assisting a customer, what should be done if another person charges to the head of the line, demanding to be serviced immediately? Or, should our enterprising employee simply limit each customer to one minute of time, before having to rejoin the line at the end, serving them in a "round-robin" fashion?

Our little excursion into the retail world perfectly mimics the issues in scheduling work in a computer. Fundamentally, there is the need to manage several conflicting goals: all customers want service immediately and exclusively, some customers are more important than others, and interruptions will always occur when attending to someone else. The ability to service only one person at a time is acceptable only if there is no-one else waiting. As new customers arrive and are forced to wait, the ability for the store to sell merchandise diminishes. A system capable of running only one program at a time faces the same dilemma. Assigning a priority to each task (or the interrupt level) improves the situation somewhat, as some important work is dispensed with quickly, but this is little comfort to the person at the end of the line.

With several customers (programs) demanding attention at once, there is the need to invent a way of dividing the available time amongst them.

Cooperative multiprogramming
There are two different mechanisms available to implement multiprogramming, each with its particular advantages and disadvantages. One of the earliest and most straightforward techniques is "cooperative multiprogramming". This technique is a reasonable choice for imbedded systems, where resources are small and programs are designed to strictly adhere to the multiprogramming architecture. As its name implies, the programs actively participate and cooperate to implement the multiprogramming environment. All programs periodically surrender control back to the Executive, which reviews the priorities of all pending tasks and selects the highest priority task to run. In this environment, the Executive's role is reduced to little more than creating, tracking and terminating jobs, and dispatching their work. Dispatching work is the act of turning control of the processor over to a program, allowing that program to begin or resume execution. In a computer that has only one processor, this program is now the only work that is executing. All other programs, including the Executive, are idle, awaiting their turn at the processor. There are many sophisticated algorithms available for dispatching work. One is to presume that the dispatching decisions are based only on priority.

In priority-based multiprogramming systems, there are no assurances that all work waiting to run will in fact be serviced within any given time. As a result, there is the very real possibility that a low priority job will wait indefinitely for service. A practical solution is for the active job to voluntarily lower its priority after its critical processing has completed, leaving itself available for execution once any backlogged work is processed. This type of multiprogramming has additional advantages for constrained systems like the AGC. Hardware and software complexity, and by extension, costs, are usually minimized by the sparse design of the system. Very little time is spent in executive routines, thus improving overall throughput. Despite its simplicity, there are advantages of a design where priority is the sole determinant of whether a job will be run. Multiprogramming systems have the responsibility for managing access to shared resources, such as memory areas containing especially critical data. While executing, a program may be using such a resource, and cannot allow other programs access to it for a period of time. Only after the higher priority work has finished (or lowers it priority) will a lower priority job start and have access to the resource. In effect, the Executive's priority scheduling acts as a lock on the resource. The ability to safely use a shared resource without a locking mechanism exists only if programs are carefully prioritized so that a job using that resource has the highest priority of any other job that might also be scheduled at the same time.

There are significant constraints to using cooperative multiprogramming. Unsurprisingly, the technique is completely dependent on a program's willingness to cooperate and return control back to the Executive on a regular basis. Applications must be designed with this cooperation in mind, and must be written with the perspective of how it will interact with other programs. The consequences of a problem occurring in cooperative multiprogramming are serious. If a program is overly busy working on a problem, or if a software bug prevents it from returning control to the Executive, no other work can reliably run.

Preemptive multiprogramming

Preemptive multiprogramming is the second major strategy for managing work in a computer. Unlike cooperative multiprogramming, where the application actively participates in the scheduling of work, programs in a preemptive system do not play any part in determining how or when they will run. Rather than depending on the application to make timely calls to the Executive, a timer generates an interrupt that suspends whatever process is running at that time. These timer interrupts are the basis behind the phrase "time-slicing", where processing during an interval is spread across several tasks. By design, the application programs are blissfully ignorant of being alternatively scheduled and then interrupted. Each time an interrupt occurs, the state of the currently running program is saved and the Executive scans the list of waiting work. Preemptive systems usually try to give all jobs access to the processor even if they are of low priority. Because a job may be interrupted at any time, there are no provisions to proactively save data or otherwise put the job in a known state before the interrupt occurs. Also, since any job can be dispatched at any time, a process usually cannot have an expectation of completing by a specific deadline.

Strictly speaking, the AGC does not have a formal preemptive multiprogramming Executive. Still, a number of timer interrupts do make calls on the Executive to start jobs. When this occurs, the entire job queue is scanned for the highest priority work, as in a purely preemptive Executive. In fact, the AGC's Executive is best considered as a mix of both cooperative and preemptive multiprogramming techniques.

THE EXECUTIVE

When discussing the features and functions of a computer, our impressions are usually defined by the capabilities of its operating system, or OS. Operating systems are designed to provide a rich set of services to user application programs, such as memory management, scheduling and prioritizing work, error recovery and I/O processing. Upon receiving a request to run a program, a modern OS gathers together the large number of resources necessary to create an execution environment. A block of memory, or an entire virtual address space is allocated, data files are located, and security parameters are identified. The program is loaded from secondary storage into memory, with the carefully built environment ready to receive it. When the application has terminated, it is up to the OS to clean up the mess, releasing the allocated resources for reuse by the next program. Managing these functions requires manipulating a staggering amount of internal tables, queues and buffers, creating a hugely complex system. Of course, none of these functions are relevant in the AGC. Unlike a 21st century general purpose computer, the AGC has the huge advantage of operating in a tightly defined world of hardware and software. Here, rules for application behavior are well understood and rigidly enforced, and reliability is ensured through rigorous testing. All resources required by the software

are anticipated during its development, and are already available when program execution begins. External hardware comprises a few basic device types, all of which use very few instructions to interact with. Such careful design results in software that can be "trusted" (for lack of a better word), eliminating the need for elaborate mechanisms to isolate the hardware and OS from the application. By eliminating or limiting the scope of services that the AGC provides to the mission software, a very small and efficient operating system is quite practical.

This system, simply known as the Executive, is the heart of the AGC. It provides capabilities that are advanced even by modern standards, especially in view of the limited amount of hardware available. In this section, we will focus on the major services provided by the Executive:

- Job initiation, termination and scheduling
- Memory allocation
- Restarts and error recovery (program alarms)
- User interface (DSKY)
- The interpretive language and its "virtual machine"
- Event timing
- I/O.

Much of an operating system's design centers on two basic assumptions: the OS must manage the execution of application programs and provide them with necessary resources, and the OS should protect itself and other applications from any errors or faults that might occur. Large commercial computers, with their abundance of processing power and memory, achieve these goals with considerable effort. However, the AGC is especially notable in its ability to effectively manage the work running in the computer, and its sophisticated restart and recovery design. The lack of hardware resources in the AGC placed a practical limit on what capabilities were available to the Executive. Still, the Executive provided services for the programmer that were sophisticated enough to manage multiple programs, tasks and hardware devices.

Scheduling and dispatching in the AGC

When executing, the representation of a program to the Executive is through data structures containing the essential information about the process. A modern operating system might include data structures for process identification numbers, references to allocated resources, execution priority, entry point addresses and so forth. Creation of this data is the foundation for defining the existence of a process – without it, by definition, there is no program to run. In the AGC, the 12-word data area known as the Core Set contains all of the information necessary to manage the execution of a program. Very little information needs to be placed into the Core Set area. Process information occupies only five words of each Core Set, with the remaining seven words available to the job for temporary variables. The five words used for process management contain the program's priority, entry point address, a copy of the BBANK register, various flags and a pointer to the Vector Accumulator (VAC) area, if one has been requested. The seven words in the temporary area have a

MPAC	00_8
MPAC+1	01_8
MPAC+2	02_8
MPAC+3	03_8
MPAC+4	04_8
MPAC+5	05_8
MPAC+6	06_8
MPAC Mode	07_8
LOC (Program Ctr)	10_8
BANKSET (BBANK)	11_8
PUSHLOC	12_8
PRIORITY / VAC Ptr	13_8

Figure 30: Core Set layout

dual functionality, depending on the type of code executing. When basic AGC instructions are running, the seven free words in the Core Set are unstructured, allowing the programmer to use the storage as desired. Interpretive programs use this area for arithmetic and logical operations. Now a highly structured memory area, these seven words become the Multipurpose Accumulator (MPAC), which performs duties similar to the AGC's hardware accumulator. For convenience, when a non-interpretive program is running, these seven words are still referred to as the MPAC.

Requesting a basic job: NOVAC

Five words of memory in the Core Set contain the basic information of a program executing in the AGC. A Core Set represents the program to the Executive, and contains all the status information necessary to determine if the program is ready for dispatching or not. Several Core Sets are stored in a contiguous table in unswitched erasable memory, with six available for the Command Module and seven for the Lunar Module. As entries in a table, Core Sets are numbered, with Core Set 0 as the first entry in the table, followed by Core Set 1 and so on. An additional Core Set is reserved for the "DUMMY" job, which "executes" only when there is no other work available to run. Available Core Sets are identified by a negative-zero (77777_8) in the priority field, making it easy for the Executive to scan for an available entry. Small efficiencies, such as organizing Core Sets into "available" and "in use" queues are not practical as the length of the table is so short that there is little benefit from the additional complexity.

To schedule a basic job (that is, one that does not call on the Interpreter) for execution, the starting address of the new job and its priority are passed to the NOVAC executive routine. NOVAC then scans the lists of Core Sets to find one that

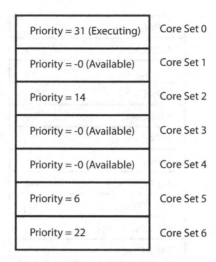

Priority = 31 (Executing)	Core Set 0
Priority = -0 (Available)	Core Set 1
Priority = 14	Core Set 2
Priority = -0 (Available)	Core Set 3
Priority = -0 (Available)	Core Set 4
Priority = 6	Core Set 5
Priority = 22	Core Set 6

Figure 31: Core Sets and job scheduling

is available. Careful design and exhaustive testing of the AGC software demonstrated the maximum number of Core Sets to be sufficient for all mission phases, so NOVAC should never find the Core Set table full.[3] On finding an unused Core Set, NOVAC stores the starting address and banking information, job priority and other data, leaving the job ready for scheduling. Now that the new job is ready to execute, the next step is to determine if its priority is higher than the currently running job. If so, the Executive must suspend the currently running work and allow the new job to begin processing.

Core Sets are very dynamic data structures, and are created and deleted as programs start and end. With one exception, there is no ordering to the jobs in the Core Set table, and executable work does not have to reside in consecutive Core Set table entries. The exception is the first entry of the Core Set table. Core Set 0 is always assigned to the job that is currently executing, with the other Core Sets representing jobs that are not ready to run or are of a lower priority. Since this property is fixed in the Executive's architecture, any job that wishes to run must swap its Core Set with the currently running job. Information necessary to resume a new or suspended job is stored in the Core Set, so dispatching requires little more than swapping the Core Sets of the new and current job. Note that during a job's lifetime, its process information may be stored in any of the Core Set areas when it is not active. As jobs with higher priorities start and end, it is not possible to know in advance where an inactive job's Core Set will reside.

---

[3] And yet, it did happen. A full Core Set table caused the 1201 program alarms during the Apollo 11 lunar landing. The effect of this alarm, and the equally nefarious 1202 program alarm (which indicated that there were no Vector Accumulator areas available), is discussed in the restart/recovery section.

Requesting an interpretive job: FINDVAC

Seven words of temporary storage might be viewed as ludicrously small, yet work is broken down sufficiently so that the Multipurpose Accumulator in the Core Set is sufficient for small jobs. This is not the case when using interpretive instructions in the program. As presented later in this chapter, the Interpreter creates a "virtual machine" with an architecture and instruction set that is unrelated to the underlying AGC hardware. The Interpreter has far larger space requirements for temporary variables than programs running in native AGC code. To satisfy this demand for storage, a second area in unswitched erasable storage, the Vector Accumulator area is available. Similar to Core Sets, the five VACs are also arranged in a table and each provides 43 words of storage to the job. The first word of each VAC area is reserved to indicate whether the VAC area is in use or not, leaving 42 words available to the application. When scheduling a job containing interpretive instructions, a call to the FINDVAC Executive routine is used rather than NOVAC. FINDVAC scans the list of Vector Accumulator areas, looking for one that is available. Once found, the VAC area is marked as "in use", but a job cannot run with only a VAC area defined. Now that a VAC area has been acquired, FINDVAC's second task is to use the NOVAC code to obtain a Core Set area for the new job. The address of the VAC area is placed in a word within the Core Set just allocated, along with the other job information, and the job is ready to be scheduled.

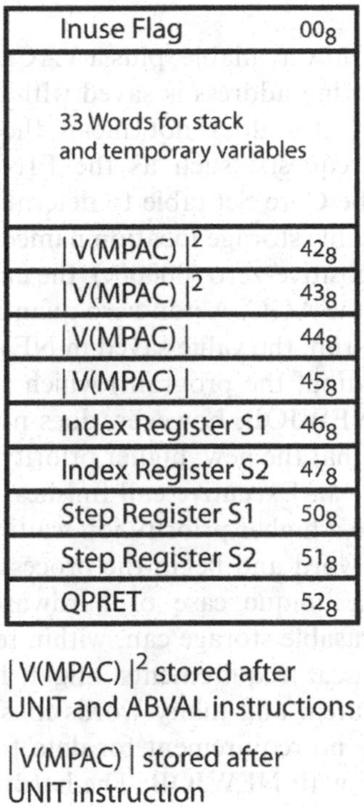

Inuse Flag	00_8
33 Words for stack and temporary variables	
$\mid V(MPAC) \mid^2$	42_8
$\mid V(MPAC) \mid^2$	43_8
$\mid V(MPAC) \mid$	44_8
$\mid V(MPAC) \mid$	45_8
Index Register S1	46_8
Index Register S2	47_8
Step Register S1	50_8
Step Register S2	51_8
QPRET	52_8

$\mid V(MPAC) \mid^2$ stored after UNIT and ABVAL instructions

$\mid V(MPAC) \mid$ stored after UNIT instruction

Figure 32: Vector Accumulator area layout

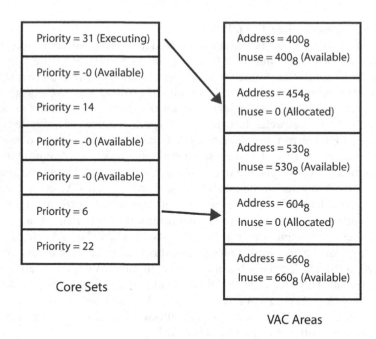

Core Sets

VAC Areas

Figure 33: Allocated Core Sets and VAC areas

At this point, a Core Set is now available, plus a VAC area if needed. The priority, along with the program's starting address is saved within the Core Set, making the job ready to run. However, this does not mean the job will begin executing immediately. All Executive requests, such as the FINDVAC and NOVAC just described, require scanning the Core Set table to determine the highest priority job. The result is placed in an erasable storage location named, appropriately, NEWJOB. When NEWJOB is equal to positive-zero (00000_8), the currently running job has the highest priority of any job in the AGC. A non-zero quantity indicates not only that a higher priority job is ready to run, the value saved in NEWJOB is also the address of the Core Set (and by implication, the program) which should be run next. Simply having a non-zero value in NEWJOB, however, does not mean that the active job will change immediately, nor that the new, higher priority job has been swapped into the Core Set 0 location. After an Executive call finishes, it resumes the current job, despite the fact there might be a higher priority job waiting. It is up to the active job to interrogate the NEWJOB word and begin the process of starting the new job.

The NEWJOB word is a unique case of hardware and Executive software integration. Most words in erasable storage can, within reasonable limits, be located anywhere in memory. Practical requirements might force some data into the unswitched erasable storage area, and many words must be in contiguous storage locations, but there is usually no requirement for data to reside in specific memory locations. This is not the case with NEWJOB. The hardware passively monitors the NEWJOB storage word, which is strictly defined as location 00067_8. Although the Executive uses and references NEWJOB no differently than it does any other memory location, the hardware maintains a special vigil on its use. For cooperative multitasking to work, a regular check for other jobs waiting to run is necessary, which requires a reference to the NEWJOB word. Assuming the goal of testing NEWJOB every 20 ms is maintained, all is well with the system. However, if there is a problem in the software, the error might prevent the regular interrogation of NEWJOB. In this case, there would be no checking for a new job, and the current job would run forever, preventing any other work from gaining access to the processor. The hardware allows 640 milliseconds to pass before it acts to clear this situation by invoking a restart of the entire system. Called the Night Watchman by the engineers, it presumed that the software "punched the clock" (e.g., checked the NEWJOB word) indicating it was still awake and operating normally.

Checking for new work

Now that the new work is scheduled and ready to run, control returns to the currently active job. In using cooperative multiprogramming to dispatch new work, the new job must wait until the active job queries NEWJOB to possibly relinquish control to a higher priority job. The timing for the test of NEWJOB is somewhat flexible, merely having the requirement that the check is performed at least every 20 milliseconds. Such flexibility allows the job to choose the point where control is potentially surrendered, providing an orderly transition from one job to the next. Of course, an exception exists with interpretive programs, where the advantages of interpretive instructions are offset by their long execution times. The interpretive instruction set, designed to implement an entirely different architecture than the native AGC, cannot easily test NEWJOB directly. The option of switching between the basic and interpretive instructions is not practical, owing to the additional overhead and disruption to program logic. A compromise is reached by having the Interpreter system itself check NEWJOB after executing each interpreted instruction.

Testing NEWJOB requires only a two instruction sequence, beginning with a Count Compare and Skip (CCS) instruction. Remembering from the instruction set section, CCS first decrements the contents of the accumulator, then skips to one of the locations following the CCS based on the value referenced by the instruction operand. When testing NEWJOB, we are interested in a positive-zero indicating that the current job is the highest priority job and no further action is necessary, or a value greater than positive-zero, defining the Core Set address of the higher priority job waiting to run. In this second case, the active job relinquishes control by calling the CHANG1 Executive routine. The coding sequence is:

```
CCS     NEWJOB      IS HIGHER PRIO WORK WAITING?
TC      CHANG1      > +0, CHANGE ACTIVE JOB
                    = +0, WE ARE HIGHEST PRIO
```

The elegance of the Executive's design, and how it interacts with the CCS instruction is apparent. The decision making for cooperative multiprogramming condenses into two instructions, only one of which executes if no higher priority work is waiting. This sequence is not immune from problems, as there is no sanity checking of the contents of NEWJOB before it is tested. Conceivably, it could hold negative-zero or less than negative-zero, both illegal values that would cause unpredictable results. The FINDVAC and NOVAC routines initially test for invalid data in NEWJOB, but this cannot guarantee the data in NEWJOB has not been overlaid in the meantime. Despite this exposure, the instruction sequence for testing NEWJOB is a fast and efficient way to switch to the highest priority work in the system.

Suspending the current job, and starting the new job

The CHANGJOB Executive routine (whose entry point is CHANG-1) is called if the CCS test of NEWJOB recognizes there is a higher priority job waiting. CHANGJOB's first task is to save the active job's return address and banking information into its Core Set in preparation for suspending it. The actual suspension of the active job, which occupies Core Set 0, and starting the new job, requires only the swapping of their respective Core Set contents. After this is done, NEWJOB is set to positive-zero, as the highest priority job is now active. Finally, the starting address and banking information are loaded from Core Set 0 into the Z and BBANK registers. With the Z register pointing to the new program and the banking registers defining the program's memory environment, the transfer of control to the new job is complete.

Changing the job's execution priority

Jobs do not always have to run continuously at their original priority. During the lifetime of a program's execution, higher priority work may be scheduled, or a critical routine in the job might have finished. As such, more efficient scheduling of work is possible when jobs dynamically adjust their priority to reflect the relative importance of their current work. Voluntarily altering execution priority is a useful element in cooperative multiprogramming, especially when recognizing that the current work does not merit monopolizing the processor. Changing priority is an effective tool only if all jobs are "well behaved"; that is, a job should alter its priority only with an awareness of its impact on other work in the system. Assuring this sort of good behavior is usually possible only during the design phase of software development, when establishing the priorities, sequencing and interrelationships of work running in the system. This requires that the developers of all the mission programs be aware of other programs that will be running at the same time. The benefit of allowing programs to manage their own priority is that elaborate job management schemes in the Executive are no longer necessary. However, the importance of tight controls on priority changing routines cannot be overstated. Where there are no restrictions, users could make inappropriate assumptions on the priority of their programs, often to the detriment of overall system performance.

Changing the priority of an active job is useful as a scheduling tool. There are instances where a job might need to run without being preempted and suspended by higher priority work. Inhibiting interrupts is one possible approach to this situation, but this brute force solution could easily prevent other desirable processing from occurring. Conversely, there are often occasions where there is no harm in letting a waiting job run. A job lowering its priority to below that of the waiting job is effectively suspending itself in favor of other work. A call to the PRIOCHNG Executive routine sets the priority of the job to its new value, and then performs a scan of the Core Set table for the highest priority work. If a higher priority job is found, a call to CHANGJOB suspends the current job and dispatches the new work. Note that only the priority of the active job is changeable. One job cannot change another's priority, nor can other work, such as waitlist tasks or other Executive routines change the priority of a job.[4] Jobs are the only type of work in the AGC that need the ability to change their priority. Interrupt routines and waitlist tasks, by definition, run with interrupts inhibited, and consequently have complete control of the processor.

End of job processing

All good things must come to an end, even a job in the AGC. The end can come in various ways; through an explicit termination by a DSKY input, a restart, or by the most common means: the program itself calling the Executive routine ENDOFJOB to say "My work here is done." As there is very little overhead in setting up a job to run, likewise only a small number of operations are needed to terminate and clean up after a job, with most of those tasks centering on preparing memory for reuse. Cleaning up the Core Set and VAC area (if applicable) is straightforward. A negative-zero (77777_8) is placed in the priority field of the job's Core Set, indicating that it is available. If a VAC area was reserved, the address of the VAC area is saved into its VACUSE field, which flags that memory area as available. After the job's memory resources are released, the job has effectively terminated. As the AGC does not have any other resources for a job to allocate, no further cleanup is necessary.

All of this work creates a small dilemma. In a cooperative multitasking system, it is up to the active job to test the NEWJOB location to see if a job is waiting, and then call the Executive to schedule the execution of the next job. As the active job has terminated, there is no longer any program executing to test NEWJOB, creating a situation where it is impossible to dispatch another program. To resolve this, the Executive itself performs a scan of the Core Set table and dispatches the highest priority work.

JOBSLEEP and JOBWAKE

Waiting for data input from the crew, or a hardware operation to complete, presents a problem in scheduling work in the system. When a process needs to wait for an extended period of time (a large fraction of a second or longer), simply looping and executing pointless code in an effort to "kill time" is inefficient at best. In cases such as these, it makes little sense for a program to remain active, and is far more practical to "put the job to sleep", suspending it until the operation or data input completes.

Consider the case where a job requires the crew to enter data in all three DSKY registers. Even quickly punching in the data can take up to a minute – an eternity in computer time! The job waiting on the data could suspend itself, with the expectation that another task will awaken it at the end of the data entry sequence to continue processing. This requires a cooperative effort by another program; in this case, the

---

[4]   Waitlist tasks are short, high priority programs that run with interrupts inhibited.

program monitoring the keystrokes and waiting for the input to complete. Without an idea of how long it will take to input the data, simply setting a timer to check the status of the input is inefficient. Hardware operations such as torquing gimbals present another example as they do not generate an interrupt when their operation has completed. In this case, a job can arrange for its awakening when the gyros have repositioned.

Although the JOBSLEEP routine does not provide a facility for waking a job at a specified time in the future, another routine, DELAYJOB does provides this capability. When called with the desired amount of time to wait, DELAYJOB puts the job to sleep and schedules a waitlist task using the delay time to schedule starting the waitlist task. When the waitlist task finally executes, it calls JOBWAKE with the Core Set address of the sleeping job. Up to three jobs can be sleeping with delay times in the Lunar Module, and up to four in the Command Module.

With the basics of JOBSLEEP and JOBWAKE established, it is possible to probe deeper into how these operations are implemented. A sleeping job is defined as having a negative priority, so JOBSLEEP will complement this field in Core Set 0. In one's complement notation, complementing is the same as multiplying by -1, so a job whose priority is 33 will sleep with a priority of -33, providing the benefit of preserving the priority's original value. Then the address of the next instruction to execute after awakening (the instruction following the JOBSLEEP call) is placed into the job's Core Set along with its banking information. With the job now asleep, it is no longer an active job, and so the Executive scans the Core Set list for the highest priority job. When one is found, its Core Set address is placed in NEWJOB, and the new work is dispatched through CHANGJOB. If no other jobs are ready to run, the DUMMY job is placed in Core Set 0, and the system waits for the sleeping job to be awakened.

CHANGJOB, as seen earlier, swaps the contents of the new and current jobs in the Core Set list, sets NEWJOB to positive-zero and transfers control to the job referenced at Core Set 0. Waking up a job is similar to scheduling in CHNGPRIO. First, the address of the sleeping job's Core Set is placed in the accumulator before the call to JOBWAKE. JOBWAKE complements the priority of the job back to a positive value, indicating the job is awake and ready to execute. At this point, the Executive scans of the Core Set table for the highest priority job, which is followed by a call to CHANGJOB. Note that putting a job to sleep is not the same as suspending a job. A suspended job can execute at any time when it becomes the highest priority job in the system, whereas a sleeping job will become active only when explicitly awakened by another job or waitlist task.

A job can theoretically run forever, if its priority is high enough to prohibit any other work from running. This is not necessarily an undesirable behavior. Consider a critical job that must run to completion before any other work can start. Here, the goal is to give this job a high priority to prevent it from being suspended or delayed by other work. As the program's importance is enough to justify a high priority, deferring the execution of other tasks is an acceptable outcome. The Executive does not take any action on a program that runs for a long time – as no accounting of processing time is ever done, it has no way of knowing if a job has been running for a second, a minute, or an hour. The only requirement is that the job continues to check the NEWJOB storage location on a regular basis in order to ensure that there are no other waiting jobs.

Major Modes in the AGC

But how is a program in the AGC directed to run in the first place? Short running work such as an interrupt routine or waitlist task is dependent on timers or external events to begin its processing. Restricted to a few milliseconds at most, it must schedule a job if more lengthy processing is necessary. An active job can also schedule other jobs to run, which execute as soon as their priority becomes the highest in the system. The third and final mechanism for the astronaut to start a job is through DSKY commands, using a Verb sequence to change the "Major Mode" of the computer. A Major Mode is best defined as a very long running program (minutes to hours) that controls an important mission function such as launch, lunar landing or rendezvous. In fact, few attributes differentiate a Major Mode from other jobs, and they are generally managed the same as other jobs in the system. Perhaps surprisingly, jobs running as a Major Mode are scheduled at a very low priority, but the rationale for this becomes apparent on closer inspection. Major Mode programs cannot monopolize the system resources, and frequently schedule other jobs to perform small but important functions. Each check of the list of ready-to-run programs in the Core Set might reveal a new, higher priority job, and this suspends the Major Mode program until the new job completes. If the Major Mode program always maintained a high priority in the system, the work that it has just scheduled would never have the opportunity to run. Maintaining a Major Mode program at a low priority ensures that all the work that it schedules will always be given an opportunity to execute.

The waitlist

Control computers must regularly schedule work to perform time-critical functions. Checking panel switches to see if changes have been made, scheduling the Digital Autopilot to run, and even the mundane task of flashing a display off and on are mandated to run at predetermined times. The AGC maintains a "waitlist", which is a pair of tables containing references for "tasks" that are scheduled to run in the future.

Waitlist tasks run in interrupt-inhibited mode, where no other work in the system can disrupt its processing. Since these tasks effectively prevent other programs from using the AGC, they must necessarily limit their execution time. An informal time limit of five milliseconds is imposed on the tasks. There is no mechanism to interrupt or otherwise enforce this limit – only careful design assures that the task will not exceed its allotted time. Five milliseconds is equivalent to executing 150 to 200 machine instructions, which is certainly not enough to perform complex operations. When a significant amount of processing is required, the task will schedule the execution of a larger job to perform the bulk of the work. The limited number of erasable storage locations available for temporary variables further restricts waitlist tasks. A few words are allocated in erasable storage for programs that run with interrupts inhibited. These are only valid during the task's execution, and there is no expectation of their contents being preserved after interrupts are reenabled. That the temporary areas are valid during waitlist processing is assured only because no other program can take control of the processor and reuse the storage while the task is running. Once the task has completed, the temporary storage is free to be reused by the next task in any way that it sees fit. Additionally, the A, L and Q special registers may be used, with the caveat that they are preserved (in the ARUPT, LRUPT and QRUPT areas) and restored before the task has terminated.

Two waitlist tables are identified by their contents. Table LST1 contains the time intervals used in scheduling the start of the tasks. LST2 holds the starting address and memory banking information. The LST2 tables are paired one-for-one with the entries in LST1, with the exception of the first LST2 entry. Normally contained in LST1, T3RUPT contains the time until the next waitlist task, which is counting down until the next interrupt. This exception leaves LST2 containing the address of the next waitlist task to run. The waitlist is aptly named, being just a list of tasks waiting for their scheduled time to run. As seen in the tour of the special registers, the TIME3 counter is used exclusively to schedule tasks in the waitlist. A time interval is placed in TIME3, which counts down one "tick" every ten milliseconds. After reaching zero and overflowing on the next clock tick, the T3RUPT interrupt is raised, transferring control to the waitlist management routine.

A waitlist task is scheduled by providing the start time and address with banking information. Again, "time", as used by the TIME3 counter, is not in the absolute "time of day" sense. Rather, the time is defined as an interval to a moment in the future, measured in increments of ten milliseconds. Waitlist tasks are ordered chronologically, with the task next scheduled to execute at the top of the list. Each interval defined for a request is relative to the point of its scheduling. That is, if a task is to start 140 milliseconds from now, the first LST1 entry will not necessarily contain the value of 140 milliseconds. The time stored in the LST1 interval table is calculated as the sum of the contents of the TIME3 counter, plus all the intervals in the requests that are ahead of it. For example, a task needs to run 180 milliseconds from now and two entries already exist in the waitlist, both scheduled to run before this new request. A check of TIME3 shows that the next task will run in 30 milliseconds, and the second task will run 40 milliseconds after the first. Properly setting the execution time for the new task, 180 milliseconds from now, requires adding the 30 milliseconds remaining in the TIME3 counter to the 40 milliseconds that will elapse between the first and second tasks. Subtracting the 70 milliseconds of waitlist time from 180 milliseconds results in an interval of 110 milliseconds. The new entry is appended to the end of the waitlist, and will run 110 milliseconds after the task ahead of it executes.

With the tasks in the waitlist organized by start time, a more complicated case occurs where a new task is scheduled to run between two existing tasks. In this example, TIME3 contains a value of 20 milliseconds, after which it will start Task A. Task B is also in the waitlist, and will start 100 milliseconds after Task A. Finally, Task C needs to start in 60 milliseconds, meaning 40 milliseconds after Task A starts. To retain the proper sequence, task B is moved down one entry in the waitlist and

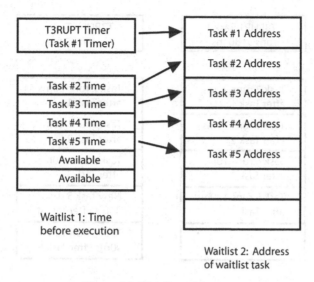

Figure 34: Waitlist tables

Task C is placed in the resulting gap. A problem now exists with task B, as its start time is now invalid and needs to be recomputed. With Task C now starting 40 milliseconds after task A, the new interval for Task B is 60 milliseconds.

Managing the waitlist is a straightforward matter of inserting and deleting entries in tables. After the active task has started, its entry in the waitlist is no longer needed and it is removed from the table. All of the remaining entries are moved up one position, and the TIME3 counter is set to the interval for the next request. A waitlist task is scheduled only once. If there is a need to run the same task according to a periodic schedule, the task is responsible for rescheduling itself for its next execution.

Phase tables part 1: restart and recovery

In complex systems such as spacecraft and aircraft, computer failures can have catastrophic consequences. The range of potential failures is broad, from a missed opportunity for a scientific observation to the destruction of the vehicle and loss of the crew. The need for a real-time system to maintain control of the vehicle is paramount, especially during critical mission phases. Failing this, the system must react to problems in a controlled manner, and if necessary degrade to a known and benign state.

The restart capability in the AGC is quite flexible, and ensures that processing can continue in all but the most extreme situations. If a job requires restart protection, it calls the PHASCHNG routine in the Executive. One or more parameters are passed to this routine to specify which of the predefined restart routines to use, along with any variable data the might be appropriate to that restart option. A restart uses one of several different strategies to recover from a program error. First, the job having difficulty can be terminated with no intention of restarting it. A second option is to

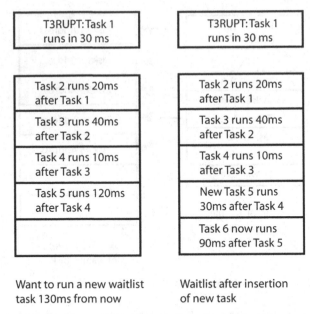

Want to run a new waitlist task 130ms from now

Waitlist after insertion of new task

Figure 35: Addition of a new waitlist task

restart the job from the beginning, hoping that the problem will not recur, and performing little in the way of cleaning up data from the previous invocation of the program. A third and far more complex option, is to schedule work that performs data cleanup and reenters the program at a previously designated restart point. Such flexibility is necessary due to the complexity of the software used in the AGC, and the realization that not all program errors are necessarily fatal. By closely matching the recovery option with the problem, it is possible to create the level of robustness that is essential in a real-time system.

In a system as complex as the AGC, a wide range of problems might occur. Some are the result of innocent procedural errors or minor hardware problems, while others arise from unexpected logic errors or unforeseen interactions of hardware and software. While it is not possible or practical to analyze a problem in real time to determine the root cause of the failure, it is reasonable to create a general set of rules and procedures for recovery that are appropriate for each executing program. In cases where a routine has few interdependencies on other programs or data, it might be easier to restart the program from the beginning. A simplified example might be a routine that starts, reads distance information from a radar, displays it on the instrument panel, and terminates. If an error occurs before the processing completes successfully, it is usually easier to try to run it again. A brief moment of data loss should not be a serious situation. In other cases, restarting from the beginning of a program that performs a lengthy computation is not always practical, such as in the case of calculating a targeting solution for a major maneuver. A more appropriate restart process might be to reenter the program to restart the calculations from the last successful iteration and reuse the existing data wherever possible.

These scenarios are simplified examples of some of the possible approaches to software recovery. Given the large number of programs that are usually running at a given time, a single restart solution is usually unworkable. For a word processor to recover from a spelling error by rebooting the PC is certainly absurd; and it is equally inappropriate to simply report that the hard drive in your laptop has failed, and blithely continue on. To address the wide range of recovery requirements, the AGC's Executive provides a restart mechanism that is dynamic and highly flexible. Note, however, that waitlist tasks are not defined with restart protection in mind.

Complexity in the restart system is managed by assigning programs to one of six "restart groups". Once defined, a restart is performed regardless of whether a job is active, suspended or sleeping. Each group manages a logically related collection of jobs, and each group can only restart one job. As an example, restart group number four in the Command Module is responsible for both SPS burn programs and also for entry software. Clearly, no conflict exists because these two situations are mutually exclusive events – by the time of atmospheric entry, the SPS will have been jettisoned. Other groups are organized in a similar fashion, such as with Servicer programs handled by restart group two, and group five restarts assigned to IMU management. The absence of restart protection for the Digital Autopilot is notable. The DAP, executing every 100 ms, is an example of a program that is run without any recovery protection. A failure during one DAP cycle is not necessarily a danger, since it will execute again within a fraction of a second. If a serious problem in the DAP were to create a control problem, it would be manually disabled through a switch on main instrument console.

The restart table

The restart table is located in unswitched fixed storage and is organized into six restart groups. Within each group, one or more restart "phases" define the specific restart procedure within the group, each assigned to a unique restart location within the program. As a program progresses, and its restart requirements change, it informs the Executive through the PHASCHNG routine that a different restart phase is now in effect. As several different programs share the same group, an individual job might only use one, or perhaps a few, restart phases within the group. Each entry in the table defines a single recovery process, which is associated with only one program. There is no universal or all-encompassing restart table, and the contents in the Lunar Module version of the AGC software are considerably different from its Command Module counterpart. Four different restart options are available, and are defined through the attributes of the restart phase number. For clarity, it is easier to use the notation "G.P" to describe the restart point, with G as the group number, and P, the restart phase. Using this notation, the list of restart options are:

G.0	Inhibit group restart
G.1	Restart last display
G.Even#	Execute two restart routines
G.Odd# (> 1)	Execute one restart routine

In this notation, "even" phase numbers and "odd" phase numbers are exactly that; G.Even# include phases such as 2.2, 2.8, 4.6, and so forth. As G.1 is already assigned, the "odd" phases must begin with 3. An important effect of this convention is that both restart phases G.0 and G.1 are common to all groups, and neither requires defining a unique process to run. During a restart, each entry in the phase table is checked for a zero entry. If so, that restart group is not active and no additional processing occurs. Restart phase one redisplays the data on the DSKY, but does no other recovery processing. Only the remaining two types of restart phases, G.Even# and G.Odd# (> 1) will attempt to restart the program itself. Phases using odd numbers greater than one perform only one operation during that group's restart processing, such as scheduling either a job or waitlist task. Two restart operations occur in even-number restart phases, and any combination of jobs or waitlist tasks are possible.

Each restart operation requires three words to define, with its contents depending on whether the restart action requires scheduling a job or a waitlist task. Odd-number restart phases performing only one restart operation will only occupy three words, while even-numbered phases require six words to reflect the two operations scheduled for that group. Entries for a job in the restart table require three words: one word for the execution priority, and two words to define the entry point address. Waitlist tasks have a word dedicated to their scheduled start time, and also use two words for their entry address. Although this format appears sufficient, it does not contain all of the information required to schedule its work. In the quest to minimize the amount of storage occupied by the phase table, the several flags necessary for specifying other program attributes are not stored in dedicated memory areas.

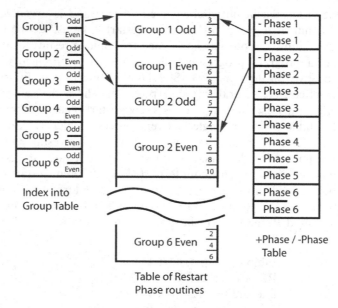

Figure 36: Restart and phase tables

Rather, these flags are implemented as encodings of the three restart table words passed to PHASCHNG. In the case of a job, the first decision is to know if a VAC area is necessary or if only a Core Set will suffice. Using the priority word in the restart table entry, a positive priority value indicates that a VAC area is necessary. If the priority is complemented, a Core Set is obtained for the job but no VAC area. The entry point address is used in a similar manner. When complemented, it indicates that a waitlist task must be scheduled (not a job) and that the word used for defining the job's priority should be used as the waitlist scheduling time. This encoding scheme, which may appear on the surface to be needlessly complex, actually saves several words in the unswitched fixed storage area.

Phase tables and restart processing

The number of options available with restart processing creates a dizzying array of data formats and parameter lists to process. However, the saving of storage space in the AGC is always a high priority, and the concise format of the PHASCHNG parameter lists reflects this. Each program protected by restart code has its own unique set of requirements, all of which must be passed to the PHASCHNG routine. At a minimum, the call to PHASCHNG must include the group and phase number, and flags to indicate important properties of the programs used for the restarts.

Two broad classes of parameters are used when calling PHASCHNG. Calls to PHASCHNG of "Type A" define a fixed, single-word parameter that references an explicit restart routine. "Type C" parameters include additional data words to describe variable data used by the restart routine. Such variable data might be the scheduling time for the defined recovery waitlist task, or the entry point address of the restart code. PHASCHNG also allows a third, "Type B", format, which is a hybrid that combines one fixed and one variable parameter. Highly condensed, the format of the data word following the call to PHASCHNG contains essential data such as the parameter type and group number. For those parameter lists that support variable data, bits also define what additional parameters are present.

The design of the phase tables

Dynamically altering the contents of the restart table is not possible, as it resides in fixed storage. This is reasonable, as the data in the restart table must be protected against any occurrence of a program failure that alters memory. However, restart routines frequently require data that is specific to the recovery needed, and this is saved in the phase table. The restart phase table contains the parameters necessary for recovering each restart group. The parameter strings passed to PHASCHNG are saved in the table, together with their bitwise complements. This odd means of storing data is an essential part of the sanity checking of the system during the recovery sequence. Saving the inverse of the data parameter is an effective means to assure that the restart tables have not suffered from corruption by a program error. Phase tables, like so many other critical data areas, are particularly vulnerable because they reside in unswitched (common) erasable storage. Without any storage protection mechanism in the AGC, damage to the phase tables can occur at any time and without any warning. Storing a second copy of the phase table might be useful, but there is always the chance of having several words overlaid by identical garbage. For example, if an errant process wrote zeros throughout memory, how would this be detected without extensive logic? In a more general case, how do we determine if two identical copies of data are invalid? While the technique of saving both the data and its complement is far from foolproof, it is trivial to implement and requires only a small amount of additional storage. Upon entering the restart processing, the phase table is checked to see if it is still valid. If a test of the phase table fails, erasable storage is presumed to be corrupted and program recovery stops. With a phase table that is longer valid, it is no longer possible to restart the program and so a hardware "fresh start" is initiated.

A check of the phase table's health is driven by more than a conservative programming philosophy. When performing a restart, no effort is devoted to attempting to diagnose the root cause of the problem; rather, the restart process begins immediately. While this allows for a fast recovery, so important in a real-time system, it leaves open the question of how extensive the problem is. The most conservative assumption is that *any* routine could be at fault, so it is necessary to cancel all active jobs and tasks to ensure that the failing component is terminated. Logic errors are always likely suspects, and might generate a restart for any number of reasons. Of course, bad logic usually begets bad data, and it is highly likely that corrupted erasable memory resulted from the problem. Regularly examining several areas of erasable memory is part of the self-checking and restart processing. Unlike the phase table entries, testing of these data areas centers on checking for known values or ranges. If a discrepancy exists, it must be presumed that erasable storage is corrupted, and a fresh start is the only option to clear the error.

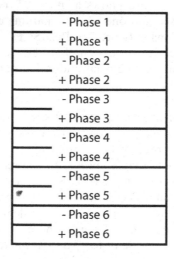

Figure 37: The + phase / –phase tables

Phase tables part 2: control of Major Modes

A secondary function of phase table management routines is maintaining the program number of the currently executing Major Mode. This quantity, saved in the MODREG erasable storage location, is displayed on the DSKY as the program number. Maintaining the Major Mode number is little more than a housekeeping task, as a change in MODREG does not result in scheduling a new program. On the other hand, a number of routines check this value, and alter their logic based on the program that is currently running.[5] In a few critical mission phases, the Major Mode progresses to a point where it starts to run significantly different execution logic. This change is presented to the crew as a change in the Major Mode number, making it appear as if a new program has started. For example, the progression of Major Modes during a lunar landing (Program 63, then Program 64, and finally Program 65) is entirely automatic, saving the crew from the distraction of manually entering DSKY commands during a time of high workload. While the follow-on programs might appear as standalone Major Modes, their execution is dependent on the successful completion of the earlier stage.

A bad day at work: program alarms and restart processing

The AGC uses three different mechanisms to handle problems with the software or its inputs, recognizing the fact that not all situations require the same level of recovery. Program alarms, aborts[6] and fresh starts are all possible options in the case of hardware or software problems, and each provides a different level of recoverability. Minor issues such as procedural errors may present themselves as program alarms, but do not require restart processing. In these cases, the program alarm is little more than an error message, such as an indication that the landing radar is not in the correct position, or that a program wishes to use the IMU while that hardware happens to be turned off. Other program alarms or aborts are a more serious matter. For a system resource shortage or a non-recoverable software error, the only recovery option is a restart of the AGC. Restarts from program aborts take on two different identities. A software restart terminates all running work and resets many of the system variables before performing recovery processing using the restart table. The objective of this type of restart is to preserve the execution environment and continue processing with a minimum of disturbance. It is important to realize that a software restart is not the same as "rebooting" a computer. Although work is thrown away, some restarts automatically reschedule programs for execution. More serious problems can result in cases where most of erasable memory is preserved but processing is halted in a so-called "P00DOO" abort. A far more drastic response to errors is the hardware restart, or fresh start. A fresh start clears all the work in the system, reinitializes the hardware and system variables, and places the system in an idle state. A fresh start is the equivalent of "rebooting" the AGC, with all work being lost and erasable memory reinitialized.

---

[5]   Exploiting this behavior is discussed as part of the Apollo 14 abort switch workaround.

[6]   Confusingly, the terms program alarm and program abort are often used interchangeably.

Each type of restart has a four-octal-digit code number associated with it that uniquely identifies the problem. A prefix digit to the code describes the type of recovery that is done: a 2 indicates that a P00DOO abort is called, and a 3 in the first digit signifies a software restart. Using this coding scheme, a 1202 alarm will be displayed as 31202 on the DSKY. Invoking a program alarm requires calling the ALARM executive routine with the alarm number. ALARM requests a priority display of the error code on the DSKY, overriding any existing display, and also illuminates the PROG[ram Alarm] light. The crew brings up Verb 05, Noun 09 to display the current alarm number, together with those for any other program alarms that might have preceded the restart. Finally, a number of important variables are set back to their initial state. The integrity of the restart phase table is verified, and restart processing for each active phase is performed. Restart groups are handled in a sequence, from group six backwards to group one. After the restart sequences are complete, the previously executing programs are running again and critical data such as IMU orientation and the state vector are preserved. All told, the entire restart process completes in seconds. After acknowledging the alarm by pressing the Reset key, the astronaut can reenter the request, terminate the program, or take other actions.

Program abort: P00DOO
Program aborts represent more significant problems than program alarms, and result in a P00DOO abort. For example, if a program encounters an unrecoverable logic error such as attempting to take the square root of a negative number, it immediately terminates without invoking the restart processing routines. Erasable storage is not cleared, thus preserving critical system data that is required even with the system idling. With the Major Mode no longer executing, the AGC enters an idling state, displayed as Program (Major Mode) 00. Program 00 reflects the state where no Major Mode is active, yet all other routines (attitude control, telemetry downlink, etc) are actively running. As in the case of program alarms, when the error is detected, the P00DOO Executive routine is called with the abort number. Again, the PROG light is illuminated, and Verb 05 Noun 09 displays the abort code to the astronaut. Program 00 appears as the Major Mode after the Reset key is pressed, and the computer is in a state ready to start a new program.

Program abort: software restarts
Both a fresh start (hardware restart) and a software restart (often confusingly called simply a "restart") begin with similar actions to quiesce virtually all system activity.[7] Software restarts focus on bringing previously running program back up, while a fresh start resets the AGC back to an initial idle state, ready to start new work. In both cases, all jobs are canceled and all tasks in the waitlist queue are deleted. Once the system is down, the hardware and software restart routines

---

[7]   The entry point label in the program abort code is BAILOUT; a phrase not lost on pilots.

continue, but with different objectives. For the recovery process to succeed, the Executive must perform several tasks without distraction, therefore early in the restart sequence all interrupts are inhibited. Timers and flags for the T3RUPT through T6RUPT interrupts are set to prevent their routines from executing until their environment is cleaned up. The three tables associated with units of work – the Core Sets, VAC areas, and the waitlist queue – are cleared and reinitialized. DSKY routines are reset to their initial state, and the display is cleared. Warning lights are turned off, with the exception of the PROG and IMU error lights if these are illuminated. Important system variables, possibly containing clues to the nature of the restart, are gathered and downlinked to the ground for analysis. A full dump of erasable storage would be both time consuming and unlikely to provide much additional diagnostic information.

Rebooting: the AGC fresh start
A fresh start, on the other hand, bypasses the program recovery process and continues shutting the system down to an idle state. Shutdown commands are sent to the SPS, DPS or APS engines if they are firing. Much of the hardware is reset, including the IMU. The loss of the inertial platform requires that the Digital Autopilot be placed into an idle state and all thruster activity terminated. Without an attitude reference, computer control during thruster firings is impossible.[8] Since the AGC is progressing to an idle state, no restarts to recover running programs are run, and the restart tables are reinitialized. Finally, with the hardware and software fully reset, Major Mode 00 (Program 00) is displayed on the DSKY.

THE ASTRONAUT INTERFACE: THE DISPLAY AND KEYBOARD

The language of the DSKY
For just a moment, return to the early 1960s and the early design phase of the AGC. Virtually all computing was batch, using punched cards and paper tapes for data storage. Hard disks, or disk drives as they were known, were slowly making their way into commercial computing. Magnetic drums, the precursor to hard disks, offered high speed (for its day) random access to data. All of these units were large, weighed hundreds of pounds and used prodigious amounts of electricity, making them completely impractical for spaceflight computing. The concept of engaging in an interactive dialog with a computer was just starting to emerge, yet it would still be years before systems with practical timesharing capabilities became commonplace. With such a background, consider the dilemmas faced early in the AGC design process. What would an interaction with a real-time system look like? What would an astronaut need to know during a particular mission phase? What kind of data would require to be entered by the crew? Most importantly, what would the "language" of such a dialog look like?

---

[8]   Fully manual control of the thrusters is always available.

A small thought experiment is revealing in how people communicate their needs outside the world of computing. Assume you are a tourist arriving in New York with a very limited English vocabulary. It is nearing lunchtime, and you are hungry for a quick meal. Looking for advice, you find a police officer and say one of the few words you know: "eat". With a few gestures, you can convey the idea you are looking for a place for lunch. So far, so good. But, New York prides itself as a "city of restaurants"; the question now becomes, "what would you like to eat?" Knowing that New Yorkers seemingly thrive on a single dietary staple, the answer is obvious. "Pizza". The police officer now has all the information he needs to help you find a pizzeria for lunch, and after a few more gestures you are on your way. From this simple exchange, the first step is to analyze the words "eat" and "pizza" on a grammatical level. "Eat" is a verb, describing an action or an operation on an object. "Pizza" in a noun, and is the subject of the verb. The verb-noun phrase, while grammatically awkward due to its brevity, conveys enough information to be understood. Returning now to the world of computing with this knowledge, this structure is recognizable as the basis for almost every interaction. Hardware operations such as ADD [accumulator with the contents of] LOCATION, or text oriented commands found in systems like Unix or DOS (EDIT FILENAME.TXT), both are composed of a verb followed by a noun. Even modern WIMP interfaces (Windows, Icons, Menus and Pointers), despite their graphical presentation, simply

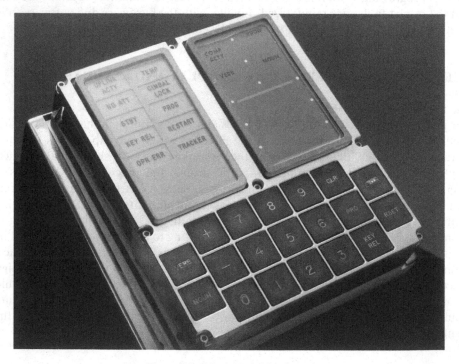

Figure 38: Display and keyboard

change the way this notation is commanded. Under "File" in a typical pull-down menu, there is "Print" (a Verb), which displays a pop-up box for specifying the filename (the Noun). Within every level of computing, almost every action can be expressed in terms of the humble verb and noun.

Therefore, it should come as no surprise to learn that two primary keys on the DSKY are labeled "Verb" and "Noun".

Using the language of Verbs and Nouns makes interaction with the AGC remarkably easy. A dialog with the computer begins with pressing the Verb key, which defines the operation to be performed, followed by a Noun code specifying the data that is to be operated upon. The primary operations performed through the DSKY are:

- Display, monitor or update data used by the AGC
- Execute and manage programs to control major mission phases
- Execute commands to control spacecraft hardware or to modify running programs.

With this background as introduction, it is time to begin the discussion of the Display and Keyboard, or DSKY. The DSKY is the visible component of the AGC, and is centrally located on the instrument panel for easy viewing by the crew. As the astronauts' primary interface to the AGC, the DSKY provides both data input and output services, plus status indications on other guidance and navigation components. Approximately 7 inches (17.8 cm) by 8.5 inches (21.6 cm) tall, the DSKY includes a 19-key keyboard, a display section for Verbs, Nouns and program numbers, three data registers, and numerous warning lights. Digits are composed of the usual seven-segment arrays, but are green electroluminescent displays, unlike the LEDs or liquid crystal displays so common today. The DSKY is assembled into a single unit, and is physically separate from the computer. A single DSKY is in the Lunar Module, but two DSKYs are in the Command Module. In the CM, the first is located on the main display console, midway between the leftmost and center positions, with a second at the optics station in the lower equipment bay at the foot of the crew couches. To illustrate the overwhelming importance of the computer to the mission, consider the locations of these units in the spacecraft. All vehicles, regardless whether they are a car, an airplane or a spacecraft, place the most important information directly in front of the driver or pilot. In the CM, the DSKY is placed squarely in the astronauts' field of view, and is one of the four primary flight instruments along with two FDAIs (attitude indicators) and the Entry Monitor System. In the Lunar Module, the DSKY is located between the Commander and the LMP at approximately waist height. In both spacecraft, ease of monitoring and inputting commands drove the DSKY's installation location.

Displays for the program number, also known as the Major Mode, and the Verb and Noun displays occupy the upper right-hand section of the DSKY. An electroluminescent light labeled CMPTR ACTY is to the left of the program number and illuminates when the computer is actively processing data, and is extinguished when the DUMMY task is running. Below the Verb and Noun displays are three, five-digit data registers used for displaying data specified by the Noun.

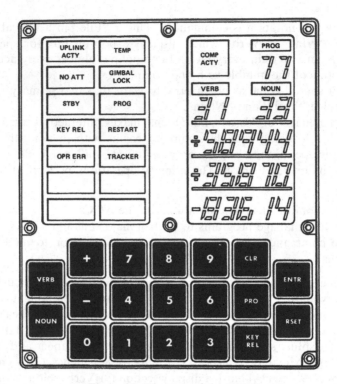

Figure 39: Diagram of the DSKY

Preceding the digits is a sign that is used only with decimal data to indicate whether the quantity is positive or negative. Nouns operate on a wide variety of data formats, from integers to octal digits, to real numbers with varying degrees of fractional display. A surprising characteristic of the display registers is the lack of an explicit decimal point. Editing numbers to determine where the decimal point falls would require computing resources, extra keystrokes by the crew and computer time, but add only a minor amount of additional readability. Octal data, of course, does not use a decimal point in its notation. In some cases, more than three data items need to be displayed by a Noun. For example, if the time before an event is displayed in minutes and seconds, both of these two-digit quantities can fit in a single register. To ensure that the data is readable, minutes will occupy the first two digits of the five-digit display, with seconds in the last two digits and the middle digit blank.

Three data registers present an excellent compromise between engineering realities and human factors, as many of the important concepts in spaceflight are expressed using three values. Attitude uses three quantities for pitch, roll and yaw, and both position and velocity vectors are defined by three values. Time is perhaps the most familiar case, using hours, minutes and seconds to express itself accurately. During the times of highest cockpit workload, the crew will be constantly scanning their instruments, looking for any indications of trouble or reassurances that the systems are performing nominally. To efficiently scan a large number of displays, and integrate that information into a coherent picture, is perhaps the most important and challenging task for a pilot. Three pieces of information, such as the DSKY register display, is the most a person can comfortably absorb and process after a quick glance. Overloading the pilot with data that might be interesting, but not essential to the task at hand, quickly becomes counterproductive.

Keys on the DSKY reflect the numeric language used in the dialog between the crew and computer. Programs, Verbs and Nouns do not use "names" in the conventional sense; they are identified by a numeric code.[9] Of the 19 keys on the DSKY, ten are for the numeric digits, plus two for the + and – signs. In addition to the Verb and Noun keys, five keys are used to interact with the software:

ENTR [Enter]: Enter provides two different functions to the crew. First, Enter indicates that the Verb, Noun or register entry is complete and ready to be used by the software. In some procedures, the computer will request that the astronaut perform a task by displaying Verb number 50, known as a "Please Perform" request. Pressing Enter acknowledges that the requested operation is completed.

PRO [Proceed]: The Proceed key is most frequently used to indicate that the program can continue using its previously calculated data. Many programs, especially those performing complex maneuvers (engine burns, rendezvous, etc) display a large number of parameters that the crew can either accept or change. In these dialogs, the computer flashes a display with the Noun code for data that is specific to that phase of the maneuver. If these numbers are acceptable, the crewmember commonly presses PRO, telling the software the values are acceptable. If they are not, they are changeable using the data entry Verbs. Proceed is also entered as a response to a flashing Verb 99, "Please enable engine ignition". By pressing the PRO key, the astronaut gives the AGC its final permission to fire the engine automatically. In conjunction with Program 06 (AGC Powerdown), pressing and holding PRO places the computer in Standby mode, where it consumes significantly less power. Bringing the AGC out of Standby mode also requires pressing the PRO key.

CLR [Clear]: Operating just as it implies, CLR allows the crewmember to clear out an entry on the display. When keying in a Noun or Verb code, or data into one of the registers, pressing CLR will blank that part of the display to allow data to be reentered from the beginning.

RSET [Reset]: A program alarm has a wide range of effects on AGC operation, ranging from a comparatively benign error message to a complete system fresh start. When these alarms occur, the PROG (Program Alarm) light illuminates and the DSKY blanks. Pressing RSET acknowledges the alarm, allowing the crew to display

---

[9]   An alphanumeric keyboard is totally impractical for a spacecraft flight computer, where rapid and concise data entry is paramount. One could easily visualize the near-comic problem of an astronaut, faced with a traditional keyboard, trying to "hunt and peck" commands while wearing bulky gloves.

the error code and continue if possible. If an error made while entering data generates an OPR ERR light, the RSET button clears the condition and allows the crewmember to reenter the key sequence.

KEY REL [Key Release]: Although the AGC is a fully multiprogrammed system, the DSKY can present information from only one program at a time. There are often occasions where an astronaut is using the DSKY to work with one program, and another routine is demanding attention with an important display request. That program will announce its need for service by illuminating the KEY REL light on the DSKY, and wait for the crewmember to press the KEY REL key. Once KEY REL is pressed, the KEY REL light extinguishes and the waiting program's data replaces the current DSKY contents. Note that there is not a requirement for the crew to immediately stop work to respond to the KEY REL light. The astronaut can finish his work on the DSKY, and then bring up the waiting program's display. Interrupting work on the current program in order to attend to another is a confusing and unworkable approach, as "toggling" the DSKY between two programs is not possible. Once the waiting program takes control of the DSKY, the astronaut must complete the dialog before returning to the other program.

Several status, warning and error lights complete the DSKY display area, and they are controlled directly by the software, not by external sensors. Not all lights are common to the Command Module and Lunar Module computers. For example, the landing radar ALT[itude] and VEL[ocity] lights are found only in the LM. A full list of the lights follows:

UPLINK ACTY [Uplink Activity]: Ground controllers frequently perform remote updates of the AGC for tasks as mundane as updating the system clock, to providing the latest guidance and navigation data. Since the crew cannot use the DSKY during the uplink, illumination of the UPLINK ACTY light advises the crew that the ground has control of the computer. Once the light has extinguished, the crew is free to resume operating the DSKY. UPLINK ACTY has two other, more obscure functions in the Command Module. During rendezvous navigation, UPLINK ACTY illuminates when a large (> 10 degrees) attitude change is necessary for tracking, yet the AGC has not requested that the crew maneuver to a new attitude. Also, in the minutes before Translunar Injection, the Launch Vehicle Digital Computer (LVDC) in the Instrument Unit of the Saturn V enters a mode called Time Base 6, which provides the guidance and control routines for the maneuver. When the LVDC begins Time Base 6 the CM's AGC is notified, and it illuminates the UPLINK ACTY light for ten seconds to notify the crew that the Translunar Injection sequence has started.

TEMP [IMU Temperature]: TEMP indicates that the Inertial Measurement Unit temperature is out of limits. Maintaining its temperature within a narrow range is important for the precision of the IMU, and accurate regulation requires elaborate heating and cooling equipment. Ambient air at a pressure of 1 atmosphere is circulated throughout the sealed spherical case, and a water-glycol loop extracts waste heat. The TEMP warning light illuminates any time the internal temperature is outside the 126° F to 134° F (52.2° C to 56.7° C) operational range. Failure of the IMU is assumed if the temperature is not within these limits, and the AGC will "cage", or reset, the IMU to prevent further damage; an intervention that results in the loss of all attitude reference information.

NO ATT [No Attitude]: For the AGC to perform any attitude control maneuver, the IMU must be powered up, within temperature limits, and aligned to a known orientation. Additionally, its interface electronics, the Coupling Data Units (CDUs) must be functioning within normal limits. Failure to meet any of these criteria results in the assumption that the IMU is unable to provide valid data, and no attitude reference is possible. This does not mean that the IMU has suffered a mechanical or electronics failure. NO ATT will illuminate when the IMU begins its power up sequence, and whenever it loses its orientation for any reason (such as gimbal lock). Here, the expectation is that the IMU is physically in good condition, but needs manual intervention before it can be used.

GIMBAL LOCK [Gimbal Lock]: Gimbal lock occurs if the middle of the three gimbals exceeds a critical angle, rendering it unable to rotate freely and maintain its orientation. This is the gimbal on the Z, or yaw, axis. Once entered, a gimbal lock situation results in the IMU losing its orientation, so realignment is necessary for recovery. Crew awareness of the IMU orientation is a key element of gimbal lock prevention, so the GIMBAL LOCK light will illuminate when the middle gimbal angle reaches ±70 degrees, still 15 degrees from actually locking up. The light provides a warning that gives the crew time to prevent further yaw motion towards gimbal lock. Only when the middle gimbal reaches ±85 degrees will the IMU actually enter a locked state. At this point, all orientation is lost and the NO ATT light will illuminate.

STBY [Standby]: To conserve power, a standby mode is available in the AGC. This reduces electrical consumption to 15 watts from the 70 watts used during normal operation. To enter Standby mode, the crew selects Program 06 and presses the PRO key until the STBY light goes on. If desired, the AGC and IMU can be powered down at this point.

PROG [Program Alarm]: Program alarms are generated by software errors or hardware failures, but are not necessarily fatal and often do not require extensive recovery. The crew normally presses the RSET key to acknowledge the alarm, and use Verb 05 Noun 09 to display the alarm code. More serious program alarms result in a variety of automatic restart options, which will illuminate the RESTART light.

KEY REL [Keyboard Release]: An issue posed by multiprogrammed systems is how to alert the crew that another program which is running in the background needs attention. When one program is using the DSKY, another program needing service will light the KEY REL lamp, requesting the astronaut to complete his work and release the DSKY for the waiting program to use. There is no information on the name of the routine requesting the DSKY, but the crew's knowledge of the currently running programs usually allows them to make a reasonable assumption of which program or routine is making the request. When the crewmember is ready to work with the waiting program, pressing the KEY REL key brings up the waiting display.

RESTART [Restart or Fresh Start]: Some errors in the AGC are serious enough to warrant restarting the failing program or perhaps even the entire computer. The type of error determines the extent of the restart, but terminating programs is usually required. As part of the restart and recovery sequence, the RESTART light will illuminate, notifying the crew that a problem exists and the computer will attempt to recover. As restarts often follow from program alarms, the crew can press the RSET key to clear the PROG and RESTART lights, and then use Verb 05 Noun 09 to determine the cause of the restart.

OPR ERR [Operator Error]: Although the "language" of the DSKY is not complex, strict rules apply to the entry of requests and data into the DSKY. Not all combinations of Nouns and Verbs are valid, and data entered must match the format specified by the Noun. Any input that is not valid will generate an OPR ERR light after pressing the Enter key. Operator errors are usually trivial problems, and incorrectly keying data is often the cause. Clearing the error requires pressing the RSET key and reentering the Verb or Noun sequence.

TRACKER [Tracking Error]: Originally used to indicate an error in the star tracker, this warning light kept its name after the star tracker hardware was deleted from the Apollo specifications. In the Command Module, errors in the optics assembly CDU electronics or problems in reading valid VHF ranging data cause the TRACKER light to illuminate. In the Lunar Module, it is illuminated if the rendezvous radar CDU electronics fail, or if landing radar data is not reasonable. In this last case, the TRACKER light may be accompanied by the ALT and VEL lights.

DAP NOT IN CONTROL [Digital Autopilot not controlling the LM]: Found only in the Lunar Module, this unlabeled light illuminates any time the Digital Autopilot is not controlling the spacecraft. Such a situation occurs when the DAP is off, or in idle or minimum impulse mode.

PRIORITY DISPLAY [Priority Display waiting to use the DSKY]: Found only in the Lunar Module, this unlabeled light is controlled by Program 20 (P20, Rendezvous Navigation). It indicates that important information is ready, and the crew should not delay in attending to it. P20 runs in the background during rendezvous maneuvers while other targeting programs run as the Major Mode.

ALT and VEL [Landing radar Altitude and Velocity lights]. During the Lunar Module's powered descent, the landing radar continually measures the vehicle's altitude above the surface as well as velocity components in all three axes. For the landing radar to report valid data, a number of operational constraints must be satisfied, such as staying within the acceptable attitude range and the data being valid for a minimum amount of time. Violating any of these constraints generates a signal to the AGC that indicates the data from the landing radar is not valid, and this in turn will cause the appropriate light to illuminate. Loss of radar altitude or velocity data can be a serious situation, especially during the later phases of the landing, as the IMU cannot provide altitude data that is sufficiently accurate for a safe landing.

DSKY operation
A dialog with the AGC usually takes the form of a Verb, followed by a Noun. Both Verbs and Nouns are identified by unique two-digit codes. After the Verb and Noun are loaded into their registers, pressing the Enter key indicates that the entry is acceptable and ready for processing. One example of a DSKY request might be:

Verb 06, Noun 33, Enter

This sequence is somewhat tedious to write, and a shorthand notation is used extensively throughout the documentation and flight procedures. In this notation the Verb and Noun sequence above is abbreviated to V06N33E.

When the Verb or Noun keys are pressed, the entry in that display flashes as the digits of the code are entered, and continues to flash until the Enter key is pressed. Overwriting a previously accepted entry is possible by pressing the Noun or Verb key again and reentering the digits. If a mistake is made when entering data into a data register, the CLR key blanks the register and the data can be rekeyed. There are a few exceptions to this syntax. Extended Verbs, which do not require a Noun, use only the Verb code followed by the Enter key. Verbs that recycle or terminate a program also do not require a Noun, and only require pressing Enter to start the processing. Some Verbs cannot be keyed into the DSKY, as they are requests made by the software for the crewmember to perform some action. For example, Verb 50 (Please Perform a Checklist Item) might tell the astronaut to perform an attitude maneuver, align the inertial platform, or reposition the landing radar. Verb 99 is displayed in the last five seconds before engine ignition, and requests the final authorization to continue and fire the engine. If either of these Verbs is keyed into the DSKY, the OPR ERR light will illuminate.

The notation of Verb ##, Noun ##, Enter is a convention, but not a hard requirement. A sequence of Noun ##, Verb ##, Enter is equally valid. A displayed Verb is also "reusable" by keying in another Noun value, and bypassing the need to punch in the Verb code again. For example, entering Verb 16, Noun 95, Enter on the Command Module DSKY displays parameters related to an engine burn. The monitor Verb 16 is chosen because it displays the Noun as it changes, counting down the time until engine ignition. While this display is active, the astronaut can also monitor the AGC's clock time by requesting Noun 36, without having to reenter Verb 16.

Regular and extended verbs
Two major types of Verb exist. "Regular" Verbs (numbered 00 through 37) are used to display, monitor and update data. Included in this range are Verbs that control overall AGC operation, such as starting, terminating and recycling programs, or restarting the entire system. "Extended" Verbs (numbered 40 through 99) run programs that perform a single discrete operation, such as setting flags, configuring panel displays, or altering the logic of the Major Mode. Most

extended Verbs run without requiring additional input, and in those cases where specific data is needed, it presents several Verb-Noun combinations to the astronaut. A typical example is initializing the Digital Autopilot through Verb 48 (Request DAP Data Load Routine). Several pieces of data are necessary before the DAP is started, including spacecraft configuration and vehicle weights. After several Verb-Noun dialogs between the initialization software and the crew, Verb 48 begins DAP operation. Extended Verbs monopolize the DSKY until their operation is complete, and a second extended Verb cannot be entered until the first has completed.

DSKY operation – entering and retrieving data

Verbs for viewing or manipulating data are broken down by data type (decimal or octal), and whether the data is loaded, monitored in real-time or simply displayed. Monitoring Verbs allow the crew to observe data such as velocity or altitude changing over time, updating at a rate of once per second. Before choosing the Verb to display data on the DSKY, the first task is to consider the Noun used with the Verb. This defines the type of data and the number of DSKY registers that are required. Most of the data is in a decimal format, such as times, velocities or angles. Computer-specific data, such as memory addresses and their contents, or data expressed as bit patterns use octal digits. Nouns reference a wide variety of data, and require one, two or all three data registers in order to display their contents, and each Noun component is assigned a register where it is displayed. For example, the hours, minutes and seconds components of Noun 33 to display the time are assigned to registers 1, 2 and 3, respectively. These assignments are fixed by the Noun format and cannot be rearranged into a different sequence.

Seven display Verbs are available, five of which are variations on displaying octal data in different registers. Arguably, the most frequently used display Verb is Verb 06, which presents decimal data in any or all of the three registers. Verb 06 understands the register requirements of the Noun it is operating with, and adapts the display accordingly. Regardless of the number of components that a Noun requires, Verb 06 will display only those that are requested. In the case of Noun 92 (Command Module Optics Shaft and Trunnion Angles), Verb 06 knows that only registers 1 and 2 are needed for the display, so register 3 will remain blank.

Monitoring data changing in real-time, such as altitudes, velocities or angles, gives the crew essential situational awareness of the state and performance of their spacecraft. While it is possible to repeatedly call up fresh data by rekeying the display request, this option is impractical during a high-workload phase of the flight. Perhaps the most visible need for monitoring real-time data is during the lunar landing. The crew references the DSKY constantly, and also cross-checks the computer's values against a cue card with pre-computed values for each point of the descent. As the descent progresses, the crew can verify their progress as they reach important milestones, and take corrective action if a problem emerges.

The list of monitoring Verbs follows a similar pattern to those used for displaying data. Most are for monitoring octal data in various registers, but Verb 16, the real workhorse, continuously displays decimal data. Updating can be suspended by pressing the PRO key or Verb 33 (Proceed Without Data). Pressing the Enter key instructs the AGC to resume continuously updating the display.

The five data load Verbs are organized by the registers they update. Each accepts the Noun's components, either singularly or in combination. Verbs 21, 22 and 23 load one Noun component into data register 1, 2 or 3 respectively. As entering a Verb and Noun code each time a component is loaded into a data register is tedious at best, multiple component loads are available by Verb 24 and Verb 25. Verb 24 loads components into registers 1 and 2. It switches automatically to Verb 21 when loading register 1, then to Verb 22 when loading register 2. Verb 25 operates similarly, using Verbs 21, 22 and 23 for loading all three registers. Decimal data is identified by a + or − sign preceding the digits as they are entered. When entering data, all leading zeros must be entered as well. Failing to enter all five digits in a register will illuminate the OPR ERR warning light, and the entry will have to be rekeyed.

DSKY error checking
A few basic rules for DSKY operation are necessary to prevent entering invalid data. First, leading zeros are necessary for all entries, with two digits required for Verb and Nouns and five digits for registers. For any given Noun, all components will use either decimal or octal notation; mixing the two types in one Noun is not allowed. The explicit datatype of each Noun must always be used; entering an octal for a decimal quantity is not allowed.

Input punched into the DSKY requires a 'sanity check' before it is passed along to the program.[10] With so many keystrokes used during flight, it is inevitable that a miskeying will occur. Only basic checks are performed on the input Nouns, and little effort is taken to assure that the data is appropriate for the Noun. Mistyping an angle 086.00 as 008.60 will not be recognized as an error, as both are valid numbers. An invalid entry will cause the OPR ERR light to illuminate, but there is no other indication of the specific problem. The crewmember must review the entry, identify the problem and reenter the request. The following situations are not legal, and will illuminate the OPR ERR warning light:

- Entering the digits 8 or 9 into a register without a preceding + or − sign. Not providing a sign indicates octal data, which only uses the digits 0 through 7.
- The software does not implement the Verb or Noun code entered.

---

[10] A note about the use of the rather archaic phrase "punched", as in "punching in the data". Throughout the early decades of computing, the dominant form of input was the punched card. Data was represented using holes punched into the card, which were read by mechanical or optical scanners. Few artifacts of computing have endured as long as the lowly but ubiquitous punched card. Although they have long disappeared from the world of data processing, the nostalgic and curious can still get a glimpse of what they were like. Airline flight coupons of the type printed at the check-in counters use the same card stock, and are virtually the same size as the venerable punch card.

- Providing decimal data when entering a machine address. Only octal data is allowed for addresses.
- A value out of limits for the component, such as specifying more than 360 degrees for an angle.
- Mixing decimal and octal data within a Noun.
- Attempting to load data into a Noun that is not loadable, such as a checklist code for the crew to act upon.
- Entering more than two digits for a Verb or Noun code, or more than five digits into a data register.
- Pressing the Enter key without a Verb and Noun already loaded (except for extended Verbs, which do not need a Noun).
- If a Priority Display is waiting, and the "mark" button is pressed on the optics assembly.

DSKY input processing

A group of routines under the formal name, "Pinball Game – Buttons and Lights" (more informally simply called "Pinball") handle entering data, interpreting the Verb and Noun codes and dispatching commands. Such a whimsical name is actually quite appropriate, as Pinball dedicates much of its code to data punched into the keyboard and managing the lights and digit displays. Although the movement of data from the output buffer to the DSKY display is driven by the T4RUPT interrupt processing, all the display contents are managed by Pinball routines. Input through the DSKY begins when an astronaut presses a key on the panel, raising the KEYRUPT1 keyboard interrupt. The keycode, identifying the key just pressed, is placed in the lower five bits of input channel 13. The interrupt routine runs quickly to schedule the Pinball routine CHARIN to process the data. CHARIN categorizes the keystroke as a Verb or Noun, a digit, or any of the other operation keys, and determines what processing must take place. Pressing the Verb key, for example, sets flags indicating that the subsequent characters must be interpreted as a Verb code, and sets bits in the display output buffer, DSPTAB, to flash the Verb and Noun fields. Essential sanity checking on the input is performed first. Invalid keystrokes, such as pressing a non-digit key when numeric data is expected, will terminate CHARIN and illuminate the OPR ERR light. As numeric data is punched into the DSKY, it is first displayed in the leftmost digit of the Verb, Noun or register, and each new digit appears successively to the right. This is different from the familiar case where each keystroke shifts the existing display left, with the new digit inserted in the rightmost position. Only after pressing the Enter key does the interpretation of the data begin.

Extended Verbs (numbered 40 through 99) are quickly dispatched by transferring control to the appropriate routine. A new job is not scheduled for performing extended Verbs, rather the Verb is processed as part of the CHARIN routine. Often an extended Verb requests additional data, as in the case of Verb 48 (Load Digital Autopilot Parameters), which presents several Verb-Noun combinations to the astronaut, requesting information on the spacecraft configuration and desired autopilot characteristics. After collecting the necessary data, Verb 48 schedules the Digital Autopilot tasks and terminates itself. Other extended Verbs can proceed without additional data, and set flags or schedule other work before ending. As extended Verbs execute as part of the CHARIN routine, only one extended Verb can execute at a time. In the event that a crewmember tries to execute a second Extended Verb before the first one has finished, the OPR ERR light illuminates and prevents a second extended Verb from executing concurrently.

Regular Verbs (when combined with Nouns) display, monitor or load data in a wide variety of formats. Regardless of the operation requested, an elaborate set of tables and routines convert the data entered into a format usable by the software routines. As seen in the hardware description section, most data is stored using a fractional notation. Additionally, the units employed internally in the AGC are exclusively metric, yet human factors demand that the crew uses English units. These requirements, plus the different numbers of components for the Nouns, result in a complex design for referencing the appropriate scaling and conversion routines.

As an example of data loading, we will use the case of the three components of Command Module Noun 89,[11] which specifies the location of a lunar surface landmark. Noun 89 is a mixed Noun, whose data components use the following format:

Landmark Latitude:	XX.XXX Degrees
Landmark Longitude/2	XX.XXX Degrees
Landmark Altitude	XXX.XX Nautical miles.

Each component of the Noun uses a different input format and scaling. Here, latitude is saved as a double precision value, as is the Longitude/2. Longitude normally is expressed as a range 0 to ± 180 degrees, but, as its name implies, Longitude/2 is scaled so that the data entered is less than ± 90 degrees. Landmark altitude, expressed as the distance from the center of the Moon, is also double precision, but scaled quite differently than the other two components. Parsing each Noun and its individual components requires a complex assemblage of tables and routines. The two types of Nouns, ordinary and mixed, require that those tables alternate between two different formats, further increasing complexity. Ordinary Nouns are numbered 01 to 40, use direct memory references to the data, and the values are all of the same data type (angles, velocity or octal). Mixed Nouns, those which are numbered above 40, are much more complex. Each component can be a different datatype from the others, and referencing the data for the display requires an extra step called indirection. Ordinary Nouns are defined by referencing data located in contiguous storage, and use the same scaling and conversion resources for each component. They require only two data structures to reference the data and routines to manipulate the DSKY data. Mixed Nouns present a completely different situation, as their numerous variants require no fewer than eight tables for processing the data.

---

[11] Many Nouns and extended Verbs are unique to the Command Module or Lunar Module. Because there is so little commonality, it is important to qualify which spacecraft the code is for.

DSKY output

Getting data out of the AGC and onto the DSKY is, in many ways, more difficult than getting the keystrokes into the system. Consider the problem of managing the data flow to the display. It has 21 digits and three $+/-$ signs (one pair for each register), plus several status and warning lights, each of which is individually addressable. Tiny mechanical relays act as switches to turn each of the segments of the display on and off. Implementing separate output channels makes programming relatively easy, but also requires an exorbitant amount of hardware and cabling. Using multiple channels for sending data to an I/O device has the huge appeal of speed, but this is unnecessary for a device like the DSKY. Humans can process data at only a fraction of the speed at which a computer can generate it, so it is best to forgo fast software and expensive hardware in favor of more appropriate technology. This exchange of complexity and cost is possible through multiplexing the data over one output channel between the computer and the DSKY, allowing the single channel to carry data intended for different destinations. For this technique to function, some overhead is necessary in the form of additional bits appended to the output data, which are used to identify the targeted DSKY field. Each digit, sign and light in the DSKY is associated with a unique identifier, which is loosely analogous to a memory address. Data intended for DSKY output is buffered in an 11-word table in erasable storage named DSPTAB. Two 5-bit fields, each corresponding to a decimal digit, reside in bits 10 through 1 in the word. Each such field contains a relay code used to display numeric data, rather than the actual number itself, simplifying the task of selecting and driving the relays that display the digit segments. Bit 11 in the DSPTAB word is employed for other purposes, such as the $+$ and $-$ sign in the data registers, and the indicator to flash the displays as data is entered. Bits 15 through 12 have a dual role. DSPTAB entries in memory use bit 15 (the sign bit) to indicate whether an update to the display is required. A positive value indicates that the data reflects the DSKY display, and no updates are necessary. When bit 15 is set to one, the word is interpreted as negative and indicates that the DSKY requires an update from the data buffer. When a data word is ready for output, a display addressing code replaces bits 15 through 12.

The display itself is an electromechanical device, using over 100 miniature latching relays to create the number segments on the panel. Using latching relays in the DSKY display creates a valuable dual benefit. First, traditional relays are susceptible to vibration and acceleration forces, where unrestrained relay contacts may open and close unexpectedly and produce meaningless displays. Additionally, mechanically latching the contacts open or closed eliminates the need to maintain power to the hardware, thus reducing the electrical demands of the device. Latching the relay open or closed does not happen instantaneously, however. Power must be applied for at least 20 milliseconds to assure that the relay has settled into its new state and can permanently maintain its configuration.

Sending data out to the DSKY occurs during the 120-millisecond T4RUPT processing cycle. DSPTAB is scanned for any negative entries, which indicate that new data is ready to be sent to the display. Recalling that the DSKY output channel

(channel 10_8) multiplexes its data, a destination address code is necessary to define which display elements are being updated. As with all channel output operations, the data word is copied from DSPTAB to the accumulator. After each data word has been loaded, bits 15 through 12 are replaced with an identifying code for the corresponding display digits. Decoders in the DSKY hardware interpret this code and route the update to the appropriate set of relays, which in turn drive one or more display elements. After each write to the output channel, the routine waits 20 milliseconds for the relays to latch into place before continuing with the next data word. During the execution of a monitoring Verb, where all 15 digits in a data register might change, an update of the entire display can take as long as 160 milliseconds. Updating a large number of digits in the display becomes a relatively slow, mechanical process, and the display perceptibly "ripples" as each new digit is placed in the display.

Noun tables

Managing the various Noun formats requires a large number of tables to define each type. Eight tables in fixed storage describe Noun characteristics such as the number components in each Noun, the datatypes used for each Noun component, and information on the proper internal scaling of the data. Ordinary Nouns are the most straightforward, as their data resides in contiguous memory and all components use a common datatype. These factors combine to require a small subset of the Noun tables. The large number of mixed Noun variants is responsible for the bulk of the complexity seen in the table structures. Combining both ordinary and mixed Nouns in the three main Noun tables reduces the overall complexity somewhat, but at the price of maintaining two different data layouts within each table.

The first group of Noun tables includes NNADTAB, NNTYPTAB and RUTMXTAB, which are respectively, the Noun addresses, type and mixed Noun tables. Entries in these tables have the important property of being referenced using the Noun number as an offset into the table. The remaining tables, IDADDTAB, SFOUTAB, SFOUTABR, SFINTAB and SFINTABR are addressed using indirect addresses located in one of the Noun tables. To completely describe how the Noun tables are used for DSKY input and output, a short introduction to each of the structures is necessary.

NNADTAB: As the first of the three main Noun tables, NNADTAB supplies references to the data referenced in each Noun component. The first 39 words of NNADTAB point to the starting address of the three ordinary Noun components. The remainder of the table uses a different format for mixed Nouns and contains two pieces of data: a component code, which defines the number of Noun components and some of their characteristics, and a pointer into another table, IDADDTAB, which in turn points to the individual mixed Noun components in storage.

NNTYPTAB: The second of the main Noun tables, NNTYPTAB also uses different formats based on the type of Noun. NNTYPTAB complements NNADTAB by providing information on the Noun components' data representation. Entries for ordinary Nouns include the component code data, an index into tables used to perform data scaling operations, and datatype information. Like NNADTAB, the information is different for mixed Nouns, and only datatype information is saved for these entries.

RUTMXTAB: The last of the main Noun tables, RUTMXTAB is used by mixed Nouns only. An entry in this table is organized into three fields, corresponding to a Noun component, each of which contains an index into the data scaling tables. Four tables are indexed by the index field, two each for DSKY input and output scaling factors.

SCALING TABLES: All Noun components entered or displayed on the DSKY have a scaling factor associated with them. Two pairs of tables are used for the scaling process, and are dedicated to either input or output. The first pair of tables (SFINTABR and SFOUTABR) contain the scaling routine addresses used for each data type. A second pair of tables (SFINTAB and SFOUTAB) contain scaling constants that are applied to the data during the conversion process. Indexes into these scaling factor routine and constant tables are located in NNTYPTAB or RUTMXTAB.

IDADDTAB: Unlike ordinary Nouns which reference data in contiguous storage locations, data for mixed Nouns can be anywhere in erasable storage, eliminating the possibility of using only a single address to point to the data (as is the case with ordinary Nouns). Additionally, mixed Nouns can have one, two or three components, in contrast to the standard three components of ordinary Nouns. Each entry in the IDADDTAB table contains the address of the variables referenced by a mixed Noun. The variable number of entries in each mixed Noun precludes using an index calculated from the Noun's number, as is the case with NNADTAB and NNTYPTAB. Instead, an offset into IDADDTAB is part of each Noun's entry in the NNADTAB.

DSKY specific scaling

In the earlier discussion of data formats, the concept of fractional notation for data was introduced. Here, maintaining the magnitude, or scaling factor, was a burden imposed upon the programmer. Consistency in applying data scaling simplifies calculations, but does not always result in the most intuitive representation of the data. For example, the value of pi is 3.1415926..., but in the AGC it might be implemented as 0.31415926 using a scaling factor of 10^{-1}. At the same time, the altitude of the spacecraft in lunar orbit might be 1,800 km above the center of the Moon (approximately 100 km above the surface). A scaling factor of 10^{-4} is necessary to convert this to its fractional value of 0.1800. While this value is completely acceptable, calculations become complex when including pi, scaled at 10^{-1}, and other variables with yet other scaling factors. Although keeping track of the mix of scaling factors is not particularly difficult, extraordinary care is necessary to prevent overflows or other computational errors. Managing scaling factors is especially important for input and output, and

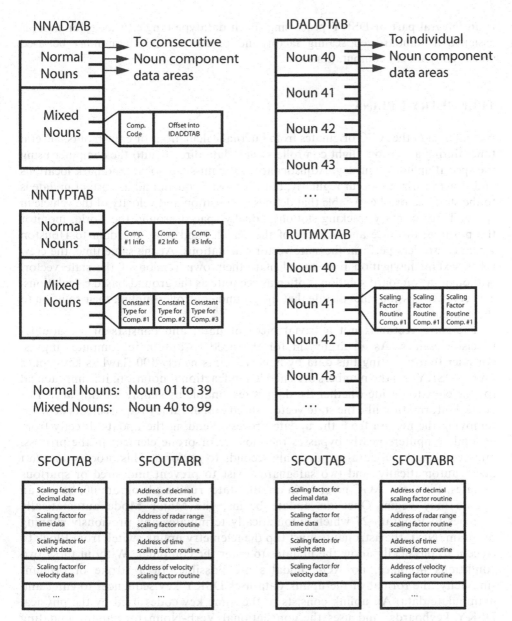

NNADTAB

Normal
Nouns

To consecutive
Noun component
data areas

Mixed
Nouns

| Comp. Code | Offset into IDADDTAB |

IDADDTAB

Noun 40

To individual
Noun component
data areas

Noun 41

Noun 42

Noun 43

...

NNTYPTAB

Normal
Nouns

| Comp. #1 Info | Comp. #2 Info | Comp. #3 Info |

Mixed
Nouns

| Constant Type for Comp. #1 | Constant Type for Comp. #2 | Constant Type for Comp. #3 |

RUTMXTAB

Noun 40

Noun 41

| Scaling Factor Routine Comp. #1 | Scaling Factor Routine Comp. #1 | Scaling Factor Routine Comp. #1 |

Noun 42

Noun 43

...

Normal Nouns: Noun 01 to 39
Mixed Nouns: Noun 40 to 99

SFOUTAB	SFOUTABR	SFOUTAB	SFOUTABR
Scaling factor for decimal data	Address of decimal scaling factor routine	Scaling factor for decimal data	Address of decimal scaling factor routine
Scaling factor for time data	Address of radar range scaling factor routine	Scaling factor for time data	Address of radar range scaling factor routine
Scaling factor for weight data	Address of time scaling factor routine	Scaling factor for weight data	Address of time scaling factor routine
Scaling factor for velocity data	Address of velocity scaling factor routine	Scaling factor for velocity data	Address of velocity scaling factor routine
...	...	...	...

Figure 40: Noun tables

is an integral part of DSKY processing. Each datatype (angle, velocity or time) is assigned a standardized scaling factor and precision, and these must be used consistently for all computations.

TELEMETRY UPLINK

Not all data in the AGC originates from internal calculations or crew inputs. Several times during a mission, flight controllers enter data directly into the computer using the special uplink frequency. Adjustments to the mission clock, landmark locations and system variables occur regularly, but the most frequent and important update is to the state vector, the variable that defines the position and velocity of the vehicle in space. Three primary tracking stations, equally spaced around the world, monitor the position, distance and velocity of the vehicle, and relay the data to Houston where controllers perform the state vector calculations. At the same time, the crew takes sextant navigation fixes to calculate their own version of the state vector. Although the onboard solution is often as accurate as the ground-based calculations, conservative mission rules dictate that the ground controllers uplink their solution to the computer.

The state vector is not a trivial piece of data, and consists of six double-precision values. As an argument for the need to automate computer inputs, consider that updating this data by hand requires nearly 100 flawless keystrokes on the DSKY. As the numbers are encoded in fractional notation, it is impractical for the crew to decide whether the data is reasonable by simply looking at it. For critical information like the state vector, speed and accuracy requirements demand removing the human from the update process. Sending the update directly from ground computers neatly bypasses the most error-prone element in the process, and even lengthy updates require only seconds to perform. This process does not occur automatically, and two safeguards exist to prevent undesired or spurious updates. First, the AGC must be in an idle state, running Program 00, before an update may begin. Once an update begins, the Major Mode automatically switches to Program 27 which automatically terminates any previously running program. Next, the astronauts must flip the telemetry uplink switch from Block to Accept to physically allow the updates to enter the computer. Without these two conditions in place, no AGC updates are possible. In a stroke of brilliant simplicity, the format of the uplink data uses DSKY key sequences to enter and manipulate data. An uplink consists of the same key codes used by the physical DSKY keyboards, and uses the conventional Verb-Noun format for updating memory.

During the mission, tracking stations maintain nearly continuous transmissions to both the Command Module and Lunar Module. Both communication links transmit in the S-Band radio frequencies, with 2106.40625 MHz used for the CM and 2101.8 MHz for the LM. A 70 KHz subcarrier channel contains the digital data, which the LM's Digital Uplink Assembly, or the CM's Updata Link electronics extract and pass to the AGC. Assembling the individual bits into a 15-bit word from this serial data uses two nonprogrammed instructions in the AGC. The uplink data is placed in a special register named INLINK, which is located at location 45_8 in storage.

At the beginning of the uplink transmission, INLINK is all zeros, with the exception of bit one, which is one. Uplinking a zero bit to the AGC triggers a SHINC instruction, which shifts the uplink register left by one bit and inserts a zero in bit 1 of the word. SHANC, like the SHINC instruction, shifts the uplink register one bit to the left, but places a one in the rightmost bit. Successively shifted left by SHINC and SHANC as new bits are received, the original bit 1 in INLINK eventually arrives in position 16, triggering the UPRUPT interrupt. Processing of the current work is suspended, and control transfers to the DSKY's Pinball routine to process this character.

Bad data (data not received correctly) is unacceptable, as a bad IMU reference or state vector has serious consequences for guidance and navigation. Transmitted data is often noisy and error prone, and a sufficiently long run of static could easily be interpreted as valid data. Two tests for data validity are performed to assure the correctness of the uplinked data. First, each 5-bit keystroke character is transmitted three times as part of the single 15-bit uplink word. Comparing all three copies against each other is reasonable insurance against corrupted data, but an additional check gives an even greater level of confidence in the data. Of the three copies of the uplinked key code, the second 5-bit string is the complement of the key code. To illustrate this, consider the keyboard code for "Verb", which is binary 10001. The INLINK register is shown in Figure 41.

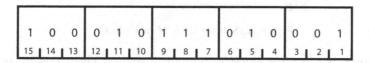

Figure 41: Data in INLINK register

The first five bits, in positions 15 through 11, contain the Verb code, 10001. The next five bits, in position 10 through 6 are an inverted copy of the Verb code (binary 01110), and the final bits repeat the normal representation of the Verb code. Checking the code against its complement, and then another, non-complemented copy of the code reliably eliminates the possibility that random radio noise would generate a valid input sequence.

After validating the uplinked character, it is passed to the CHARIN routine of Pinball, where it is processed in the same manner as all other DSKY keyboard input. From here, key sequences sent via remote control from the ground are no different from having a crewmember physically press the keys in the spacecraft. Most uplink operations modify data in erasable storage, but many updates are to ranges of erasable storage words that might not have a Noun associated with them. Solving this issue requires the introduction of a special set of extended Verbs. Designed to function only with the uplink software, Verbs 70 through Verb 73 are not available for the crew to use. Upon recognizing one of these Verbs in the uplink data stream, the Major Mode display changes to Program 27 to indicate that the update is in progress.[12]

Each of these Verbs performs a specific update function. Verbs 70 and 73 update the AGC's clocks, using the uplinked data to adjust the time up or down. Invoking Verb 71 allows the update of several words of contiguous storage. The first entry is the count of the number of words to follow, the first of which is the starting address parameter, with the update data trailed by Verb 33 (Proceed Without DSKY Inputs) to finalize the update.

Verb 72 is similar to Verb 71, except that the updates are not necessarily in contiguous storage. After the number of words to be updated has been passed to Verb 72, pairs of storage addresses and data are sent to the computer.

Note that in the cases of these four extended Verbs, data is not loaded into any of the three display registers. Freed of this physical limitation, Verbs 71 and 72 can accept between 10 and 20 words of data per update. Verbs 70 and 73 adjust the timings in the AGC. Verb 70 updates the liftoff time, and Verb 73 updates the AGC clock.

V71E	Begin updates
5E	5 entries to follow
01016E	Data will be stored at location 01016_8
14000E	First word of data
01603E	Second word of data
16067E	Third word of data
V33E	Update complete

Figure 42: Sample Verb 71 in Uplink

At this point, it is not yet apparent how these update Verbs prevent problems with slow and noisy data transmissions from the ground. The most important safeguard is to buffer the updates in memory during the update process, leaving critical AGC data untouched during this period. Only after the uplink is complete and all the data is validated will the update process begin. At this point, the ground sends a Verb 33 sequence to indicate that the uplink is complete (no additional data is expected), and the software can now copy the buffered data over the AGC's existing data. First, a restart phase is defined so that any interruption or failure will insure that the changes are implemented. Next, the actual update of the data is performed. Because it is running as an extended Verb, which in turn is running as a Pinball routine, it has the luxury of running with interrupts inhibited. Under this

---

[12] Here is an excellent example of how the Major Mode does not necessarily correspond to an explicitly running program. The update routines are only extended Verbs, but the update process is such a change to the normal state of the system that a new Major Mode is justified. Although the DSKY's "PROG" display switches to Program 27 from Program 00, a new job is not scheduled to run.

configuration, there is no need to worry about losing control of the processor and spoiling the update, even for a short time. After copying the update data from the buffer to its destination, the software returns to get the next set of DSKY commands from the ground. When all the updates are complete, the ground sends Verb 34, which terminates all update operations and returns the Major Mode back to Program 00.

SYNCHRONOUS I/O PROCESSING AND T4RUPT

Communicating with the hardware external to the AGC is a major responsibility of a control computer such as the AGC. In addition to the wide range of hardware that the computer interfaces with, the AGC must also adapt to the responsiveness of each device. At one end of the spectrum, software routines identify an action for a piece of hardware and send a command to that device with the expectation that that action occurs immediately. For example, thruster firings are important I/O requests which occur frequently and demand timings on a millisecond scale. Thrusters must fire the moment that the data is placed in the I/O channel, and they need not respond with any status information. Other devices can be managed satisfactorily without such near-instantaneous responses. Only a few of these asynchronous operations exist in the AGC, and they are usually limited to initializing a device prior to its first use. For example, powering up and initializing the inertial platform requires about two minutes to complete. Such a device, with its low latency requirements, eliminates the need for high resolution timers or dedicated interrupt vectors and their associated software. Perhaps the most familiar instance is the DSKY, which responds quickly to software commands, but in turn, must interface with the slowest component of all, the human crew.

Another group of devices are characterized by needing periodic attention, whether to check their status, to update their displays or to monitor their physical movement. The handling of these tasks is performed by the T4RUPT routine, named for the interrupt that drives its execution. Devices managed by the T4RUPT do not raise an interrupt or otherwise alter the AGC's overall processing. Rather, T4RUPT runs according to a fixed schedule, interrogating I/O status bits and directing processing based on their settings. Unlike thruster timers or DSKY keyboard interrupts, T4RUPT manages I/O synchronously, forcing each device to wait its turn to have its status bits polled. Outputting data for the DSKY display also operates under a similar fixed schedule.

T4RUPT is scheduled through the T4RUPT clock, one of the four event timers in the AGC, and it schedules eight distinct operations, or "passes", 120 ms apart, repeating the sequence every 960 ms. Breaking T4RUPT into several passes has two key advantages. Many devices are serviced more than once during the T4RUPT 960 ms cycle, which improves the responsiveness of the computer to the devices' needs. Also, using several small tasks necessarily limits the amount of processing that is performed during each pass. As all timer interrupts are entered in interrupt-inhibited mode, all work must necessarily be very brief. It might appear that the timings for T4RUPT are oddly precise, especially when a change of a few milliseconds would make little difference in handling the devices. That is, surely rounding to a full second between the time T4RUPT starts its first pass would be more intuitive? The answer lies in the limitations of the TIME4 clock. At centisecond (i.e. 10 ms) resolution, it simply cannot operate at timings of exactly 1/8, or 0.125 second.

Prior to determining which 120 ms phase is scheduled next, T4RUPT performs two tasks. First, the DSKY output table is checked to see if any output is waiting to be displayed. If an update to the DSKY display or its status lights exists, the output word and its relay codes are assembled in the accumulator and the data is written to output channel 10_8. Output channel 10_8, like all output channels, is seen as a set of "flip-flops"; digital devices that maintain their data until a change is commanded. All the while data is in output channel 10_8, the AGC maintains the signal in its "flip-flops". But keeping the DSKY relays endlessly energized is unnecessary. As the relays are designed to latch into position within a short while, power can be safely removed after the latches are secure. To apply power for much longer than 20 milliseconds would not only result in unnecessary power consumption but also risk overheating the relay coils. With these constraints in mind, T4RUPT will schedule another interrupt to clear the output channel and remove power from the relays. When this special case of T4RUPT processing occurs, the next interrupt is scheduled for 100 ms in the future to maintain the 120 ms T4RUPT processing cycle.

Checking the status of the Proceed key is the second DSKY-related task T4RUPT performs before any 120 ms cycle. Unlike all the other DSKY keys, Proceed does not generate a KEYRUPT interrupt. Rather, it sets bit 14 of input channel 32_8, which is interrogated at the beginning of each 120 ms pass. When a crewmember presses Proceed, T4RUPT will detect this state and schedule a Pinball job to process this keystroke.

So far, IMU processing has taken up only two of the eight available passes through T4RUPT, leaving several cycles available for servicing other devices. Up until this point, the implied assumption in this discussion of T4RUPT is that processing is similar for both the Lunar Module and Command Module. However, this is true only for IMU and DSKY processing, which are the only TRUPT4-managed devices that are common to the two spacecraft. The remaining operations performed in the T4RUPT processing cycle are tailored for each spacecraft's environment. In the Lunar Module, for example, the remaining T4RUPT processing passes focus on controlling the radars and converting their data for analogue displays, and for monitoring changes in the RCS configuration. Lacking radars and computer-selected thrusters, the Command Module's AGC uses T4RUPT to manage the optics assembly. Data from other AGC programs will request an orientation of the sextant and telescope assembly, and the T4RUPT routine controls the movement of those devices.

Although eight passes are scheduled during each 960 ms T4RUPT cycle, only four different routines are executed. To improve I/O responsiveness, each routine is scheduled twice, always 480 ms apart. For example, the IMU's monitor executes during the second pass of T4RUPT, and will also run 480 ms later during its sixth pass. A check of the RCS configuration in the LM is scheduled for the fourth pass, and again during the eighth and final pass, 480 ms later.

T4RUPT processing in the Command Module is significantly different than its Lunar Module counterpart. IMU processing, such as health checking, gimbal lock detection, power-on and power-off, are essentially identical for the two spacecraft. However, T4RUPT in the CM has no radars to command, nor input changes defining the RCS configuration. Rather, T4RUPT processing centers on the optics system, its health and status checks, and driving the sextant to its desired position.

LM T4RUPT passes 0 and 4: RCSMONIT

In the lunar descent, vehicle controllability is of paramount importance, especially during the final landing phase. Attitude control is managed by the Digital Autopilot, creating a responsive "fly-by-wire" system that takes commands from both the targeting software and the astronaut inputs. Given that the nature of a lunar landing allows "one attempt, no second approaches", a failure in any spacecraft system must be addressed immediately. In the case of the Reaction Control System, the crew can quickly shut down failing attitude control thrusters by closing their propellant valves using switches on the LMP's panel. These shutoff valves are organized by thruster pairs, and are designed to minimize controllability problems if a thruster has to be disabled. This intervention is communicated to the AGC by setting a bit in input channel 32_8. Eight bits, one for each thruster pair, are checked during each 480 ms T4RUPT cycle. When a thruster pair shutoff is detected by the T4RUPT routine, the Digital Autopilot's understanding of the RCS configuration is updated, and the DAP adjusts its control laws to prevent selecting the disabled thruster jets. In contrast, the Command Module has no comparable set of T4RUPT routines and the computer is not notified of any changes in the RCS configuration. If a thruster is disabled, the crew must manually update that spacecraft's DAP configuration using Noun 46.

LM T4RUPT passes 1 and 5: RRAUTCHK

The Rendezvous radar antenna dish is located above the "face" of the LM ascent stage, and can rotate on two axes to point directly at the CSM transponder. The radar antenna is mounted on the end of a shaft assembly and gimbals allow the dish to rotate left and right. The radar shaft is then mounted perpendicularly on the trunnion assembly, which in turn is mounted on the ascent stage. Gimbals on the trunnion are oriented parallel to the Y-axis, and allow rotation of the shaft/antenna assembly in the vertical axis. Resolvers in the shaft and trunnion gimbals send angular data to the CDUs, generating counter interrupts in the AGC.

The rendezvous radar tracks the CSM target vehicle in two different ways. First, its internal electronics can acquire, lock onto, and automatically move the radar antenna dish to follow the relative movement of the CSM. The second mode is for the AGC to compute the CSM's position and drive the radar antenna to where it expects the CSM to be. As the LM maneuvers, or if different orbits cause the relative positions of the two vehicles to change, the AGC commands a new position for the antenna to point. During T4RUPT passes, the AGC sends commands to the shaft and trunnion gimbals to reposition the antenna to its new orientation. This motion can occur slowly, at 1.3°/sec, or more rapidly, at 7°/sec. In all likelihood, several T4RUPT passes will execute during this repositioning operation. T4RUPT monitors the antenna's progress, and stops driving the gimbals when the desired orientation is reached. If the radar is tracking automatically and is not relying on the AGC for pointing control, T4RUPT will monitor the antenna's shaft and trunnion angles and if they exceed a critical value then the AGC will command the antenna back within limits.

During a T4RUPT pass, the radar's health and status is checked. Like the IMU, a power-up request requires special handling to initialize the electronics properly, and to prevent jobs that wish to use the radar from referencing it during this time. Powering down, or a failure in the radar electronics, forces a check for any work that might require the radar. A program alarm terminates any work using the radar, and the TRACKER warning light illuminates on the DSKY. T4RUPT also performs basic checking of the landing radar, verifying that the altitude and velocity sensing

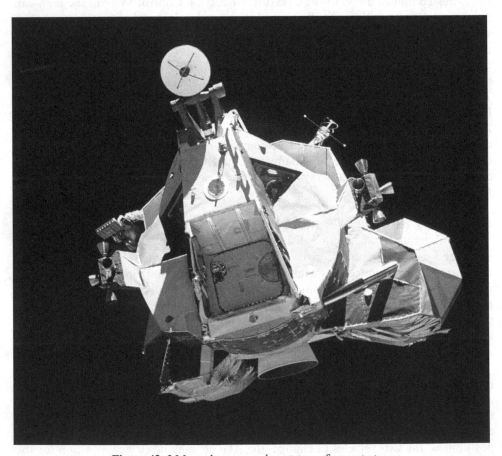

Figure 43: LM rendezvous radar at top of ascent stage

electronics have not failed. A problem in these electronics will not cause a program alarm or terminate any active work, as to do so might cause a loss of data during the critical landing phase of the mission. Note that this is different from the "radar data not good" discrete, which is often due to an antenna positioning error and is not the result of a hardware failure. Although programs continue to run, the TRACKER warning light illuminates to warn the crew the radar program is not able to follow its target.

LM and CSM T4RUPT passes 2 and 6: IMUMON
Two T4RUPT passes are dedicated to checking the status and health of the Inertial Measurement Unit. Applying power to the IMU and its related systems will set the POWER ON request in bit 14 of input channel 30_8. T4RUPT tests this bit every 480 ms, and if it remains set for one full T4RUPT cycle (960 ms) then a 90 second delay in any IMU processing is begun. By that time, the IMU should have completed its power up and initialization sequence, and the gyros should be up to speed. If no errors are detected, T4RUPT performs additional failure checks, and if everything is in satisfactory order will initialize the CDUs and direct the gimbals to move to their coarse align orientation. The NO ATTITUDE lamp is illuminated on the DSKY, informing the crew that the platform is not usable until it is aligned to a known reference.

After the testing of IMU state changes is complete, the T4RUPT pass continues by checking other bits in input channels 30_8 and 33_8. Failure indications such as temperature out of limits, or failures in any of the PIPAs, CDUs or the IMU itself, are presented in these channels. When any of these conditions arise, the IMU is not considered usable, and is placed in a coarse align mode, losing its attitude reference in the process. When T4RUPT detects such a failure, it illuminates a warning light on the DSKY panel to alert the crew to the loss of the IMU. In other cases, the IMU and its associated electronics may be operating normally, but other health checks might reveal issues. A problem in the PIPAs, identified in input channel 33_8 bit 13, is a serious fault and generates a program alarm. Used by the IMU to measure changes in acceleration, and, over time, velocity, a PIPA failure results in the loss of velocity data to the AGC, but does not cause the IMU to lose it attitude reference. While critical, the IMU is still operable in a degraded mode in which it provides only attitude references.

If AGC telemetry uplink or downlink is too fast, an error in channel 33_8 is raised and T4RUPT schedules a program alarm to alert the crew and clear the error. An additional test checks whether the crew is requesting the IMU be caged. Never performed during normal operations, caging the IMU forces the platform to a known, fixed orientation, and resets its associated electronics. All attitude and velocity data is lost and the vehicle must rely on a backup system (the SCS in the CM, and the AGS in the LM) for attitude control. Only after the platform is realigned and an updated state vector is either calculated or uplinked, is the platform usable for guidance, navigation and control. Once T4RUPT detects the caging request, it resets the software and illuminates the NO ATT warning light.

The IMU hardware itself does not detect or report on a potential gimbal lock situation. Only during the T4RUPT processing is the middle (Z, or yaw axis) gimbal tested against gimbal lock conditions. When the middle gimbal moves into the range between 70 and 85 degrees, the GIMBAL LOCK light is illuminated. At this point, the light is only a warning, although a very stern one, that the spacecraft will lose its attitude reference if rotation around the Z-axis continues. In this region, the attitude reference is still valid, and no harm results if the rotation is reversed. If the middle gimbal continues past 85 degrees, gimbal lock occurs and all attitude references are lost. The IMU is again placed into a coarse align state, which is unusable as an attitude reference, but is ready for the crew to perform realignment procedures.

LM T4RUPT passes 3 and 7: DAPT4S

For the Inertial Measurement Unit to provide an attitude reference, it must be oriented to a known, fixed orientation in space. This orientation might be the launch or landing site, or an attitude required by the spacecraft for a major maneuver. As the LM rotates, the gimbals in the IMU allow the platform to maintain its fixed orientation in space. DAPT4S takes the gimbal information and makes two sets of single-precision matrices for use in attitude computations. The first translates the gimbal angles into the coordinate system used by the spacecraft body axis, which are recognizable by the pilot as roll, pitch and yaw. Second, it rotates that matrix back to the coordinate system used by the gimbals to produce an easily referenced, single-precision matrix representing the gimbal angles. DAPT4S is unique in that it executes four times during T4RUPT processing. In addition to being called explicitly in T4RUPT passes 3 and 7, RRAUTCHK also transfers control to DAPT4S in passes 1 and 5, causing the attitude data to be calculated every 240 ms. Not all functions of the Digital Autopilot require highly precise values for the spacecraft attitude, making the single-precision version of this data perfectly acceptable. None of the data calculated during these passes is used by T4RUPT routines, the data is saved in an area of memory ready for use by other routines.

CM T4RUPT passes 0 and 4: OPTTEST/OPTDRIVE

During OPTTEST/OPTDRIVE T4RUPT passes, the AGC first determines if the optics need to be moved to another orientation. If so, the optics are driven to the position commanded by the tracking or IMU alignment program or through a DSKY entry supplied by the crew. The sextant trunnion gimbals are limited in the range it can move – while the sextant shaft may rotate a full 360 degrees, the trunnion is limited to 90 degrees. Within this trunnion movement, the sextant's view is only 67 degrees before the spacecraft structure obscures the view. When driving the optics to their new location, an output register is set with the number of "pulses" requested to move the optics assembly. Each pulse translates to a discrete amount of motion, but the two optics axes do not move at the same rate for each pulse transmitted. One pulse sent to the trunnion drive motor will move the sextant 40 arc-seconds, yet the same pulse will command the shaft to move the sextant 160 arc-seconds. These translate into a trunnion rotation rate of $3.8°/sec$, and a shaft rate of 15.3°/sec. The two CDUs are limited to 165 pulses in a single T4RUPT pass, or 1.83 and 7.33 degrees of rotation in the trunnion and shaft, respectively. This constraint assures that the previously commanded rotation is completed by the time of the next T4RUPT pass. During the repositioning process, the software checks the orientation of the optics not only to assure that the target orientation is achieved but also to prevent a potential hardware problem. Electrical and mechanical limitations of the optics assembly require avoiding movements near the trunnion's limit of travel. If the optics were to come near this zone, the T4RUPT will halt any further commands to the trunnion.

CM T4RUPT passes 1 and 5: OPTMON
The optics assembly comprises a wide field of view, unity-power telescope and a 28-power sextant.[13] Eyepieces for the telescope and sextant are mounted on a panel in the lower equipment bay, and observe through a window in the hull of the CM. In general, the AGC controls only the sextant, monitoring the angle at which it is pointing and driving it to a new orientation. Only when the telescope is explicitly slaved to the sextant will it be under the control of the computer. The sextant is used to identify the proper orientation for the inertial platform, and to obtain navigation fixes by taking sightings of the Earth and Moon.

Managing the optics system in T4RUPT begins with a check of the CDUs, which convert the sextant's angular orientation into counter pulses that are usable by the AGC. A failure in this hardware sets a bit in the AGC, which T4RUPT senses and then raises a program alarm to alert the crew that the sextant is unusable. Next, the optics mode is tested to see if the sextant is being operated manually, or if it is commanded automatically by the AGC. A manual mode, of course, will inhibit any software requests for driving the sextant to a new position and will generate a program alarm if this is attempted. A third possible state is for the optics to be in the process of being "zeroed". Zeroing is performed before every sighting operation, and drives the sextant to a known position and resets the CDUs.

CM T4RUPT passes 3 and 7: RESUME
Unlike the Lunar Module AGC software, the Command Module version does not use all of the T4RUPT cycles: passes 3 and 7 simply return control to the T4RUPT processor without performing any operations.

---

[13] A rather cynical comment says that only the United States government could be fooled into purchasing a 1x power telescope. This, of course, ignores the fact that the telescope provides an exceptionally wide field for locating the navigation stars, and can be slaved to the sextant. After centering the star in the telescope, attention turns to the sextant with its much finer resolution for the actual sighting.

HIGH LEVEL LANGUAGES AND THE INTERPRETER

Since the start of the computer era, there has been the need to express problems for the computer in a form that is understandable to humans, yet easily mechanized for execution in a processor. Early systems placed the burden of effort wholly on the programmer, forcing them to structure their problems in steps that corresponded one-for-one with the basic machine operations. One can only imagine the difficulty in translating symbolic mathematical equations into the arcane language of basic arithmetic and procedural operations. In the section on AGC instructions, Figure 24 showed an example of using basic instructions to add a list of numbers. While quite useful as a demonstration of several interesting AGC instructions, it also showed the level of detail necessary for performing even trivial tasks. Expending a large amount of effort to develop such a routine is acceptable if the code remains static. However, if there is a change in the AGC instruction architecture, or if the routine is ported to an entirely new architecture, we are forced to rewrite the code to reflect the new hardware characteristics. The time and expense of such a recoding effort is often so high that it negates using a newer, more flexible architecture. A not completely implausible example would be implementing the routine to sum a table of numbers in the Space Shuttle computer. The Shuttle's General Purpose Computers are a completely different design than the AGC, so our program must be rewritten to reflect the different architecture. Figure 44 shows how this might look in the computer's native, low-level language.[14]

   While the Space Shuttle code is arguably more compact and efficient, there is still the issue of rewriting the summation program to use multiple registers, indexes and a

```
          SR     R2,R2           SET SUM IN REGISTER 2 TO ZERO
          SR     R3,R3           STARTING INDEX IS ZERO
          LA     R4,4            INCREMENT IS 4 BYTES IN THE TABLE
          LA     R5,20           FIVE TABLE ENTRIES * 4 BYTES/ENTRY
LOOP      A      R2,NUMBRTAB(R3)      ADD THE NEXT ENTRY IN THE TABLE
          BXLE   R3,R4,LOOP      INCREMENT THE INDEX, CHECK IF
*                                WE ARE DONE, IF NOT, CONTINUE
NUMBRTAB  DC     '2'
          DC     '5'
          DC     '7'
          DC     '4'
          DC     '6'
```

Figure 44: Summing a table of numbers in the Space Shuttle computer

---

[14] Some readers will recognize this code as IBM 360 assembler. The Shuttle General Purpose Computer, the AP-101S, is based on the classic IBM System 360 architecture. Although the AP-101S is not completely compatible with the S/360 series, it is sufficiently similar that using S/360 assembly for this example is acceptable.

complex branch instruction. Clearly, the goal is to express the problem in a form that is as independent of the underlying architecture as possible. At the same time, the notation should more closely reflect how the user would express the original problem. High level languages are a huge improvement over machine level coding, as they reflect the problem statement more closely, and are far easier to read and maintain. Literally hundreds of languages exist, each with its own target audience, but some of the most familiar are C, Java, and yes, even Cobol. The power of a high level language is easily demonstrated by the trivial assignment of the value "1" to the variable "X". Coding this in the AGC might look like this:

```
        CA    ONE     COPY THE VALUE OF 1 INTO ACCUMULATOR
        TS    X       SAVE THE VALUE OF 1 IN STORAGE
        ...
X       DEC   0       STORAGE AREA CALLED "X"
ONE     DEC   1       VALUE OF ONE
```

The same expression can be coded in the C language:

```
integer x = 0;
x = 1;
```

While this small code fragment is not a particularly dramatic improvement over the AGC code, mathematicians will immediately understand the intent. We now expand the idea of high level coding, using our number summation program written in the C language:

```
integer numbrtab[5] = {0,1,6,4,2,3};
integer sum = 0;
integer i = 0;
for (i=0; i++; i<5) {
    sum = sum + numbrtab[i];
}
```

While not completely intuitive, this code is far more readable than our attempts with machine coding, and at first glance appears to be the solution to our coding needs. While the program is more "human friendly", it cannot execute in its current form without a conversion to the processor's native machine instructions. A software program known as a compiler performs this task, taking each of the high level statements and generating the equivalent machine instructions. Unfortunately, there are important limitations to the quality of code that a compiler can generate. In particular, early compilers did not produce efficient code, compared with the best hand-written machine instructions. The fact that both execution and memory requirements are larger for compiled programs, made high level languages impractical in the resource-limited AGC.

A compromise between the high efficiency of machine code and the easy programmability of a high level language is realized in the AGC's Interpreter. As implemented in modern computers, interpreters perform a similar function to that of a compiler. Both begin with the source code written as a high level language. After taking the source statements and generating machine instructions, a compiler will never need the original source code again. Interpreters do not generate machine level instructions for the hardware to execute, but use a different process to convert high level statements into a form that can be processed. The simplest interpreters begin with the original source code, and execute the program directly without converting the program into a more efficient representation. Each source statement is reparsed and reinterpreted each time it is encountered, regardless of how many times the statement has been executed previously. As expected, this approach is very slow and inefficient, and there are no intermediate results from the interpretive process to save and reuse.

The output of more sophisticated interpreters is not machine instructions as produced by a compiler, but an intermediate language, often called "pseudo-code" or "P-code". This interpretive code is the "language" of the interpretive system, and is analogous to a processor's machine instructions. An imaginary, or "virtual", processor is written in software, and it reads and executes the pseudo-code as its "basic instructions". A huge benefit of such a virtual machine is the ability to create any desired processor characteristic, bypassing the limitations inherent in the underlying hardware's architecture. In the description of the basic design of the AGC, the lack of support for index registers and sophisticated datatypes is apparent. With an interpreted system, many of the limitations of the AGC's architecture are neatly eliminated. Interpreters are targeted for developing applications such as the mission programs, rather than system-level coding in the Executive. Restricting the interpretive software to mission software eliminates the complexities involved with managing hardware operations such as interrupt handling. This is usually not a problem, since Executive programs rarely use complex datatypes or complex indexing, and cannot afford the greatly increased execution time that interpreted programs require.

Unlike contemporary languages like Java, whose statements are converted to "P-code" and then executed, there is no comparable high-level language for the AGC. The task of efficiently converting human readable code to an internal format requires sophisticated optimizing compilers and interpreters, which are fiendishly difficult to develop, and were simply not available to the AGC developers. To create the most efficient programs possible, the AGC developers wrote the intermediate code directly, bypassing the troublesome conversion from a high level language.

Parsing high level language statements

To understand the AGC's interpretive system, it is essential to return to a fundamental issue. What is the process of converting the high level logic and equations into executable machine instructions? Beginning with a simple equation:

$$a = \frac{(b^2 + c^2)}{(b^2 - c^2)}$$

The limitations of typing on a computer require equations to be coded on a single line:

a = (b**2 + c**2) / (b**2 – c**2)

While perhaps not as graphically satisfying as our original expression, the new representation is equivalent and has its own advantages. The sequence of operations is more apparent, using the convention of reading from left to right. Parentheses are used to group operations together and eliminate any ambiguity in identifying which terms should be executed first. Because superscripts are not possible in this simple single-line notation, exponents are replaced by the characters "**". Using this notation, b^2 is now rewritten as b**2. Starting with the expression immediately following the equals sign, the value of b is squared, as is the value of c, and the two are added together. Now that the numerator has been evaluated, it is time to begin work on the denominator. However, both the numerator and denominator must be evaluated before the division operation can proceed. To preserve the work already done on the numerator, the intermediate value of $(b^2 + c^2)$ is set aside in a temporary location that we will call T1. Next, the value of c^2 is subtracted from the value of b^2. Like the first half of the equation, this intermediate value is saved as well, this time in a location called T2. The only remaining operation is to divide the temporary value of T1 by the temporary value T2, and save the result in the variable named a.

Processing an equation requires reading and parsing the entire statement, and analyzing the result to determine the proper execution sequence. The notation used has its roots in computer science, and is called a "tree structure". Many problems in computer science lend themselves to being expressed as trees, especially those with complex interactions. In its simplest form, a tree structure has "branches" and "leaves", akin to its physical namesake. Operands such as variables are the leaves, are connected to branches, and are joined by a single computer operation. In addition to benefiting the programmer with a graphical description of the problem, translating a tree structure into basic computer instructions is quite straightforward. An expression, 3 + 2 = 5, has two numbers added together to create a sum. In a tree structure, this expression is described in Figure 45(a). Read from the bottom-most leaves upwards, we take the two operands, three and two, and add them together to create five as the sum. Extending this form to an expression such as X = 3 + 2, we have the variable X assigned the sum of three plus two in Figure 45(b); and 45(c) creates a generalized expression which retains the same structure, yet now is usable for any set of input variables.

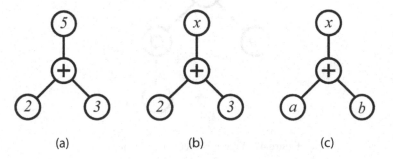

(a)   (b)   (c)

Figure 45: Expressions in tree structures

The full range of binary operations are available to the programmer, including addition, subtraction, multiplication and division. Unary operators also exist, and include not only negation, but also mathematical operations such as square and root plus trigonometric functions.

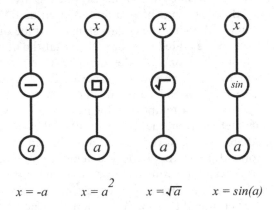

$$x = -a \qquad x = a^2 \qquad x = \sqrt{a} \qquad x = sin(a)$$

Figure 46: Unary and trigonometric functions

As with all building blocks, each such individual element appears barely worthy of even passing attention. It is only when they are combined that they become truly interesting. Consider a slightly more complex expression, $E = mc^2$. This familiar equation requires both a unary and a binary operation. Traversing the tree upwards, the first step is to square the value of c. The result is multiplied by m, and the final value placed in E.

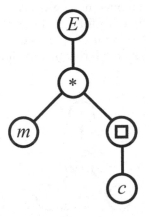

Figure 47: Tree representing $E = mc^2$

The ability to assemble basic operations into tree structures to define an equation is now becoming apparent. A final, more complex example allows us to address the more important issues that arise in real world problems. Consider the polynomial expression $ax^2 + bx + c = 0$. Solving for the value of x requires the quadratic equation:

$$x = \frac{-b \pm \sqrt{b^2 - 4ac}}{2a}$$

Two values for x are possible, so we must break it into two separate expressions.

$$x = \frac{-b + \sqrt{b^2 - 4ac}}{2a} \text{ and } x = \frac{-b - \sqrt{b^2 - 4ac}}{2a}$$

The trees differ only by the plus or minus term in the numerator, so our next example works equally well with either expression. A tree structure of the first equation looks like:

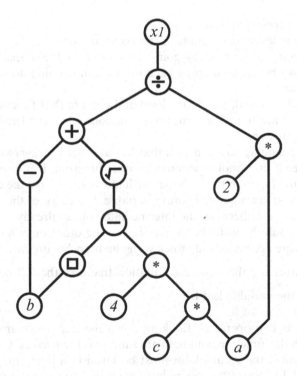

Figure 48: Quadratic equation tree

The quadratic equation uses two variables (a and b) more than once, which complicates our tree somewhat. The variable b is both squared and negated, and a is in both the numerator and denominator. The equation is already in its simplest form, so it is not possible to combine or cancel out either variable.

Traversing the quadratic equation

We return to one of the possible roots of the quadratic equation:

$$x = \frac{-b + \sqrt{b^2 - 4ac}}{2a}$$

The traversing process starts in the lower left-hand corner of the tree, with the goal of traversing all the branches to the right and upwards. Only a few rules are necessary for navigating the tree, and they reflect the natural process of evaluating an expression from left to right.

- Start at the lowest, left-most leaf on the tree. If there are multiple branches on a leaf, start with the left-most branch.
- Variables, constants and temporary variables are all handled in the same manner.
- A branch is traversed only once.
- All functions leave their results in the accumulator.
- Upon reaching a function by going up one of its branches, take the other branch down before continuing back up. Continue going down the tree until a leaf is found.
- When a leaf is found, go up one level and execute that function.
- If two leaves are found underneath a function, select the left-hand leaf first, then the right one.
- When traversing up to a function that has another untraversed branch, look ahead to see if that branch contains another function. If it does, and we have arrived at the function through the left-hand branch, save the contents of the accumulator in a unique temporary variable. Once saved, the accumulator is cleared to zero. Otherwise, the intermediate values already calculated up to this point would be lost when traversing to the other branch.
- Operations are performed only when going up branches on the tree, never down.

The process of traversing the quadratic equation tree uses the following steps:

1) Start with the variable leaf b.
2) Negate the value of b.
3) Addition has two operands. Looking down the tree, there are two branches underneath this function, and both are functions themselves. Coming up from the left branch, the accumulator must be saved in a temporary variable that we will call T1. Next, the accumulator is cleared to prevent interference with the next set of arithmetic instructions. The next branch is down and to the right, to the Square Root function.

4) Square Root is a unary function. No special handling is necessary, and we continue down to determine its operand.

5) Subtraction, like addition, requires two operands. Both are functions, requiring that the intermediate variable be saved when we finish traversing the left branch.

6) The first function is Squaring. Like Square Root, Square is a unary operation, so we continue down the next branch and see the variable b. This is squared, placed in the accumulator, and we move up the left branch of the Subtract function.

7) Back at the Subtract function, all of the operations on the left-hand branch are completed. Noticing that the right-hand branch is another function, we will again save the accumulator, this time in temporary location T2.

8) Moving down the right-hand branch of the Subtract, we encounter the Multiply operation. On the left branch we have the constant 4, so no further progress is possible here. The branch to the right will provide the second operand.

9) The branch takes us to another Multiplication function, with its two branches leading to the variables a and c. There is now enough information to perform the Multiply, so the product of a and c is placed in the accumulator.

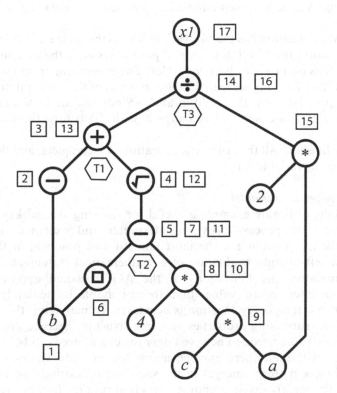

Figure 49: Labeled quadratic equation tree

10) Moving up with a value in the accumulator, the next step is to perform the Multiplication of the accumulator and the constant 4. Once the Multiply is complete, the traverse up the tree can continue.

11) Returning to the Subtraction operation, there is the left-hand value stored in the temporary value T2, and the value left in the accumulator from the just completed traverse of the right-hand branch. The contents of the accumulator are subtracted from T2, the result saved in the accumulator, and the traverse continues upwards.

12) Back to the Square Root, now with a value that that is usable for that function. Take the Root of the accumulator contents, replacing the existing value in the accumulator.

13) We now find ourselves in a similar situation as we were in step 11. The value for the left-hand side (-b) is saved in temporary variable T1. The addition operation takes T1 and adds it to the accumulator.

14) Calculations for the numerator are now complete, and attention moves to the denominator of the equation. The division operation has a value provided through its left-hand branch, but the right-hand side is a Multiplication function. We will save the intermediate result in another temporary variable, T3, and begin the traverse down the right-hand branch. Again, the accumulator is cleared when creating the temporary variable.

15) The Multiplication has the two variables available as leaves on the tree, so we can perform the Multiplication and put the result in the accumulator.

16) The Division now has both variables. The numerator from the left branch in variable T3. This is divided by the contents of the accumulator built from the operations on the right branch. Since the denominator is in the accumulator, we need to exchange it with T3 before the Division takes place.

17) We're finished! All the arithmetic operations are complete, and the final value is saved in variable X1.

Stacks and temporary variables

This admittedly elaborate example is useful for showing several key concepts of how equations are processed by an interpreter and executed. Parsing and generating the tree structure is the most familiar and practical methodology to represent the left-to-right evaluation of a mathematical statement. With only a single accumulator, the limitations of the AGC become apparent. Finding ourselves with intermediate values that are not usable immediately, saving the accumulator in a temporary location is necessary. In managing these temporary variables, a few surprising properties of a tree structure become apparent. First, an intermediate value needs to be saved only once, and needs to be retrieved only once as well. Although there may be many intermediate values saved at one particular time, a pattern emerges in the sequence of storing and retrieving the data. The most recently saved temporary data is always the first brought back into the equation. Any data saved previously is recalled only after more recently saved data is itself retrieved, a process called "first in, last out". To see this property demonstrated, assume three intermediate values T1, T2 and T3 are saved in sequence: T1 is saved first, then T2 and lastly T3. Only after the point where T3 is retrieved will T2 be needed for its part of the equation. Continuing further, only after T2 is used will T1 become available in the expression.

A restaurant kitchen offers a perfect physical analogy of how the temporary variables are referenced. Plates are stacked in a dispenser cart, which contains a spring to keep the top plate visible and easy to grasp. Whenever a clean plate arrives from the dishwasher, it is placed on top of the stack, pushing down whatever other plates might already be in the dispenser. When the chef needs a new plate, she takes it from the top of the stack and the remaining plates "pop up". With this analogy, we have described the most common and powerful data structure in computer science, the stack. When data has to be stored in an ordered sequence and retrieved in reverse order, stacks are used to store and keep track of the data. The stack is especially appropriate where one program calls another, which in turn, calls a third, which calls a fourth, etc. Local variables and parameters are pushed onto the stack at each successive program call in the same manner that plates were placed onto the dispenser. As each program completes processing and control returns to its calling program, the data is "popped" from the stack.

A particularly important characteristic for the AGC is that a stack uses memory very efficiently. Consider the case if, instead of a stack, we explicitly assigned a word of memory for each temporary variable that we might use. Programs generally do not simultaneously use all temporary variables at all times; yet without a stack, all of the storage for temporary variables must be allocated in advance. Memory is wasted since only a few temporary variables are needed at any one time, leaving many words of storage sitting idle. Granted, it is possible to develop elaborate schemes to reuse specific memory areas, but these are cumbersome and error prone. In contrast, a stack "grows" in storage only as a program requires it.

Stack implementations follow the physical analogy of the plate dispenser only so far. Each time a plate is added or removed, the remaining plates physically move up or down in unison. In designing a software version of the dispenser, there is no need to move the data back and forth in memory. Rather, we reserve enough storage to hold the maximum amount of data that might be expected in the stack. The first item is placed in the first storage location in the stack memory area, and is (by definition) on the "top". When the second item of data needs to be placed on the stack, the first data word is not moved out of the way. Instead, the new word is placed in the second storage location in the stack. This action creates a dilemma: specifically, how is it possible to keep track of the location where the "top" of the stack is? Solving this problem is done by introducing a new variable that is an integral part of a stack implementation, the stack pointer. Stack pointers, as their name implies, always reference the most recent item in the stack. "Pushing" a new item of data onto the stack requires that the stack pointer change its reference to the next available location, into which the data is copied. "Popping" data off the stack operates in the opposite way. After retrieving the data at the top of the stack and passing it to the requesting program, the stack pointer is updated to reference the next oldest entry. Note that only the value of the stack pointer changes during the push and pop operations, and the data in the stack remains fixed. Clearing out the locations that previously held the "popped" data is not a mandatory housekeeping task, but it is often good practice to do so. All stack operations must use the stack pointer to reference the data, and only the top element is legitimately available. From this set of assumptions, any storage not within the scope of the stack at any given moment is irrelevant.

Double and triple precision datatypes

The notion of "datatypes", that is, whether a variable is single or double precision, integer or fractional, has had little relevance until this time. In the task of analyzing an equation and creating a process tree, the exact characteristics of the datatypes were not important. Only now, as the process tree is converted into machine instructions do these characteristics need to be considered. The example for finding one root of a quadratic equation used single precision values, but double or even triple precision variables are possible simply by generating a few extra machine instructions for each arithmetic operation. Generating an extra instruction or two might be acceptable for multiple precision values, if done sparingly. However, when complex datatypes (such as vectors and matrices) are used, repeatedly generating the same instruction sequences consumes large amounts of memory. Using a traditional subroutine library for complex operations is one alternative, with parameter lists of input and output for exchanging data with the main program. Unfortunately, this approach also consumes memory at a prodigious rate, and incurs the overhead of moving data in and out of the parameter list.

At this point, extending the AGC architecture further only results in marginal functional improvements. Even if the new system extensions are desirable, the additional level of system complexity is expensive in terms of the development and testing effort.

Thus, we conclude that the AGC is ripe for a radical departure from its currently implemented architecture.

THE INTERPRETER

Thus far, this chapter has focused on the system level software found in the Executive. System routines are characterized by their own type of complexity and operate in a world of interrupts, timers and process management. Logically complex, the programming focus of the Executive software is hardware, and it uses simple datatypes such as single and double precision variables. While the Executive routines are essential, they are only the necessary overhead required for the most important software: the mission-specific programming. Mission software is very different from that in the Executive. Each mission program aims to achieve a specific objective, whether it be navigating to the Moon, the landing itself, or the rendezvous after the lunar module ascends from the surface.

Developing the mission software to solve these problem is a huge undertaking.

Ideally, mission programmers want to concern themselves with the nuances of the problem, and not with the idiosyncrasies of the underlying system. In these cases, a high level language would be preferable, but their relatively inefficient code and excessive memory requirements would be impractical in the AGC. Despite the desire to remove developers from the nagging details of the machine, the AGC's resource constraints are so severe that only hand-coded and optimized programs are practical. The obvious need to extend the AGC architecture and the inability to do so, forced the designers to consider a radical departure from the evolutionary steps used up to this time. If extending the existing hardware is impractical, the alternative is to invent an entirely new system architecture and implement it in software. Freed of (most of) the limitations of the underlying AGC hardware, a new architecture can implement features to ease the development of mission software. The Interpreter, as this new system architecture is named, supports several important features:

- Single, double and triple precision variables
- Complex datatypes, such as vectors and matrices
- Vector operations, dot and cross product, and shifts
- Transcendental functions, such as square root and trigonometric functions
- Two index registers, plus index increment registers
- A large stack area and stack pointer
- Simplified addressing – banking registers are not explicitly used.

The Interpreter complements rather than replaces basic AGC coding, as not all the mission software programming requires the capabilities of the Interpreter. Additionally, mission programs remain wholly dependent on Executive routines for process management, I/O, interrupts and other system level functions. As such, while the Interpreter has a number of features in common with a hardware CPU, it is less a machine emulator than it is an extension of the programming environment. Programs may begin with basic AGC instructions and then enter the Interpreter to evaluate a complex expression, only to return to executing basic instructions when the calculation is complete. Fundamentally, the role of the Interpreter is to extend the AGC's architecture using software written in the AGC's native code. Elements such as specific registers or status bits that are not found in the physical machine are implemented in erasable storage, using software logic to define the rules of their operation. In the AGC's interpretive system, the goal of a rich and comprehensive instruction set and the desire to hide details of memory banking from the user forces an unpleasant reality. With only a 15-bit word available, interpretive instructions and their operands cannot fit into a single word. The Interpreter's instruction set must be extended into a second word to hold all the necessary bits. However, by breaking through the psychological barrier of "one word equals one instruction", a wide range of possibilities present themselves.

Two issues deserve immediate attention when designing an interpretive environment: the number of operation codes and the characteristics of memory addressing. The experiences with the basic AGC instruction set and its three bit opcode proved that an interpretive instruction must have sufficient bits to represent the entire instruction set without resorting to elaborate tricks in order to extend an inadequate design. But how many bits are truly necessary? Obviously, three bits for an opcode is unacceptably small, five bits is only marginal, but using the entire 15-bit word (for more than 32,000 unique instructions) is absurd. At the same time, using only a fraction of the word for the opcode is wasteful, presenting the question of what to do with the remaining bits. Addressing in the AGC is limited to 15 bits, so using the leftover bits in the opcode word for addressing does not provide any additional capability. The solution for maximizing the usage of the interpretive instruction word has profound consequences for writing and storing a program in memory. Choosing seven bits for the length of the interpretive opcode provides a generous 128 unique instructions, and also allows two opcodes to fit comfortably within a single word. By squeezing two opcodes into the same word, we create one of the unique properties of the interpretive programming system. Unlike the basic AGC instruction set, instructions for the Interpreter are not in a single, fixed format, nor are the operands necessarily contiguous with their opcode. After the word that holds the two interpretive operation codes, operands for the first opcode are specified, followed by those for the second opcode. This format is complex, and neither the interpretive language nor the assembler provide the programmer with any coding assistance. Each word of the interpretive program is coded individually, with the programmer solely responsible for operand placement.

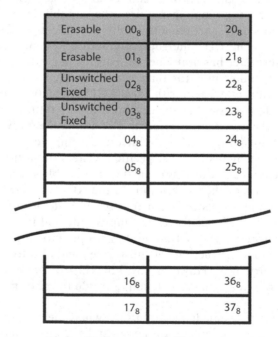

Erasable	00_8		20_8
Erasable	01_8		21_8
Unswitched Fixed	02_8		22_8
Unswitched Fixed	03_8		23_8
	04_8		24_8
	05_8		25_8
	16_8		36_8
	17_8		37_8

Figure 50: Interpreter half-memories

The Interpreter's address word uses only 14 bits to reference the location of the data, leaving bit 15 to define which (if any) index register is used. However, 14 bits can reference only 16K of storage, which is far less than even the AGC's limited memory. Circumventing this serious restriction of storage access requires an adjustment to our view of memory management in the Interpreter. By introducing the concept of "half memories" to an interpretive program we can view fixed storage as two independent 16K areas.[15] However, not all of the 16K is fully used, nor are the half-memories contiguous. All references in half-memories must be in switched fixed storage, and banks 00_8 through 03_8 are inaccessible. The first half-memory occupies fixed storage banks 04_8 through 17_8, giving 12K of addressable storage in the first half-memory. The second half-memory uses fixed banks 21_8 through 37_8, for a total of 15K words, with bank 20_8 unused by the Interpreter.

The instruction format of the Interpreter
The structure of the interpretive instruction set is quite different from the basic instructions of the AGC. Up to this point, software was converted to machine instructions using a single, self-contained instruction for each line of our source code program. Interpretive instructions depart from this convention by taking one or two instructions and coding them over several lines of source, which translates into several words of storage. Instructions begin with the Interpretive Instruction Word (IIW), which contains one or two operation codes. Each opcode is analogous to an AGC basic instruction code, and may include arithmetic, logical and branching operations. Source code instructions are written from left to right, reflecting the sequence in which they should execute, but the first opcode is stored in the lower seven bits of the IIW. The second opcode, if it exists, is placed in the upper half of the word, in bits 14 through 8.

Each opcode of the Interpretative Instruction Word requires zero, one or two Interpretive Address Words (IAW). An address word is exactly that – the address of an operand required by an interpretive instruction. The number of IAWs is determined by the operations in the instruction word. An operation may not require an operand, such as a trigonometric function that acts only on the accumulator contents, or a "vacuous" operand on the Interpreter's push down stack. In both cases, there will not be an IAW for this operation following the instruction word. Other operations can require one or two IAWs, depending on the interpretive instruction. Operands for the first interpretive opcode immediately follow the instruction word, and any IAWs for the second opcode follow next in storage. Figure 51 shows a representative layout of an IIW, followed by IAWs.

---

[15] Breaking fixed storage into half-memories might be properly thought of as a memory banking technique, not unlike how fixed and erasable storage is managed. While this observation is correct, the literature never uses the word "banking" to describe half-memories, so we will refrain from using this term as well.

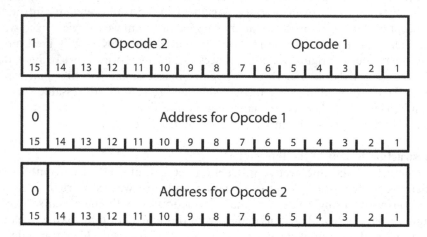

Figure 51: Generic interpretive instruction format

Interpretive opcode encoding and addressing
The encoding of the Interpretive Instruction Word is complex, and is driven partially by the need to differentiate itself from the Interpretive Address Word. Interpretive opcodes are located in bits 14 through 1 of the instruction word, with bit 15 equal to zero. When placed in storage, the entire word is complemented to create a negative value, readily identifiable from normally positive Instruction Address Words.[16] Complementing the IIW becomes the first step in the instruction fetch process. The first operation code (called OP1) is in bits 7 through 1 of the word, and is isolated by masking out the higher order bits. Subtracting one from the opcode leaves us with the final value for the interpretive opcode. A different process extracts the second operation in the instruction word, OP2, which is in bits 14 through 8. Repetitive shifting of the IIW to move these bits to the low end of the word is possible, but this is a time consuming process. A special hardware assist is available by saving the IIW into storage location 00023_8, the Edit Opcode (EDOP) special register. When data is saved in the EDOP register, it is automatically shifted right by seven bits, filling the higher order bits with zeros. After retrieving the operation code from the EDOP register and subtracting one, the second interpretive opcode is ready for decoding and execution.

Interpretive Address Words are coupled with an interpretive operation code and define the address for its operands. In most cases, IAWs use only addresses as

---

[16] There is one exception to this case. The address is complemented, and bit 15 in an IAW is set to one when indicating that the X2 index register is used.

operands, pointing to data in either erasable or fixed memory. The few special addressing cases are notable, because they create useful extensions to the Interpreter. Immediate operands, where the operand itself is the data, are available in the cases of shifting and index register manipulation. Indirect addressing, when the operand points to a word in storage that contains yet another address at which to save the data, is possible only when storing the contents of the Multipurpose Accumulator. Indexing using one of the interpretive index registers, X1 and X2, is a welcome addition to the AGC repertoire, and replaces the powerful but awkward INDEX instruction. Indexed addresses are possible only with special operation codes, where the choice of index register is embedded within the IAW. In cases where the first interpretive opcode does not require an operand, coding IAWs for the second opcode is no problem, because the Interpreter has no difficulty in determining what operation code the IAWs are associated with. However, the case where the first operand intends to use the stack as an operand (a vacuous address) can present a problem. If the second operand requires an operand, then the Interpreter requires a means to determine which instruction the IAW is associated with. The solution is through the STADR opcode, which is discussed below.

Programmers must exercise caution when coding the address words, as they do not contain identifying fields that assign the IAW to a specific opcode. No sanity checking is performed during the source code assembly or interpretive execution to assure that the IAWs are correctly associated with the operation codes. A forgotten IAW, for example, will not prevent the Interpreter from fetching a word that "should" be the intended operand for the other operation code in the IIW. Address words are not exclusively for data addresses. Branching and other transfer-of-control instructions use IAWs to specify the destination of the next instruction to execute.

Together, instruction and address words are termed an "interpretive string". This is a more accurate characterization than referring to an interpretive "program", since the "string" is not executable code in the traditional sense. The Interpreter itself is the executable unit of work in the AGC when a string is processed, and the string is simply its "data". This might appear arcane, but the difference becomes relevant when following the execution of a job in the AGC. Special registers containing addresses within the executing program (such as Z or Q) will never reference any location in the string, only in the Interpreter.

Breakdown of the instruction classes

The Interpreter presents a rich set of well over 100 instructions, plus a special "no operation" code. Of the seven bits available, the last two bits define important characteristics of an instruction and also group them into four distinct addressing schemes.

Addressing Class	Characteristic
00	Unary instructions, which do not require an operand. Operations include transcendental functions and short shifts. Unary instructions should not be confused with operations using vacuous addressing.
01	Arithmetic operations using a non-indexed address to reference storage. Addressing is limited to only local erasable or the half-memory from which the instruction was fetched.
10	Branching and index manipulation instructions. Branching instructions may use all of the interpretive storage areas (both half-memories). Addressing in index manipulation instructions can reference any area in interpretive storage, and may also use the address word as an immediate operand.
11	Arithmetic instructions that use index registers X1 or X2 modify the operand address word. Values in the index register are subtracted from the operand address, complementing the looping technique seen in the Count, Compare and Skip basic instruction. Only storage in erasable storage or the interpretive half-memory is accessible.

While a maximum of 128 opcodes appears more than sufficient to provide a wide variety of operations, closer inspection reveals that many instructions are simply variations of a single operation. As an example, consider the four types of short shift – the left and right forms of shift, and shift and round. For each type, there are four distinct opcodes for shifting one, two, three or four bits. Hence, 16 opcodes are dedicated to the short shifts. As the opcode's 2-bit suffix differentiates between indexed and non-indexed operations, two opcodes are necessary for the instruction to use both types of addressing. Organized according to function, the interpretive instruction set is broken down into the following groups:

- Store, load and push-down
- Scalar and arithmetic
- Vector arithmetic
- Transcendental functions
- Vector functions
- Shifts; short shifts, general and normalized
- Branching, sequence generation and subroutine linkage
- Switch instructions
- Switch test and branch
- Index register manipulation
- Miscellaneous instructions.

This breakdown is necessarily broad, and instruction assignments often fall into a number of categories. Vector operations are especially problematic. Shifting the components of a vector is a good example: should it be categorized as a vector or a shift operation? In this case, the instruction is assigned to the shift operations, as the operation (shifting) is not unique to the vector datatype. Other operations are equally ambiguous, and this book will defer to the categories established in the primary AGC documentation.

The Multipurpose Accumulator and the datatype mode of the Interpreter

Arithmetic and logical operations in the basic AGC hardware are performed in the accumulator, located in the lowest part of addressable memory. The wide variety of datatypes used in the Interpreter make the existing hardware accumulator woefully insufficient. Without adequate hardware support, software becomes the only means to create an accumulator that will handle a variety of datatypes ranging from single precision to vectors. Earlier, we defined the Multipurpose Accumulator (MPAC) as a seven word area in a job's Core Set. For jobs that do not invoke the Interpreter, these words are available for general use by the program. No formal structure is defined for the MPAC in non-interpretive programs, and all of the storage may be used as the programmer sees fit. When used during interpretive string execution, the MPAC assumes a strictly defined format that is under the control of the Interpreter itself. The Interpreter also establishes the concept of the data "mode" for instruction execution, representing the datatype first loaded into the MPAC at the start of the interpretive string. These datatypes are defined as double or triple precision scalar quantities, or as three-element double-precision vectors. Interpreter programs can also operate with single-precision scalar quantities and with matrices (3 by 3, double-precision elements), but such operations do not establish a new datatype mode. During the execution of the interpretive string, the mode changes only when a different datatype is loaded into the MPAC, or when the mode is explicitly changed during a mathematical operation.

As the MPAC and its data mode are determined by software, not hardware, a highly flexible architecture is possible. Consider the basic AGC hardware, where the primary datatype is the single-word integer.[17] All hardware elements, such as the special registers, I/O channels and the most basic instructions, work on this 15-bit structure. Most mathematical operations using double precision data require programmers using basic instructions to manipulate both parts of the two-word quantity. To provide native support of the full range of datatypes, the Interpreter uses the same basic AGC operations but hides the implementation details, allowing the developer to concentrate more on the program requirements.

Operations such as pushing data onto the stack or storing data in memory are not tied to a specific datatype. Taking its cue from the datatype "mode", a PUSH instruction will place a double precision, triple precision or vector onto the stack. The notion that the Interpreter supports several datatypes is significant, but a large part of its power and flexibility lies in its ability to adjust its processing based on the type of data being manipulated.

---

[17] To be sure, a few instructions do operate on two consecutive words at once, such as Double Clear and Add (DCA)

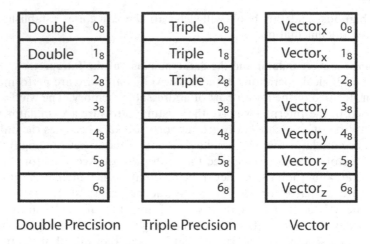

Figure 52: Multipurpose Accumulator and datatype modes

Indexing in the Interpreter

Indexed addressing in the Interpreter overcomes a key limitation in the basic AGC architecture; specifically, the lack of a hardware index register. While the AGC's basic instruction INDEX has a novel method of addressing elements in an table, the Interpreter's indexing operations are much more conventional. Many of the arithmetic instructions (both scalar and vector) can use indexing to reference their operands, as can the general shift instructions. By reserving the last two bits in the opcodes to indicate that the interpretive instruction will use indexed operands, the Interpreter now knows to look within the operand for information on which index register to use. Index registers X1 and X2 are available, and either may be used wherever indexing is possible. Two "step" registers, S1 and S2, are paired with the index registers, and are used by the Transfer on Index instruction for loop control. Unlike other registers in the Interpreter, the index and step registers cannot operate with fractional data. For an index register to modify an address, or for the step register to decrement its index, only single-precision integer data is allowed.

Loading data into an index register can take several forms, depending on the location of the data. In cases where the data to initialize the index is well defined and unchanging, Address to Index True (AXT)[18] uses the operand as immediate data and not as an address. In less well defined situations where the initial index value is not known until execution time, Load Index From Address (LXA) uses its operand to point to the data in storage. Two instructions allow us to preserve the contents of the index registers. Store Index into Erasable (SXA) copies the contents of the register into erasable storage. A two-way exchange operation is also available. Exchange Index with Erasable (XCHX) takes the contents of the index register and swaps it

---

[18] In fact, the AXT instruction actually has two distinct opcodes, AXT,1 and AXT,2.

with the word in erasable storage. Arithmetic functions are limited to two addition operations and a subtraction. Adding the contents of a fixed or erasable storage word to the index register is through Index Register Add (XAD). Index Register Subtract (XSU) operates similarly, taking the value in storage and subtracting it from the index register. Finally, Increment Index (INCR) also adds a quantity to the index register, but interprets the IAW as an immediate operand. All overflowing index operations leave their overflow corrected result in the index register and never in the operand in storage.

Unlike index registers, step registers do not have dedicated instructions available to manipulate their contents. The roles of step registers are narrowly defined, as they are a static value used only with the TIX instruction for decrementing an index. Given the limited scope of the step registers, there is little need for a rich set of instructions for moving and manipulating data within them. While the Interpreter has a comprehensive set of instructions for loading, storing and manipulating the index register contents, it does not extend this functionality to the step register. One instruction, Set Single Precision (SSP), is useful for initializing step registers. The erasable storage location specified by the first operand is loaded with the contents of the second operand. Saving the contents of a step register is rarely necessary, and using basic AGC instructions is the only practical option available for saving S1 and S2.

During the time that indexing is not used during the processing of the interpretive string, both the index and step registers are available for holding temporary data. In addition to their indexing function, index registers have capabilities similar to an accumulator. Loading and storing data, addition and subtraction, plus exchange functions are available, allowing limited arithmetic capabilities in parallel with the MPAC.

For an instruction to use an index register, additional notations must be added to the opcode and its operand. First, the opcode notation is changed with the addition of a '*' as a suffix to the opcode's name. A Double Add instruction, DAD, is now coded as DAD* when indexing of its operand is required, and the last two bits in the seven bit operation code are changed to 11 from 01. It is important to note that the basic operation of Double Add is unchanged, and only the process of obtaining the operand is different. However, this notation is not yet sufficient for completely defining an indexed instruction. When the DAD* instruction is decoded by the Interpreter, all that is known at this point is that indexing is requested. Given the limited number of bits available even in the interpretive opcodes, there are no available bits to specify which of the two index registers is to be used. The operand must specify which index register to use by placing a comma after the operand address, followed by the register number. The indexed Double Add instruction now looks like this:

DAD*   Double Add the indexed table entry
ADDTABLE,1   Address of table, using X1 for indexing

Note that the Interpreter does not know of the choice of index register until it fetches the IAW from storage. Saving the index register specification in the operand is feasible only because 14 bits are used for the address and bit 15 of the IAW is reserved to indicate which index register is used. Index register X1 is specified when bit 15 of the IAW is set to zero. Setting bit 15 to one indicates X2 is desired, but has additional consequences. Since Bit 15 is also the sign bit, the entire quantity is seen as a negative value. To simplify the decoding process after fetching, the entire IAW is saved in a complemented form when the X2 register is requested.

The notation for interpretive instructions that perform operations directly on the index registers is similar to that used for operation codes. Dedicating an operand word to the single bit required to specify the index register is wasteful, and the relative abundance of interpretive opcodes permits using a unique opcode for each index register. The opcode explicitly identifies which index register to use and only one operand is necessary to complete the instruction. The resulting notation for instructions operating on index registers is similar to that for indexed operands. A comma follows the name of the instruction, which is then followed by the index register number. Not surprisingly, operations that manipulate the index register are themselves not indexable.

SXA,2   Store index X2 into erasable memory
X2SAVE   Address of Index X2 save area

Figure 53: Interpretive instruction directly operating on an index register

The most familiar datatype used with indexing is a table, often called an array. A table is a set of related data organized in a list format, occupying contiguous memory locations. A familiar example is the AGC's star catalog, containing the celestial coordinates of the 37 stars used for navigational fixes. Each star has its position in space defined by a vector, making each entry in the table six words long. The need to examine tables in memory is so prevalent that the final indexing instruction, Transfer on Index (TIX), combines modifying the index register, testing its value, and branching based on its contents. Arguably one of the more sophisticated interpretive instructions, it replaces several frequently used programming sequences. More than simply an assist to indexing, TIX provides capabilities found in the FOR and DO loop constructs of high level languages such as C and Java. In high level languages, looping instructions define a block of code that it executed multiple times, until a particular ending condition is reached. Typically, an index or other variable is modified each time through the loop and is tested to determine whether the ending condition has been reached.

The first task in using the TIX instruction is initializing a step and index register pair; either X1 and S1, or X2 and S2. The index is loaded with the length of the table and the step register with the amount by which the index is decremented. After loading the registers, the body of the loop is entered, where indexed instructions that reference the table are found. The TIX instruction is located at the bottom of the loop, and compares the value in the index register with the contents of the step register. TIX subtracts the value in the step register from the index register, which now points the index to the next element in the table. The index and step registers are compared, and if the index register is greater than the step register, TIX transfers control to the top of the loop. Only when the index register is less than or equal to the contents of the step register will TIX not perform any operation, and thereby exit the loop to resume with the next operation.

Using the star catalog as an example, assume that the table begins at location 01000_8. A program wishing to reference *each* entry in the catalog starts with the table reference pointing to the end of the table, and a zero value in the index register. As each entry in the table is a six-word vector, a six is placed in the step register. References to the table use the ending, rather than the starting address, owing to the fact the value in the index register is subtracted from the operand address. The example in Figure 54 uses an indexed VLOAD instruction to bring a table entry into the MPAC, with its operand pointing to the end of the star catalog table located at $00336_8 + 01000_8 = 01336_8$. The first reference takes the operand address of 01336_8 (the end of the table) and subtracts 00336_8, the contents of the index register, to point at the start of the star table at location 01000_8. After the processing of this first star entry is complete, the loop continues to the TIX instruction. TIX subtracts the six found in the step register from the index register, leaving 00330_8 in the index. Repeating this sequence for each entry, the loop terminates when the index reaches zero.

It is often useful to have instructions that negate, or bitwise complement, data as it is loaded into the index register. The complementing variations of AXT and LXA are Address to Index Complemented (AXC) and Load Index from Erasable Complemented (LXC). As with the AXT instruction, AXC uses an immediate

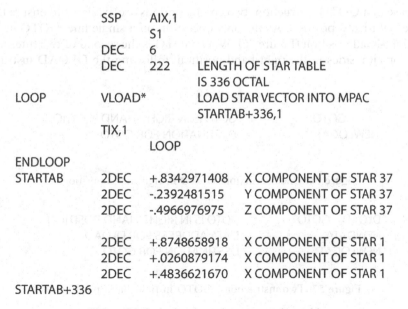

```
            SSP     AIX,1
                    S1
            DEC     6
            DEC     222     LENGTH OF STAR TABLE
                            IS 336 OCTAL
LOOP        VLOAD*          LOAD STAR VECTOR INTO MPAC
                            STARTAB+336,1
            TIX,1
                    LOOP
ENDLOOP
STARTAB     2DEC    +.8342971408    X COMPONENT OF STAR 37
            2DEC    -.2392481515    Y COMPONENT OF STAR 37
            2DEC    -.4966976975    Z COMPONENT OF STAR 37

            2DEC    +.8748658918    X COMPONENT OF STAR 1
            2DEC    +.0260879174    X COMPONENT OF STAR 1
            2DEC    +.4836621670    X COMPONENT OF STAR 1
STARTAB+336
```

Figure 54: Indexing into the star catalog table

operand. LXC follows from LXA, and its operand references a location in erasable storage.

Right- and left-handed instructions

Some interpretive instructions are designated as "right-hand only", meaning they must be either the only operation in the instruction word, or the second of two present. In both cases, the instruction is coded on the "right hand" side of the two possible instruction positions in an interpretive statement. This requirement reflects the condensed nature of the interpreted code, where two operations occupy a single instruction word. Addressing storage locations in the AGC is limited to word boundaries, and there is no mechanism in the interpretive language to indicate a branch to the "left" or "right" operation. Thus it is impossible to transfer control to an operation in the right-hand coding position. This becomes an issue when branching instructions are used. In Figure 55 an unconditional branch (GOTO) is coded in the "left hand" position, followed by a short shift to the right (SR1) in the right-hand position. SR1 will never execute since the programming flow will never reach that instruction. Ignoring the constraints of "left hand" and "right hand" for instructions creates a conflict between what appears to be coded, and the actual software that will run.

```
        GOTO   SR1      # GOTO, then Shift Right 1
        NEWLOCAT        # Destination address of GOTO
```

Figure 55: Inaccessible right-hand instruction

In the case of a GOTO instruction, two coding options are available to ensure that it is in the right-hand position. As the only operation on a single line, GOTO defaults to the right-hand position (Figure 56). When GOTO is the second of two interpretive instructions it resides in the right-hand position, following the DLOAD instruction (Figure 57).

```
        GOTO            GOTO IS IN RIGHT-HAND POSITION
    NEWLOCAT            DESTINATION FOR GOTO
```

Figure 56: One instruction, GOTO in right-hand position

```
    DLOAD  GOTO         GOTO IS IN RIGHT-HAND POSITION
    NEWDATA            DATA ADDRESS FOR DLOAD
    NEWLOCAT            DESTINATION FOR GOTO
```

Figure 57: Two instructions, GOTO in right-hand position

In cases where the right-hand opcode is the only instruction coded in the Interpreter Instruction Word, the Interpreter will fetch and attempt to execute the non-existent left-hand operation. Leaving the left-hand operation undefined is not a problem, but it does create the need to introduce a null operation code. A null opcode is fetched and decoded like all other instructions, but no operation is performed when the null code is "executed". This null instruction, often called a NOOP (NO OPeration), is placed in the operation code field by the assembler. It is important to note that the NOOP code is implied, and never written explicitly; nor does it have a formal opcode name associated with it. NOOP does not modify data, change the state of the Interpreter or alter the flow of control in the interpretive string.

Overflow processing in interpretive code

Overflows are possible during the execution of an interpretive string, and are handled somewhat differently than in the basic AGC instruction set. When an overflow (or underflow) occurs, the instruction sequence does not change, nor are interrupts inhibited. Rather, an overflow corrected result is saved in the MPAC or the storage location specified by the Interpretive Address Word, and the overflow flag OVFIND is set. Testing the overflow flag is now the duty of the programmer, as the Interpreter will continue processing despite the existence of the overflow condition. Instructions to test the OVFIND flag, and branching to another location exist to handle these cases. If no test is made, the condition will remain undetected and processing will proceed with invalid data.

Vector arithmetic

Vectors are contrasted with scalar quantities. Scalar data is only a singular value, such as temperature or energy. Vectors, in the context of physical objects, have both magnitude and direction. An automobile may be moving through two dimensional space, and have a component of motion in both dimensions. Combined, a resultant vector defines the overall motion of the car. In our three dimensional universe, we have adopted vector notation to describe the position, motion and orientation of a body. The three mutually perpendicular axes which form a vector are assigned names, which by convention are x, y and z. Starting from a given point, say a specific location on the Earth, we will also define an orientation for our vector's frame of reference. Such assignments are often arbitrary, so let's specify that the X-axis points straight up from the center of the Earth, perpendicular to the surface; the Y-axis points north; and to complete the triad, the Z-axis points east (Figure 58).

A practical example might begin with a reference point at Tower Bridge, London. An airplane is flying directly over the bridge, heading towards New York along the Great Circle route. This route initially heads slightly north of west on a heading of 287 degrees true. Thus, its position can be defined as 35,000 feet on the X-axis (height above the bridge), and zero feet north and east as the plane is directly over the bridge. Its velocity of 600 miles/hour, relative to the bridge, is zero miles/hour in the X-axis (indicating level flight), 574 miles/hour in the negative east direction (west) and 175 miles/hour northwards. From these individual components, the magnitude of the vector by applying the well-known Pythagorean theorem is:

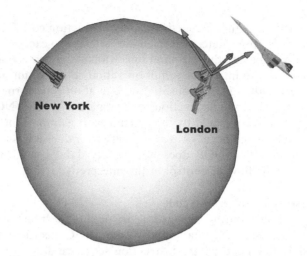

Figure 58: Orientation of vector space on the surface of the Earth.

$$c = \sqrt{a^2 + b^2}$$

In a triangle, the hypotenuse is thought of as the magnitude of a two-component vector. Extending this idea to a three-dimensional vector create three components x, y and z, with the magnitude M calculated as:

$$M = \sqrt{x^2 + y^2 + z^2}$$

To appreciate using arithmetic operations that involve vectors requires introducing new mathematical principles that extend variables into possessing more than one component. Of the sixteen vector operations available in the AGC interpretive environment, basic operations such as add and subtract are conceptually similar to their scalar counterparts. Other operations, such as dot and cross product are unique to vectors. Depending on the operation, either a scalar or vector operand is used in operations with a vector, and a scalar or vector quantity will be the result.

In the Interpreter, vector datatypes are composed of three double-precision fractional quantities. All vector functions require loading the vector into the Multipurpose Accumulator, the seven-word area located in a job's Core Set. However, the six words of the vector are not loaded into contiguous locations in MPAC storage. The first component of the vector, the x-component, occupies locations 0 and 1 of the Core Set. The second, the y-component, is stored in locations 3 and 4, and the z-component is in 5 and 6. Location 2 of the MPAC follows locations 0 and 1, creating a three-word area for triple precision scalar numbers. As there is no requirement for the vector components to be loaded into adjacent storage areas, there is no harm in separating the y and z components from the area used for triple precision data.

The description of the arithmetic vector operations begins with moving the vector into the MPAC. VLOAD places the six words of the vector quantity into locations 0, 1, 3, 4, 5 and 6 of the MPAC, and sets the MPAC mode to vector. Two vectors are added or subtracted using the VAD or VSU interpretive operations. In these operations, each of the three components is added (or subtracted) from its corresponding component in the MPAC. Thus, with a vector $V_{MPAC} = (0.1, 0.3, 0.7)$ in the MPAC, adding the vector $V = (0.2, 0.5, -0.2)$ will produce $V_{MPAC} = (0.3, 0.8, 0.5)$. If an overflow occurs, the OVFIND flag will be set and the overflow-corrected result placed into the MPAC. Vector subtraction, like its scalar equivalent, is not a commutative operation. Unlike addition, where $A + B$ produces the same result as $B + A$, A-B produces a different value than B-A. VSU has the same dilemma when taking the contents of the MPAC and subtracting it from a vector in storage. One solution would be to temporarily move the vector from the MPAC to a location in storage or the pushdown stack, load the other vector into the MPAC, perform the subtraction, and then save the result. All of these steps are eliminated through the "Backwards" VSU instruction (BVSU). In a rare example where erasable storage is the target of an arithmetic operation, this instruction subtracts the contents of the MPAC from the vector in storage, leaving the contents of the MPAC unchanged.

In the Interpreter, a division operation on a vector is defined as the division of each of the vector components by a scalar quantity. Starting with a vector $V = (0.3, 0.5, 0.1)$, division by 2 gives the result $V = (0.15, 0.25, 0.05)$. The interpretive operation Vector Divided by Scalar (V/SC),[19] performs the scalar division of the vector stored in the MPAC, where it also leaves the result. As the quotient is a fractional value with several digits of precision, a remainder is unnecessary and is not generated.

Multiplication of a vector is realized in a number of ways, depending on whether it is multiplied by a scalar quantity or another vector. Multiplying a vector by a scalar quantity in the Vector Multiply by Scalar (VXSC) instruction is similar to the V/SC vector division. Each vector component is multiplied independently by the scalar quantity, with the result remaining in the MPAC. For example, a vector $V = (0.3, 0.1, 0.8)$ multiplied by 0.4 leaves $V = (0.12, 0.04, 0.32)$ in the MPAC.

Vector operations: dot and cross product
Additionally, vector algebra provides us with two new multiplication operations: the dot, or scalar product, and the cross product. These new operations move from the comfort of intuitive operations such as adding or dividing vector components, and introduce new mathematical concepts. Rather than simply changing the magnitude of a vector, dot product yields a scalar, and cross product creates an entirely new vector. Calculating a dot product starts with two vectors, each with components x, y and z. First, the two x-components are multiplied together, then the y-components, and finally the z-components. Adding the results completes the dot product, which is a scalar value. Mathematically, we can express this as:

---

[19] Note that a slash "/" is a valid character in the name of the instruction code.

$$V1 \bullet V2 = ((V1_x * V2_x) + (V1_y * V2_y) + (V1_z * V2_z))$$

As an example, assume there are two vectors, $V1 = (3, 1, 5)$ and $V2 = (2, 7, 4)$. Multiplying together the corresponding components and adding their products gives the following result:

$$(3 * 2) + (1 * 7) + (5 * 4) = 33$$

Dot products facilitate computing the amount of work performed, where work is defined as the application of a force over a particular distance. In many cases, the forces are not acting parallel to the direction of motion of the object.

A second multiplication operation between two vectors is called a cross product. Vector cross products take two three-component vectors and create a third vector that is perpendicular to the other two. Cross products are frequently used to compute torques and other solutions involving rotating bodies. Software in the AGC also uses cross products in areas as diverse as calculating attitude maneuvers and guidance computations.

There are situations where operations on vectors require that the magnitude of the vector is exactly one, or unit length. Most vector operations involving trigonometric functions require unit vectors in their calculations, making it necessary to convert many vectors from their original scaling. Any vector can be converted to its unit form by dividing each component by the vector's magnitude. Assume there is a two-component vector whose x and y components are 2 and 3 units long, giving a magnitude equal to $\sqrt{13}$. Dividing the components by the vector's magnitude gives us $x = \frac{2}{\sqrt{13}}$, and $y = \frac{3}{\sqrt{13}}$, resulting in the unit vector. Note that the ratio of the two components is unchanged, so the direction of the vector remains the same. The UNIT interpretive operation converts a vector in the MPAC to a unit vector, and leaves the result in the MPAC.

Complements, squares and absolute values of a vector

Each component of a vector is important because it provides directional information, but many applications require only the magnitude of the vector. Two related functions provide the length of a vector, and its square. Because the magnitude is a sum of squares, it maintains the essential property of an absolute value; that is, it is never negative. Vector Absolute Value (ABVAL) and Vector Magnitude Squared (VSQ) operate in a similar manner. Both take the vector components of the MPAC, squaring each component and adding them together. When requesting a VSQ operation, this result is placed into the MPAC. Otherwise, if the programmer requests an ABVAL operation, the square root of the sum is placed in the MPAC. Both operations set the mode to double precision.

In the same manner as negating a scalar quantity, such as converting a 7 to -7, it is also possible to complement a vector. By reversing the sign of each component, we reverse the direction of the vector without altering its magnitude. The unary Vector Complement (VCOMP) function takes the vector contents of the MPAC and complements the signs, leaving the result in the MPAC.

Defining a vector

A number of important vector operations begin not with a vector, but with a trio of individually computed double precision values that must be reinterpreted as a vector datatype to be consistent with later mathematical operations. The Vector Define (VDEF) instruction begins with the first component of the vector, V_x, already in the MPAC. The remaining two vectors components V_y and V_z are popped off the pushdown stack to join V_x in the MPAC. With all three components necessary for a vector now in place in the MPAC, the mode can now be set to vector.

Matrix multiplication

Mathematical operators for guidance, navigation and control often extend beyond the relatively simple structures such as vectors. Frequently, matrices are necessary to represent the vehicle's state, or in performing transformations from one frame of reference to another. This requirement makes it essential that the Interpreter provide at least basic matrix support. Given the need for simplicity and speed in the AGC, it might come as a surprise to see support for such a complex datatype. However, in view of their widespread use throughout all areas of flight software, the Interpreter would be seriously limited without the ability to perform these operations.

Multiplication of matrices is not especially complex, but does require a number of repetitive steps. Multiplication operations are restricted by the sizes (rows and columns) of the two matrices. For matrix A of size $m \times n$ to be multiplied by matrix B, B must be of the form $n \times p$, and the resulting matrix is sized $m \times p$. Stated in a more intuitive way, the number of columns in A must equal the number of rows in B. The formal expression for matrix multiplication is:

$$C_{i,j} = \sum_{r=1}^{n} A_{i,r} B_{r,j}$$

where n equals the number of rows in A and the number of columns in B, i is the number of columns in A, and j is the number of rows in B.

A practical example is if matrix A is 2×3 (two rows by three columns), matrix B must have three rows. If B is sized 3×4, then the matrix resulting from the multiplication will be 2×4. Graphically, our example becomes:

$$\begin{bmatrix} 1 & 5 & 4 \\ 2 & 1 & 3 \end{bmatrix} \times \begin{bmatrix} 3 & 9 & 2 & 7 \\ 2 & 8 & 5 & 6 \\ 4 & 1 & 8 & 3 \end{bmatrix} = \begin{bmatrix} 29 & 53 & 59 & 49 \\ 20 & 29 & 33 & 29 \end{bmatrix}$$

Matrices in the AGC are implemented using double precision variables. Although the Interpreter limits the size of matrices to 3×3, operations on larger matrices are possible by using elaborate programming techniques.[20] Limited to only seven words,

---

[20] The most notable exception is the error transition matrix, or W-matrix. Used by the Kalman filter in the navigation routines, the W-matrix may be sized as a 6 x 6 or 9 x 9 matrix.

the MPAC is too small to hold the eighteen words necessary for a 3 × 3 matrix. Since arithmetic operations require that one operand resides in the MPAC, the Interpreter simplifies the problem by requiring that one of the multipliers be a vector, which by definition, is also a 1 × 3 matrix. The most important side effect of these two restrictions is that the result of the multiplication will have a size of 1 × 3 or 3 × 1, which are both vectors! With a vector operand in the MPAC, the matrix resides only in storage, freeing the Interpreter from the need to load, store or operate on matrices in the MPAC. However, matrix multiplication is not commutative – meaning that given two matrices A and B, multiplying A times B does not produce the same result as multiplying B times A. Given the restriction of having only one operand, a vector, in the MPAC, non-commutative multiplication forces us to introduce two distinct multiplication operations, MXV and VXM. Both instructions use the vector in the MPAC and the matrix in storage, and resolve the commutation issue by performing the multiplication in both directions. Matrix, Post-Multiplication by Vector (MXV) multiplies the matrix by the vector in the MPAC, leaving a vector in the MPAC. Matrix Pre-Multiplication by Vector (VXM) reverses the sequence by multiplying the vector by the matrix, again placing the vector result in the MPAC.

One function that the Interpreter does not implement is matrix division. Matrix division is a complex and computationally intensive operation, and in some cases a solution might not even exist. Indeed, it is usually far easier to eliminate the need for matrix division operations by restructuring the problem to avoid its use.

Flagwords

Mission software has the obvious need to adapt its processing as the different stages of the problem are computed, as well as recording the states of the onboard systems or maintaining the selections requested by the crew. A new datatype, the flagword, is used for this task. Flagwords, as their name implies, are a group of words reflecting the various states of the software, and are located contiguously in erasable storage. Each of the 15 bits of a given flagword defines a different parameter or indicator as a binary value. Limited to the yes/no, on/off nature of binary data, the flagword bits inform basic decision tests by the software. Several hundred flagword bits, also known as switches, are defined between the Command Module and Lunar Module. The CM software uses 12 flagwords, and the LM has 14 flagwords.

For example, a seemingly trivial but quite important example is the MOON-FLAG flagword bit in the Command Module software. When set to zero, it indicates that the spacecraft is in the Earth's gravitational sphere of influence; when set to one, the Moon is the dominant gravitational body. Flagwords and their bits are vital for maintaining the logical state of mission programs. Only mission software uses flagwords for managing their states; the Executive and related routines use an independent set of words to manage their execution. Knowing the state of the software is so important that flagwords are downlinked continuously to the ground for diagnostic purposes.

If the use of a word containing bits that specify various spacecraft conditions sounds familiar, it is likely because flagwords resemble the I/O channels discussed earlier. Channels in the AGC are the interface between the computer and the spacecraft hardware; one bit might indicate that the liftoff discrete has arrived, and another bit might command a thruster. In both cases, requests and information are communicated between the spacecraft and the computer through the channels. Conversely, the domain of flagword switches lies entirely within the AGC. While software will interpret the channel data, and generate commands based on it, data contained in the flagwords stays wholly within the computer.

Operations on switches are necessarily limited. With a 1-bit switch, the only possible operations are setting a bit to zero, setting it to one, or inverting whatever value is already in the switch. The implementation of flagwords also exposes an issue that arises in multiprogramming systems, and is particularly troublesome in the AGC. At any time during its execution, a program can reference or change a flagword. Unfortunately, coupled with this flexibility is a significant drawback; if there were not any means to serialize access to the flagword, two processes might attempt to access or update a particular bit at the same time. Even with the vastly increased capabilities of the interpretive system over the basic AGC instruction set, there is still no means to ensure that one process has exclusive access to the flagword. Specifically, the issue is that without a basic AGC instruction to operate on an individual bit, several instructions are needed to perform any bitwise operations. A window several milliseconds long results, in which the switch is unprotected from changes made by other processes. At any point during this window, an interrupt might be raised which in turn schedules other work that also needs to update one of the flagword switches. This creates the situation where two processes are trying to update the same location in memory, which certainly would result in untold problems. The only practical solution is to use the brute-force technique of inhibiting all interrupts during the time the flagword is processed. With interrupts disabled, it is impossible for another process to take control of the CPU and attempt an operation on a switch. After the Interpreter finishes the switch operation, interrupts are reenabled, thereby allowing other processes access to the flagwords.

The Interpreter provides a rich set of operations to set, clear and invert switches within the flagwords. There are also more complex operations, including testing switches and branching based on their state. Although the underlying operations in the switch instructions are logical AND, OR and XOR, the Interpreter allows manipulating only one bit at a time. This is in contrast to the AGC's basic instruction set, which will perform logical operations on several bits at once within the word. This is not a significant limitation, as switch bits represent such diverse conditions that there is little need to operate on more than one bit at a time. Beginning with the most basic switch operations, SET, CLEAR and INVERT are the most intuitive. A one is placed in the switch by the SET instruction, a zero by the CLEAR, and the bit is flipped by INVERT. From these three basic instructions, we can derive the other eleven flagword switch operations. Recognizing the frequent need to set a switch and then follow that instruction immediately with a branch gives us a set of three more operations, SETGO,

CLRGO and INVGO. Here, an unconditional branch follows the operation on the flagword switch.

Two instructions complement those that operate on a bit and immediately branch. BON and BOFF do not alter the contents of the switch, but conditionally branch based on its settings. The two instructions test if a switch is set or cleared and if the test evaluates to "true" then the branch is taken. The remaining six instructions combine testing the switch, modifying the switch's value, and then conditionally branch based on its original state. Using the example of BONSET and BOFSET, the original switch value is saved, and the bit is set to one. Next, the decision to branch is based on the original value of the switch. Given a switch set to one, the BONSET instruction tests the bit, obtains a "true" result, and performs the branch. With the switch bit cleared and set to zero, BONSET's test evaluates to "false", and the sequence of interpretive statements continues unchanged. Note that in both cases, the switch is set to one regardless of its original value. BOFSET operates in a similar manner, testing the switch for "off". Other instructions follow this same structure. Rather than setting the switch and then conditionally branching, BONCLR and BOFCLR will clear the switch, and BONINV and BOFINV will toggle the switch's contents to 1 from 0, or to 0 from 1.

With so many switches available to manipulate by the AGC software, there is the problem of how to efficiently address a large number of bits within a set of storage words. Only seven bits are available for an interpretive opcode, giving 128 possible combinations of values. At first glance, we might be able to develop a notation defining an opcode and the bit within the flagword it references. But the sheer number of individual switches in the Command Module and Lunar Module rules out encoding the switch number with the opcode. Accepting the necessity of dedicating an operand word that uniquely defines the switch bit within the flagword, the next task is to develop a format to efficiently reference individual switches. This is done in terms of the flagword number, and the switch number within the flagword. The 15 switches within the word can be specified using four bits. With fewer than 16 flagwords in the AGC software, uniquely identifying the specific word requires no more than four bits. From this encoding scheme, it is apparent that at least eight bits are all that is necessary for uniquely identifying an individual switch. At the same time, the number of switch manipulation instructions is worrisome. The fourteen operations consume over ten percent of all possible operation codes. An alternative is to define a single interpretive opcode for all switch operations, and qualify which of the fourteen possible operations is desired. Noting that the number of switches forces us to use an operand word, and that only eight bits of the operand are required to identify a specific switch, the six remaining operand bits are available to indicate the specific switch operation. The interpretive operation code is now common to all flagword manipulation instructions, and the specific operation (SET, CLRGO, BOFINV, etc) is defined within the operand. The final layout of the operand used is shown in Figure 59. Beginning with bits 4 through 1 of the operand, the number of the switch bit within the flagword is coded, and bits 8 through 5 define the unique switch operation code. Bit 15 is set to zero to assure a positive value for the operand, leaving six bits

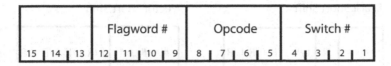

	Flagword #	Opcode	Switch #
15 | 14 | 13	12 | 11 | 10 | 9	8 | 7 | 6 | 5	4 | 3 | 2 | 1

Figure 59: Flagword operand format

available to define up to 64 flagwords. The CSM and LM software required no more than four bits to define all of their flagwords, but the switch operand layout allowed for many more if needed.

Shifting data in storage
An extensive library of shift instructions exists in the Interpreter, using two distinct implementations. Frequently, shift operations replace multiplication or division when one of the operands is a power of two. Shifting data is often necessary for scaling variables before further calculations, to avoid a loss of precision, or to isolate logical data in the word.

Short shifts, those which shift data one, two, three or four bits, are unary operators, with all the information encoded into a single 7-bit opcode. Short shifts of scalar values come in four variations: shifting left or right, and whether the result is rounded or not. Within each variant of short shift, we also have the number of bits to shift. All told, sixteen unique scalar shift operations are available. Shifts present the same problem of proliferating opcodes that flagword manipulation posed, but short shifts are designed to eliminate an operand from the instruction. As a unary operation, that is, without an operand reference, short shifts can only operate on quantities in the MPAC. In the description of the fields defined for short shift operations, it becomes apparent that nothing in the opcode states whether the shift is to operate on scalar or vector quantities. While the operation code does not differentiate between scalar and vector, a reliable means already exists to decide which operation to use. The MPAC's mode, which consistently tracks the type of data last operated on, provides the cue. When set to double or triple precision, a scalar shift is selected, otherwise a vector shift is performed.

As unary operators must contain all relevant instruction information within the opcode, the short shifts necessarily require sixteen different opcodes. To ease the effort of parsing so many operations, each attribute of the shift is encoded in a dedicated field of the interpretive opcode. Beginning with the rightmost bits of the 7-bit opcode, we assign four bits for the direction of the shift. Moving left, we use one bit to define whether the shifted result is rounded or not, and the final two bits specify the number of bits to shift (Figure 60).

Short scalar shifts perform all operations in triple precision, shifting data through all three words of the MPAC regardless of the current mode. The use of such a long datatype demands that the programmer be vigilant about unwanted values in the MPAC. Particularly in the case of unused lower order words, the unexpected bits may become part of the shifted result. When shifting to the right,

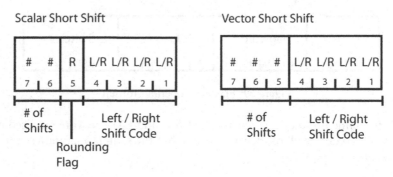

Figure 60: Scalar and vector short shift formats

zeros are placed in the highest order bits; conversely, zeros enter the lowest order bits of the triple precision value during a left shift. If the sign of the quantity changes during a left shift, an overflow condition is raised, setting the OVFIND flag. Rounding is an option available for all scalar shifts. This involves adding 0.5 to the least significant word and allowing the result to carry forward through the other two words. After the addition, the third and least significant word is set to zero, effectively producing a double precision result and leaving the MPAC mode unchanged.

Short vector shifts borrow many of their concepts from their scalar counterparts. As with short scalar shifts, the lack of an operand forces all operations to take place in the MPAC, and all three vector components are affected. Unlike the short scalar shifts, shifting is performed only on the double precision quantities found in a vector, as no space is available in the MPAC for triple precision vectors. Rounding is not an option available to short vector shifts because there is no third word to round with. This allows us to reuse the rounding bit. Combining the bit previously used for rounding with the two which specify the magnitude of a scalar shift provides a three bit field sufficient for specifying a vector shift from one to eight bits.

To shift a quantity more than four bits for a scalar or eight bits in the case of a vector requires a new set of instructions that use an operand word to contain all the shifting parameter data. Short shifts are necessarily restricted in the number of bits they can shift, but long shifts allow shifting up to the number of bits in the datatype. A double precision scalar value, for example, can accept shifts of 28 bits left or right. The operand word (Figure 61) contains four fields that define the characteristics of the shift operation. Bits 7 through 1 contain the number of bits to shift, and bit 8 defines a "pseudo sign bit" that is used to detect sign changes in indexed shifts. Bit 9 describes the direction of the shift, right or left, with the flag to specify whether rounding is desired in bit 10. Bits 15 through 11 are fixed, and are set to 01000. Note that the type of shift, scalar or vector, is not included in the fields of the operand word, as the datatype is again obtained from the MPAC mode. All long shift operations share the same opcode, plus one indexed variant, so long shifts also use the MPAC mode for specifying whether a scalar or vector shift is required. As in the

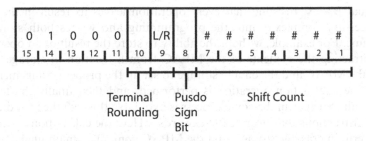

0	1	0	0	0	L/R		#	#	#	#	#	#	#
15	14	13	12	11	10	9	8	7	6	5	4	3	2	1

Terminal   Pusdo   Shift Count
Rounding   Sign
                   Bit

Figure 61: Long shift address word format

case of short scalar shifts, all operations on scalar quantities start with operating on triple precision values in the MPAC. Rounding operations are also the same as in short scalar shifts, where the triple precision value is rounded into a double word and the third word is set to zero.

Scalar arithmetic

Double precision quantities are the most prevalent datatype in the AGC's Interpreter, and, as a result, few instructions operate on single or triple precision data directly. This survey of the scalar operations began with the four basic arithmetic operations: add, subtract, multiply and divide. The four instructions DAD, DSU, DMP and DDV are their interpretive equivalents. They take a double precision operand and perform the operation on the double precision quantity in the MPAC. Any overflow that might occur sets the OVFIND overflow flag, and DAD, DSU and DMP leave an overflow corrected result in the MPAC. An overflow occurring with DDV leaves ± 0.99999999 in the MPAC. Unlike basic AGC instructions, an interpretive instruction never inhibits interrupts when an overflow occurs. Many instructions that operate on a double precision value use the MPAC's triple precision area, but these are not true triple precision operations because the MPAC mode remains double precision.

Multiplication and division require special handling, as their results often require a triple precision word to maintain an acceptable level of precision. Recalling that numeric data is represented using fractional notation, multiplying two fractional numbers together yields a smaller fractional value as a result. This situation is made worse if we start with a number that is small to begin with. However, much of the precision is retained by producing a triple precision result, even as the MPAC mode is held at double precision. After the multiplication or division operation is completed, the programmer can either strip off the unnecessary bits in the third word through the ROUND operation, or can recover the bits back into a double precision quantity through the NORM function. Using one of the shift instructions is also an option to recover bits from the third word of a triple precision result.

These four basic arithmetic operations must necessarily expand into a total of six interpretive instructions. All arithmetic operations would normally occur in the MPAC, but subtraction and division are not commutative operations; where A − B does not equal B − A. Insisting that every operation leaves its result in the MPAC forces unnecessary complexity into the programming and wastes both storage and processing time. For example, without the ability to store the result of an operation in storage, subtracting the contents of the MPAC from a location in storage requires exchanging the MPAC and the data in storage to set up the proper arrangement of the two terms. The subtraction operation is performed, and then finally the terms are exchanged again. Such a cumbersome process is avoidable through the introduction of interpretive instructions that operate "backwards". Here the calculation is performed against the term in erasable storage, and the MPAC contents remain unchanged. For backwards subtractions, the BDSU instruction is used. For dividing a value in erasable storage by the MPAC contents, BDDV is available. BDDV is unique, in that it does not behave the same as its "forward" counterpart, whose result is a triple precision quantity saved in the MPAC.

Loading the MPAC
Four instructions are available to load data directly into the MPAC. DLOAD, TLOAD and VLOAD place double precision, triple precision and double precision vector data into the MPAC, and, importantly, set the MPAC's mode to that of the datatype loaded. Operands are obtained through direct and indirect addressing to storage, and through vacuous addressing by popping the data off the pushdown stack. The Interpreter architecture provides native support for the three datatypes mentioned above, plus limited support for matrices, but the ability to manipulate single precision values is notably absent. A large part of this rationale is that the Interpreter is designed to manipulate multiple precision and complex datatypes, and single precision is arguably most efficiently handled with basic AGC instructions. Still, a significant amount of data required by the interpretive routines is in a single precision format, and so the Interpreter provides a minimal amount of support for the datatype.

   SLOAD is the fourth means of loading data into the MPAC. A single word in storage is placed in the first word of the MPAC using either a direct or indexed storage address. Using vacuous addressing to obtain the operand is not possible, as the pushdown stack does not support single precision datatypes. Since all other interpretive operators use only multiple precision or vector datatypes, the MPAC mode must be set to one of the supported datatypes, and double precision is used. When loading only a single word, the remaining two words of the MPAC's triple precision area are cleared. Without setting the second and third words to zero, any data remaining from previous computations might be confused as part of the "double" precision value. Once the MPAC is properly cleaned up and the mode set, the single precision value is now ready for processing.

Stack operations
The Interpreter's provision of the stack represents a significant improvement in function over the basic AGC hardware. Recalling the earlier description of program trees, a stack is a data structure that allows storing data in an orderly first-in, last-out sequence. Conceptually, the convention is to "push" data onto the stack to save it,

and each time data is retrieved from the stack it is "popped" off and removed from the stack entirely. The reality of the Interpreter's implementation is rather different. The image of "pushing" and "popping" data creates the convenient illusion that each stack operation physically shuffles data back and forth, with the "top" of the stack at some arbitrarily fixed location in storage. In real-world implementations, the Interpreter included, data is fixed in storage, and reference to the "top" of the stack is through a variable called a stack pointer. Push operations append data onto the existing contents of the stack and update the stack pointer to the new "top" of the stack. "Popping" removes the data from the stack by updating the stack pointer to point to the data "beneath" the top entry in the stack. Normally, it is irrelevant whether the top entry is cleared as part of the popping process, since from the stack's perspective, once the data is popped, the data is no longer accessible. However, the Interpreter does not clear the data from memory after popping the stack, leaving the data accessible to a cunning programmer.

In the Interpreter, the stack pointer is a variable named PUSHLOC which resides in the job's Core Set. The stack itself occupies the first 38 words of the Vector Accumulator Area, with the "top" of the stack originating at offset zero. Pushing a double precision variable onto the stack, for example, places the data into contiguous locations in the stack area and increments PUSHLOC by two. When an interpretive instruction requests data from the stack, the data is passed to the opcode processing routines and the stack is popped by decrementing PUSHLOC by two. A powerful aspect of the Interpreter is the manner in which it pushes and pops datatypes. Taking its cues from the MPAC mode, a double or triple precision word or a double precision vector is pushed onto the stack. However, popping the stack through an interpretive instruction is driven by the mode appropriate to that instruction. For example, when a vector load instruction requests data from the stack (as a vacuous operand) the top six words on the stack are interpreted as a vector and loaded into the MPAC. Note that the Interpreter does not check to assure that the data in the stack entry was originally loaded as a vector, or that the mode of the instruction matches the mode of the MPAC. It is perfectly reasonable to push three consecutive and unrelated double precision values onto the pushdown stack, and use them later as an operand to a vector function. While this might be perceived as a useful "feature", it also makes tracing errors extremely difficult.

With the interpretive system highly reliant on the pushdown stack, there is occasionally the need for a mechanism to manipulate its contents directly. The pushdown stack is never loaded directly from storage; rather, all data must first pass through the MPAC. The first of three stack manipulation instructions, named appropriately, PUSH, takes the contents of the MPAC and pushes it onto the top of the stack. Whether to push a double precision, triple precision or vector onto the stack is indicated by the mode of the MPAC, eliminating the need to specify the datatype. Once the data from the MPAC is pushed onto the stack, there is often the need to reload the MPAC with data for the next computation. Combining two instructions at once, Push Down and Load Double (PDDL) and Push Down and Load Vector (PDVL) first push the contents of the MPAC onto the stack and then reload the MPAC with either a double precision or vector quantity, setting the MPAC mode to the datatype just loaded. As expected, the source of the data for the MPAC may come from direct or indexed storage. With the stack being loaded from the MPAC, there is now a curious dilemma: what is the final result of executing PDDL or PDVL with a vacuous operand? This special case of PDDL and PDVL is defined as a data exchange between the pushdown stack and the MPAC. Notice that there is no requirement that the data pushed onto the stack and that loaded from the stack be the same datatype. Using the PDVL instruction as an example, assume that the MPAC mode is set to double precision. PDVL will push that double precision quantity down onto the stack, but not before the vector on the stack is popped off and saved in a temporary location. Finally, that vector is moved from the temporary location into the MPAC and the mode is set to vector.

The pushdown stack is not a complex structure, but does have strict rules on how data is placed into it, and how it is removed. Unfortunately, there is no means to verify whether the data moving between the stack and MPAC actually reflects the datatypes required by an interpretive instruction. A series of instructions operating exclusively on double precision quantities can conclude by an instruction to save a vector onto the stack, and the Interpreter will not be the wiser. In essence, there is nothing preventing a programmer from bypassing the little integrity that the stack provides, and even its internal structure is not immune from operations that alter its contents. Ideally, as a higher level data structure, the physical characteristics of the stack should be invisible to us: the number of words allocated, the direction of the stack pointer movement, and even the physical location of the stack should be hidden from the programmer. In reality, in a system as storage constrained as the AGC, the luxury of such a "black box" implementation of the stack is not practical. The amount of space allocated to the stack in the Vector Accumulator Area is a compromise of many factors, but ultimately, must be large enough to hold the maximum number of entries anticipated in an interpretive mission program. There are no program alarms or error recovery defined for stack overflows, placing the responsibility for stack management squarely on the programmer.

The manual management of the stack does have a useful side effect. Absolute knowledge of how far a stack is pushed down during a series of calculations gives rise to an obvious but overlooked fact: knowing the amount of data already pushed onto the stack gives the location of the top of the stack. If the programmer is assured that the current operations will not place more than 30 words on the stack, eight words in the pushdown stack remain unused. In the AGC, the availability of unused erasable memory locations often presents an irresistible temptation. The Set Pushdown Stack Pointer (SETPD) instruction gives us the ability to manually override the value of the stack pointer, PUSHLOC, to a known vacant location, the unused stack area can be used as a temporary storage area for intermediate results. Another more common use of the SETPD instruction is when a computation has progressed to a point where a decision is necessary on how further processing will continue. If the present calculation is abandoned in order to start another one, using SETPD to reset PUSHLOC back to zero effectively clears the stack and throws away its contents.

Store instructions

Once the thrill of a well executed interpretive string is over, there is the need to save the results in storage. Contrasted with the basic AGC instructions that are used to move data from the accumulator, the variety of datatypes and indexing options require a far more complex set of instructions for storing data. Starting with the apparently trivial objective of saving the contents of the MPAC, important decisions quickly arise on whether to save a double or triple precision variable, a scalar value or a vector.[21] Optionally, there is the ability to use the indexing capability built into the Interpreter. In all, six combinations of datatypes and indexing operations are possible. While it would be possible to create unique operation codes for each combination, the Interpreter architecture provides all the elements necessary to implement a single STORE instruction.

Flexibility in the STORE instruction requires abandoning the traditional Interpretive Instruction Word format for an entirely new layout adapted for storing data. Specifically, the STORE instruction combines both the instruction opcode and the erasable storage address into a single word of storage. It designates eleven bits (bits 11 through 1) for its address field, bits 14 through 12 for the opcode, and always has zero in bit 15. Setting bit 15 to zero, thus creating a positive number, is a significant departure from the convention used by the Interpreter, where all IIWs are saved as negative numbers. A positive value for the STORE instruction has the potential to cause confusion, as positive values are also used for operand addresses.[22] With only three bits in the opcode field, eight variations of the store instruction are possible. Perhaps the most important feature of the STORE format is that eleven bits provides addressability throughout all 2K of erasable storage, thereby allowing the instruction to save data without the need to explicitly manipulate the EBANK register.

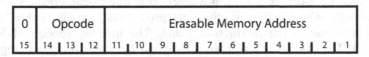

Figure 62: STORE instruction format

The prospect of eight variations on storing data seems a bit excessive, but only a few distinct addressing modes are implemented. The STORE instruction is not limited to moving the MPAC contents directly to erasable storage. Either of the two index registers may modify the destination address, creating three different addressing modes available to STORE: no indexing, indexing using X1 and indexing using X2. Left with an additional five possible variations, powerful extensions to the otherwise utilitarian STORE instruction are available.

---

[21] Storing a double word when a vector is in the MPAC is not supported in the Interpreter, allowing us to make the weak argument that the Interpreter is a "strongly typed" language.

[22] We will consider this issue in our upcoming discussion of the STADR instruction.

When writing a program, a useful observation becomes apparent: after data is stored, the implication is that the particular program fragment is complete and no further processing remains. In an interpretive string, this is seen as the STORE operation followed by exiting the Interpreter and returning to the execution of basic AGC instructions. A second case is more common, where processing continues in the Interpreter and computation begins on another piece of data. Invariably then, the STORE instruction is followed immediately by a load into the MPAC of the next piece of data. This second case occurs so frequently that it is useful to create a combined "Store and Load" instruction. Such an instruction fits comfortably in the repertoire, as it requires only two operands: the store location and the address of the next piece of data to load. The instructions STODL and STOVL first copy the MPAC contents to erasable storage and then reload the MPAC with a new operand. STODL loads a double word into the MPAC, setting its mode to double. Likewise, STOVL loads the MPAC with a vector, setting the mode to vector. The efficiency of combining two common instructions is obvious, and the benefits of these combined instructions extend beyond the notational shorthand. Savings on processing cycles are not trivial, as the interpretive process is already quite slow. The time saved by not fetching, decoding and dispatching a new instruction adds up quickly, allowing the AGC to pursue work that is more productive. In addition, extensive use of STODL, STOVL and STCALL result in the saving of hundreds of precious words of memory in both the Command Module and Lunar Module computers. Loads of triple precision values are not supported in the combined STORE/LOAD instructions, nor are loads of single precision values. Neither of these datatypes are sufficiently common to justify a separate instruction.

Although the targets of STODL and STOVL are limited to direct and vacuous addressing (that is, indexing is not possible), two additional opcodes do allow indexing. Indexing is specified as in other interpretive instructions, with a '*' being appended to the end of the instruction name. The indexed version of STODL becomes STODL*. As in other interpretive instructions, the index register is appended to the operand address, separated by a comma. The differences between these two syntaxes are seen below.

STODL STOREADDR	STODL* STOREADDR
LOADADDR	LOADADDR,1
Not Indexed	Indexed

A third variant of STORE, STCALL, reflects the observation that after storing an operand there is often the need to call to another routine to continue processing. STCALL, in a departure from the load and store model, follows the data store with a call to another location in the interpretive string. As no new data is loaded into the MPAC, its mode remains unchanged. Identical to a sequence of STORE followed by a CALL instruction, STCALL places the address of the following instruction into the interpretive return register, QPRET.

Instruction formats for STODL, STOVL and STCALL are the same as that for STORE: bit 15 equals zero, followed by a three-bit opcode and eleven bits for the erasable storage address. Coding in the source statements is also similar, where the operand address follows the operation code, all on the same source statement. Following the operation code is the operand address of the data to be loaded into the MPAC, or the address of the subroutine that will be called. When examining the instruction format of the STORE variants, a seemingly insignificant characteristic becomes apparent. With bit 15 of the instruction word equal to zero, it is now indistinguishable from Interpreter Address Words. The flexibility of the interpretive programming language now exposes an issue when using STORE instructions. Consider the situation where an instruction uses a vacuous address, taking its operand from the pushdown stack rather than fetching an operand whose address lies in the interpretive string. Vacuous addresses rely on the fact that the operand address is not explicitly coded, implying that the pushdown stack is the source or destination of the data.

When the Interpreter fetches the word in storage for what it expects to be an operand, a check of bit 15 is meant to reveal whether an operand or another instruction word is present. If bit 15 is set to one (a negative value), the word is interpreted as an operand address and vacuous addressing is not used. Conversely, when bit 15 is zero, we have encountered an instruction word and the pushdown stack is used as the operand. But what happens if an instruction intending to use a vacuous address is followed immediately by a STORE instruction? By the rules just described, the Interpreter recognizes STORE as an operand address, and uses it to fetch data. Once treated as an operand, the STORE instruction cannot execute as intended, leaving the programmer with a difficult-to-find error.

Explicit Operand	Vacuous Operand
INST1 DLOAD	INST1 DLOAD
DLOADAD	DLOADAD
STORE STORELOC	INST1 INST2
	STORE STORELOC

INST1, INST2 are arbitrary instructions

Problem: Is STORE an Instruction or Operand?	Solution: STADR tells Interpreter STORE is an Instruction
INST1 DLOAD	INST1 DLOAD
STORE STORELOC	STADR
	STORE STORELOC

Figure 63: Uses of the STORE instruction and STADR

STADR is a novel instruction used to resolve this conflict. STADR insures the intended program execution by first instructing the assembler to complement the following instruction when it is saved in storage. Thus, a STORE instruction that might resolve as 02715_8 is now saved as 75062_8. During execution, STADR instructs the Interpreter to complement the instruction one more time, and the original STORE instruction executes as expected. STADR can only exist as a right-hand operation code, as it operates only on the next word of an interpretive string.

Managing scaling

A scalar function is used to set the magnitude of a triple precision value in the MPAC to between $0.5 \leqslant N \leqslant 0.999999999$. Remember that the fractional datatype in the AGC is to simplify the number format itself; elaborate floating point data formats are not part of the AGC architecture. As calculations progress on a particular variable, it is possible that the magnitude of the data becomes less and less, as seen when multiplying small numbers together. The precision of the resulting quantity, while perfectly accurate, becomes impractically small to use with subsequent calculations, creating the risk of underflowing the accumulator in the ensuing operations. Through the Normalization (NORM) instruction, the fractional value is shifted to the left until the highest order "1" bit is in bit 14 of the high order word. For example, NORM will alter the value of $0.741 * 2^{-9}$ to $0.741 * 2^{-1}$, allowing further multiplications without the risk of underflow. As calculations in the AGC place the responsibility for knowing each variable's magnitude squarely on the programmer, he must maintain a record of how much the magnitude of the data has changed. Unfortunately, the programmer might know the exact scale of a variable in the middle of a computation only generally, perhaps at best to within only an order of magnitude. NORM solves the problem of losing scaling information by requiring an operand, into which it saves the negated value of the number of shifts performed.

UNIT is a unary operator acting only on the vector in the MPAC. In spaceflight, vectors are used in a wide variety of applications ranging from the position and velocity of a vehicle on its way to the Moon, to the small relative motions between two spacecraft about to dock. These vectors exist with widely different magnitudes, where managing the scaling is an aggravating necessity. Many calculations, especially those involving trigonometric operations, cannot easily use these values with significantly different magnitudes, and require their data in the form of a unit vector. Unit vectors in themselves are completely unremarkable, except for the essential property of having a magnitude equal to one. Converting an arbitrary vector to a unit vector preserves the relative magnitude of each component, leaving the direction of the vector unchanged. Additionally, UNIT scales the vector so that any concern about very large or small components is minimized.

To see how unit vectors work, consider the following example. Assume a car is traveling 75 km/hr; 13.3 km/hr along the X-axis, 73.8 km/hr along the Y-axis, and 0 km/hr along the Z-axis. Converting these components to a unit vector creates an X-axis value of 0.178, a Y-axis value of 0.984, and 0.0 for the Z-axis, resulting in a magnitude of one. The UNIT vector function converts the vector in the MPAC into the vector form, and leaves two other useful values in the Vector Accumulator Area. These values are intermediate quantities generated when calculating the unit vector. The square (dot product of itself) of the original vector is saved in VAC offset 34, and the original magnitude of the vector is saved in VAC offset 36. These values are available to the programmer.

Miscellaneous arithmetic instructions
Double Precision Multiply and Round (DMPR) is a variation on the Double Precision Multiply (DMP) instruction. Rather than placing a triple precision product into the MPAC, the multiplication result is rounded to a double precision value before it is saved in the MPAC. To assure a true double precision result, the most significant two words are placed in the MPAC. As double and triple precision words share the same area in the MPAC, the third and least significant word of the MPAC's triple precision data area is set to zero. Multiplication then takes place as expected, in the same manner as the DMP instruction.

Similar to Double Precision Add, Triple Precision Add (TAD) adds the contents of the MPAC with all three words of the triple precision operand. As double precision data is the predominant scalar datatype in interpretive strings, it is not surprising that addition is the only triple precision operation. This is not as restrictive as it might appear. Complementing a triple precision value, which can be used as part of subtraction logic, is available as one of the scalar functions.

Combining the testing of a double precision value's sign with an optional data load into the MPAC is possible with the SIGN instruction. This first tests whether the double precision value in erasable storage is positive or negative. If the variable is greater than or equal to zero, processing continues without any further action. If a negative value is found, the triple precision value in the MPAC is complemented regardless of whether the mode is double or triple precision. If the MPAC's mode is set to vector, all components of the vector in the MPAC are complemented.

Trigonometric instructions
With so much of the guidance and navigation expressed using angles and ellipses, it is no surprise that trigonometric functions form the backbone of their equations. Four trigonometric operations SIN, COS, ARCSIN and ARCCOS (SINE, COSINE, ASIN and ACOS are also valid names) are available as unary operations to replace the double precision value in the MPAC with a triple precision solution. SIN and ACOS are computed by the Hastings approximation, which calculates a reasonable estimate using high order polynomials. While more accurate techniques are available, methods such as a Taylor series require far more computational resources. Despite using approximations, a single trigonometric operation still requires several milliseconds to calculate: 5.6 ms for SIN, and 9.1 ms for ACOS. To economize on memory, code that calculates COSINE and ARCSIN is not written explicitly, but rather is derived from solving for SIN and ARCCOS. A COSINE, for example, is calculated from $SIN(\frac{\pi}{2}-x)$, and ARCSIN is derived from $\frac{\pi}{2}-ARCCOS(x)$.

This incurs the small cost of additional processing, but provides savings by not devoting memory to several mathematically related functions. The remaining trigonometric functions, TANGENT, SECANT and inverses are not included in the Interpreter instruction set. In the few cases they are required, they can be derived from the existing functions.

Subroutines calls and transfers of control

As in other languages, interpretive strings perform a series of computations, and are followed by a transfer of control to another location to perform another task. Frequently, these transfers of control are dependent on the state of some piece of data. Questions such as "Has the astronaut pressed the ENTER or PROCEED key?" or "Are we in the correct attitude for the next maneuver?" determine the future processing of a piece of code. This discussion now turns to three groups of instructions that are dedicated to sequence changes: subroutine calls, conditional and unconditional branches, and transfers into and out of the Interpreter.

Beginning with the most familiar type of sequence changes, the Interpreter provides a full complement of branches, with and without condition testing. In all types of transfer of control, the target address points to the first instruction of the Interpretive Instruction Word. By limiting references to operation codes on whole word boundaries, returning subroutines can only transfer control back to left-hand opcodes. While this is unsurprising in the AGC basic instruction set, the ability to store two interpretive operations in a single word presents a problem of addressing within the IIW. As seen in previous discussions, the fixed word size in the AGC demands that bit allocation is a zero-sum game. In assigning an additional bit for specifying an opcode within the IIW, another bit previously used for the address must be sacrificed. Ultimately, the inconvenience of forcing destination addresses to word boundaries is outweighed by the additional cost of interpretive addressing limitations.[23]

The most basic transfer of control, GOTO, is an unconditional branch to another IIW in the interpretive string.[24] As an unconditional transfer of control, GOTO is written as a right-hand opcode, ensuring that no instruction follows it in the IIW. As expected, an unconditional branch as a left-hand operation would make any instruction coded in the right-hand field inaccessible. A Computed GOTO (CGOTO) augments the unconditional branch with the means to dynamically modify the branch's destination address. Used to perform multiway branches based

---

[23] To reference individual opcodes within an IIW requires at least one additional bit in the interpretive address to specify which of the two opcodes is the branch destination. In the Interpretive Address Word, this forces us to reduce the addressing range of the IAW to 13 bits, or 8K words, from the current 14 bits and 16K words. Addressing would be further complicated by the fact that the reduced addressing range would impose the use of memory banking.

[24] Contemporary programmers are allowed to cringe at the sight of a GOTO instruction. It is important to remember, however, that much of the AGC software development was well under way by the time Edsger Dijkstra made his case against the GOTO statement.

on a computed value, a CGOTO instruction has two operands: the first IAW points to an erasable storage location whose contents serve as an index value for the second operand; the second operand is a destination address, usually referencing a table of address words in storage, with each address word containing a final destination address of the CGOTO. When CGOTO executes, the contents of the first operand are added to the table address specified in the second operand, and then this address is used to fetch the final destination address of the GOTO. The complexity of an instruction combining both indexing and indirection makes the following example useful:

```
       NOOP  CGOTO              IGNORE LEFT HAND OPCODE
             INDEX              ADDRESS OF INDEX VALUE
             GOTOTBL            TABLE OF GOTO ADDRESSES
             ...
             ...
      INDEX  DEC   0            IN ERASABLE STORAGE
    GOTOTBL  CADR  GOTO1        TABLE OF GOTO
             CADR  GOTO2        DESTINATIONS, ALL
             CADR  GOTO3        IN FIXED STORAGE
```

Figure 64: Computed GOTO example

Calling a subroutine is a universal concept in software. It is used as a structured programming technique for where a single routine performs functions common to several different programs. In servicing many different calling programs, subroutines must also have the ability to return to any of those programs, each residing at a different location in memory. The process begins with the calling program saving its return address in a well defined location, and then branching into the subroutine. The requirement that the calling program saves its return address is critical. When written, a called program has no idea where it will be called from, and thus no idea where it must return to. After completing its processing, the subroutine transfers control back to the calling program using the return address as the point where processing resumes. In the Interpreter, the QPRET storage area located in the Vector Accumulator area is the location where all calling programs store their return addresses. QPRET is entirely separate from, but similar in function to the AGC's Q register. Since each VAC are has its own QPRET area, interpretive strings can call subroutines without regard to the calling sequences of other interpretive jobs. Consistent with the limitations on addressing individual opcodes, the Interpreter saves the address of the next instruction word in the interpretive string, rather than of the next interpretive operation code in the IIW.

Calling a subroutine in the Interpreter is through the CALL instruction, with the address of the subroutine in the Interpretive Address Word. When executed, CALL places the address of the next word of the interpretive string into QPRET and branches to the address in the operand. A related operation called the Computed Call (CCALL) is similar to the Computed GOTO. The operand in erasable storage is added to the subroutine address to create a new calling address. CCALL performs a subroutine call to the computed address and saves the return address in QPRET. Creating a destination address calculated from this addition allows the programmer to dynamically select which subroutine is required in a given situation. CALL and CCALL, like all unconditional branching instructions, must be coded as right-hand instructions.

After completing the processing in the subroutine, it is time to return to the calling program using the address saved in the QPRET storage area. Executing the Return Via QPRET (RVQ) instruction is similar to an indirect branch, as the address stored in the QPRET location becomes the destination address. But what if the current subroutine is entered through a sequence of previous subroutine calls? Since QPRET is only a single storage word and not a dynamic structure like a stack, each called program has the responsibility of saving its return address before it makes a subroutine call on its own. Although the Interpreter implements a stack, it is used only for data computations, and not for nesting subroutine calls. Lacking any hardware or software structures to retain the calling history, it is the burden of each called program to save QPRET before it makes a subroutine call of its own. The dearth of erasable storage in the AGC limits the depth of subroutine calls. Although QPRET is addressable in the VAC area, a dedicated instruction, Store QPRET (STQ) is used to save the contents of QPRET in erasable storage. Unfortunately, no comparable dedicated instruction exists to restore QPRET. Reloading QPRET is a manual and imperfect process, requiring loading the saved return address in the MPAC and then saving it in QPRET. After the restoration of QPRET is complete, the program can return using the RVQ instruction.

Sequence changing and indirection

Program logic often requires transferring control to different locations based on the current state of its processing. For example, a timing loop may sample the clock, and then execute one of several functions based on the current time. Unfortunately, implementing a branch instruction that uses a dynamically chosen target address is particularly difficult. Multiway test and branch statements, as implemented in C and other languages, are impractical due to the large amount of storage required and the difficulty of implementing such statements within the IIW/IAW format of the Interpreter. None of the transfer of control instructions use indexing, so index registers are not available as a means of dynamically choosing a destination address. Another possible option is generating a branch address at execution time, using the Computed GOTO or Computed Call instructions, but these are not appropriate for all transfers of control. However, setting the target address to point to an erasable storage location creates a powerful new possibility for transfer-of-control instructions. All Interpreter instructions must be in fixed storage, but it is perfectly allowable to have operand references to erasable storage. Although an instruction in erasable storage cannot serve as the target for a transfer of control, this allows defining a branch to erasable storage as a new address mode, called indirection. In indirect addressing, the destination of the branch instruction is in erasable storage, which is not a legal address for an IIW. Rather than aborting the instruction due to a bad destination address, the data in the erasable storage location is used as another storage address, not an instruction word. Presumably, this address references a legitimate fixed storage address. The Interpreter, recognizing that the destination address is in erasable storage, uses the contents of the erasable word as the "real" branch destination. Not surprisingly, instructions that transfer control cannot be limited to a single level of indirection. Consider the case where the contents of the indirect word saved in erasable storage is itself in erasable storage. One level of indirection results in an invalid address, which requires us to use additional levels of indirection in the hope of finding a fixed storage address to branch to.

Conditional branches

Without a means to test data and alter processing flow based on the results of that evaluation, few programs of any sophistication are possible. Conditional branches combine into one instruction both a test of the MPAC contents and going to a new location. Ideally, the ability to test the MPAC against a value located in storage would simplify the task for programmers and improve the readability of the code. However, testing against a user-defined value requires additional overhead when fetching a double or triple precision value from storage and performing the comparison. While admittedly useful, the Interpreter avoids the processing and storage overhead of testing the MPAC against values in storage by making all comparisons a test of the MPAC against a value of zero. From testing against zero, the MPAC can be tested for a positive value, negative, or equal to zero; and by extension, for being not equal to any of these. Operating only on scalar quantities further simplifies conditional branches, and comparisons against vectors or matrices are not possible.

Six conditional branching instructions offer five different tests of the MPAC's contents. These instructions first evaluate the triple precision value in the MPAC against zero. Next, a test for a particular condition (such as positive or negative) is made, and if the result is true, the branch is taken. Branch Plus (BPL) tests if the MPAC is positive. Branch Zero (BZE) checks if the MPAC is zero. Branch Minus (BMN) tests whether the MPAC contents are negative. A fourth operation, Branch on High Order Zero (BHIZ) tests only in the higher order word in the MPAC against zero. BHIZ has two novel uses. First, if a single precision value is in the MPAC, only that value is evaluated, as the remainder of the MPAC might contain invalid data. BHIZ is also useful when generating a transcendental number, and we want to find out if the computation has reached a particular level of precision. Subtracting the result of the latest iteration with the previous one, we know we have reached at least four decimal digits of precision when the high order word is zero.

Two additional conditional branches do not test the value in the MPAC; rather, they test whether a previous arithmetic operation has set the overflow indicator, OVFIND. As discussed earlier, overflows in the Interpreter do not inhibit interrupts, and only an explicit test for overflows can identify the condition. Branch on Overflow (BOV) tests the overflow indicator, and if set, executes the branch. Branch on Overflow to Basic (BOVB) is used where processing an overflow condition is beyond the capabilities of the Interpreter. If so, BOVB exits the Interpreter and branches to a location containing basic AGC instructions. Transferring control out of and back into Interpreter is a powerful capability discussed more fully with the RTB instruction.

Addressing scope of branches and calls

Branching instructions can reference any location in the Interpreter's addressing range, in contrast to the addressing of other instructions, which are limited to the current half-memory. These transfers of control become quite powerful, but sanity checking of destination addresses does not occur during program execution. Nothing in the Interpreter prevents a branch or subroutine call from transferring control outside of the interpretive string and into basic AGC code, and the results would certainly be unpredictable. The Interpreter, unable to tell whether a word in storage contains an interpretive instruction, a basic AGC instruction, or data, will blithely execute whatever the contents of the word decode to. A run of misinterpreted instructions will create garbage data that will sooner or later prompt a program alarm, followed by a restart to clear the error.

Entering and exiting the Interpreter

Strictly speaking, an operation that transfers control within the Interpreter also allows transfers out of the Interpreter's control. So much of the AGC hardware is hidden from the programmer through the virtual machine created by the Interpreter that most Executive level operations are inaccessible. Leaving the Interpreter entirely becomes the only means to access the hardware status bits, some memory areas, or reading and setting bits in the I/O channels. A common reason for exiting the Interpreter is to access the Executive routines that perform operations on specific registers and I/O channels. Under this design, a common technique is to leave the Interpreter, call the processing routine and then return to the Interpreter's control. Leaving the Interpreter is by executing the Return to Basic (RTB) operation, which suspends the fetching and execution of IIWs. RTB combines the exit from the Interpreter with a subroutine call using basic AGC instructions.[25] Establishing a standardized call to a subroutine requires saving the return address in the AGC's Q register, which is the return address for all basic instruction subroutine calls.

A novel twist of RTB is that the Q register is not loaded with the address of the IIW immediately following the RTB operation. Returning from the subroutine directly into an interpretive string would quickly fail, as the IIWs would then be fetched and executed as basic AGC instructions. Instead, the program must reenter the interpretive system. By placing the entry address of the Interpreter itself in the Q register, the familiar "TC Q" at the end of the AGC routine brings the execution sequence back into the Interpreter, which then restarts processing of the IIW immediately following the RTB. This design is especially elegant, as the called routine is now available to both basic and interpretive programs. No special coding

---

[25] Arguably, RTB could just as easily be renamed "Call Basic Subroutine".

is necessary to adapt the routine to these two modes, creating important simplifications in the design of the code. The sole restriction for the called routine is that it must reside in fixed storage, reflecting the requirement that the Interpreter can only process fixed storage addresses. Data passed to the routine must also be independent of whether a basic or interpretive program is performing the call. This precludes using the VAC area as a location for exchanging data between the programs.

Executing RTB does not signify that the job is through with interpretive processing. Rather, RTB is executed with the intention that its time in basic mode will be very brief – a quick excursion to perform an important task before returning back to the Interpreter. No explicit provisions are made to preserve the interpretive environment for the time when the subroutine completes and returns control to the Interpreter. Importantly, no changes are made to the banking registers, forcing the basic AGC routine to use the same memory scheme as the Interpreter. The job's Core Set and VAC area are addressable when executing the basic AGC instructions, and the job is free to modify either as it sees fit. Upon returning to the Interpreter, these storage areas are not refreshed or restored in any way; they retain the changes made while in basic mode. Because the interpretive environment is preserved, there is no need to restrict RTB to the class of right-hand opcodes. As the intent is to resume the interpretive string immediately after the point it was exited, right-hand instructions must be allowable.

RTB allows us to leave the Interpreter temporarily, but always with the intention of returning to its control. A more permanent means of departing the Interpreter is through the EXIT statement. Unlike RTB, the banking registers are restored to the values that existed before the Interpreter was called. The Core Set and VAC are not cleared, but the VAC area is now available for reuse. After EXIT is processed, execution continues under hardware control with the next basic AGC instruction following the interpretive string.

3

The basics of guidance and navigation

HARDWARE UNIQUE TO SOLVING GUIDANCE AND NAVIGATION PROBLEMS

The IMU gimbals and the stable platform

Alone by itself, the computer is blind to the world around it, lacking the most basic information on orientation and the accelerations that the spacecraft undergoes. Without a source for this data, the AGC's role of a guidance computer is invalidated. Two additional components are necessary for a completely integrated guidance, navigation and control package: an Inertial Measurement Unit (IMU) to provide both attitude and acceleration data, and an optics system to initialize the platform and perform navigation fixes using stellar observations. Despite being a marvel of precision mechanical engineering, the basics of an IMU are surprisingly straightforward. The IMU performs only two fundamental measurements: attitude and acceleration. Maintaining a known attitude is the first requirement, which is accomplished by maintaining a rigid orientation with respect to a known reference, such as a set of stars. The most well-known property of gyroscopes is their ability to use their angular momentum to maintain a given orientation in space. A single gyroscope is limited for this purpose, as it will only hold itself fixed in a single axis.[1] The gyro may appear to be holding itself perfectly steady, but nothing prevents the structure attached to it from rotating around its spin axis. To truly stabilize a body in all three axes, three gyros are needed, mounted mutually perpendicular to each other. Assembled onto a common base, these orthogonal gyros create the IMU's stable member.

---

[1] International Space Station Science Officer Don Pettit created a superb video demonstrating gyroscopic stability in one, two and three axes using only portable CD players as part of his "Science on Saturdays" series.

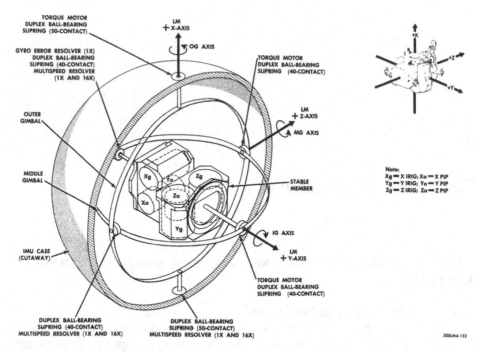

Figure 65: Schematic of the Inertial Measurement Unit

Naming conventions for the rotational axes

Each of the three gimbals corresponds to one of the spacecraft's rotational axes. Rotations about the Y-axis are recognized by the inner gimbal; likewise, rotations in the X- and Z-axis are sensed by the outer and middle gimbals. Although the IMUs are physically identical in the CM and LM, assignment of the names "roll", "pitch" and "yaw" differ between the two vehicles. These terms are foremost a convention based on the pilot's perspective, as if they were a kind of anthropomorphism for the vehicle. Ideally, the names should be associated with the same axis in the CM and LM, but the most important determinant, the orientation of the pilot relative to the spacecraft axis is completely different. In the Command Module, the crew is seated facing along the $+$X-axis, towards the front of the vehicle. Traditionally, the $+$X-axis is aligned along the major thrust axis, and this is indeed the case with the CM's Service Propulsion System. The $+$X-axis is also aligned with the thrust line of the LM's descent and ascent engines but, unlike the Command Module, the men in the LM do not face along the X-axis. Standing parallel to the X-axis to get the best view of the landing site, the LM crew faces the forward-looking $+$Z-axis. The crew perceives a rotation around the X-axis in the CM as a roll, but the crew's orientation in the LM views the same rotation as yaw. Likewise, rotations around the Z-axis have different interpretations. A yaw is a rotation around the Z-axis in the CM, but is roll in the LM. Thankfully, one axis remains the same. In both the CM and LM, the Y-axis is described as a line from left to right, and rotations about the Y-axis define pitch in both vehicles. A summary of these different orientations is in Figure 66.

Gimbal	Axis	CM	LM
Inner	Y	Pitch	Pitch
Middle	Z	Yaw	Roll
Outer	X	Roll	Yaw

Figure 66: Body axes on the CM and LM

To visualize how the axes relate to the pilot, it helps to imagine that the CM is flown like a conventional winged aircraft and the LM is maneuvered like a helicopter. In an aircraft, the pilot faces along the direction of thrust from the engine, but a helicopter's thrust is directed mostly downwards. Pitch remains the same, a rotation around the Y-axis, but roll and yaw in the LM are reversed from the CM convention. Rotations about the X-, or thrust axis, is now yaw, reflecting the left and right rotations from the pilot's perspective, leaving roll as the rotation around the Z-axis.

Cradled within the three sets of gimbals, the platform appears ready to rotate in any direction it pleases. Unfortunately, in a three gimbal system, the middle, Z, gimbal is susceptible to "locking" when the inner and outer gimbals align themselves in the same plane. Figure 67 shows the three gimbals. As we yaw in the CM (or roll on the LM), the inner and outer gimbals come into alignment, and after 90 degrees of travel the gimbals are coplanar. At this point, the gimbals cannot follow the spacecraft if it now rotates in the plane of the outer gimbal. The gimbal lock monitor routines that execute as part of T4RUPT processing are designed to give the crew ample warning when approaching a lock situation, and illuminates a warning light when the middle gimbal angle falls between 70 and 85 degrees. Beyond 85 degrees is the true gimbal lock region where the IMU cannot recover, and the AGC illuminates the NO ATT lamp and places the platform in a coarse align mode. With the attitude reference lost, the crew must realign the platform to a known reference before it can be useful again.

It is important to note that if the IMU locks, the spacecraft does not fly out of control, with its hapless crew facing certain doom. In this situation, the crew switches control to the secondary system, the Stabilization and Control System (SCS) in the

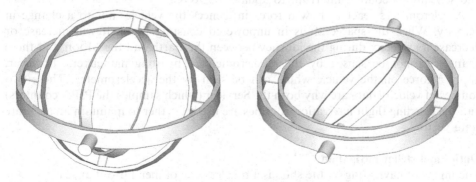

Figure 67: Gimbal unlocked (left) and locked (right)

CM and the Abort Guidance System (AGS) in the LM. Both the SCS and the AGS are perfectly capable of assuming attitude control duties while the crew realigns the platform. The choice to "sacrifice" the Z-axis is also not completely arbitrary. Pilots are completely comfortable with 360 degree rolls and loops, which are rotations about the X- and Y-axes. A complete yaw rotation about the Z-axis is not a natural act in aviation, as excessive yaw can result in spinning a fixed-wing aircraft. In the end, the ability to perform unrestricted rotations around the pitch and roll axes more than compensates for the small restrictions along the Z-axis.

The Pulse Integrating Pendulous Accelerometers
The IMU contains three accelerometers, each parallel to one of the three gyroscopes on the stable platform. Fixed in the same orientation as the gyros, which in turn hold the assembly fixed in space, the accelerometers are able to resolve motion in each axis. The term Pulse Integrating Pendulous Accelerometer is particularly descriptive of the technology used in this device. A small pendulum whose mass is known with great accuracy is immersed in a viscous fluid. The pendulum is allowed to swing along the one axis it is designed to sense. When the PIPA has an accelerating force applied to it, the pendulum moves and needs to be restored to its original position. Sensors built into the pendulum's pivot determine the amount of its motion and the amount of force necessary to bring the pendulum back to its original position. Since the mass of the pendulum is well known, as is the force required to reposition the pendulum, the acceleration is readily determined. A continuous output of the acceleration data is very useful, but the computer would be overwhelmed by the constant stream of acceleration data necessary to calculate the vehicle's velocity changes. A more useful presentation of data to the AGC is for the accelerometer to give a summary of accelerations it has experienced over a period of time. Mathematically, this is called an integration, and the effect of acceleration is computed over time as velocity. Now the accelerometer is expressing the accelerations as changes in velocity, which the AGC can use directly in its calculations. Only a short period of time is represented by the velocity data, and therefore the velocity change data is usually a small value. Finally, the velocity change data is sent to the CDU which converts the analog signal to a digital pulse and schedules a counter interrupt to update the AGC.

Accelerometers react to how a force influences the vehicle, not to a change in velocity. When the spacecraft is in unpowered coasting flight, it will increase or decrease its velocity during the journey between the Earth and the Moon, but these changes cannot be sensed by the accelerometer. Thrusting maneuvers, however, apply a force on the vehicle which can be seen by the accelerometer. These two sources of velocity data are why both the Servicer (which samples the PIPA counters) and the coasting flight navigation routines are used together to maintain an accurate state vector.

Optics and stellar navigation
The image of navigating by the stars is a timeless one of men far out at sea in fragile wooden ships, but is also a thoroughly modern technique in spaceflight. Although

the problem of longitude proved intractable for centuries, sailors long knew how to determine latitude from the stars. The star Polaris, by its lucky fate, is sufficiently close to the celestial north pole that the angle between itself and the horizon yields an accurate value of an observer's latitude. Keeping a steady gaze on both the horizon and a star (or the Sun) and measuring their separation is difficult to do in practice, especially when the heavenly object is far above the horizon. Sextant optics evolved to simultaneously observe both the celestial object and the horizon, allowing the navigator to read the angle off a scale (Figure 68). Using the telescope, the navigator views the horizon through a half-silvered index mirror. Light from the Sun or star is reflected by the index mirror, where it is collimated with the view of the horizon. By adjusting the mirror, the navigator superimposes the image of the star or Sun onto the view of the horizon. The mirror is connected to an arm that follows a scale, from which the angle between the celestial object and the horizon is read.

Despite its precision and complexity, the Apollo sextant could be easily understood by an 18th century navigator, as navigating in space uses an instrument very similar to the nautical sextant. The optics assembly in the CM consists of two instruments, a sextant and a telescope. Mounted together on a common assembly, the optics system is itself mounted on the same navigation base as the Inertial Measurement Unit. Mounting these components separately in different areas of the spacecraft could easily disturb their precise orientation as mechanical and thermal stresses create small physical displacements. Physically integrating the IMU and optics into one unit ensures that their mutual alignments are the most accurate possible, given that even small changes in orientation are sufficient to introduce measurable errors. Such attention to this level of detail might seem somewhat extreme, but is necessary when measurements are made with arc-second precision. The telescope has only a unity (1X) magnification, but its very wide 60 degree field of view performs the role of a "finder telescope", allowing the astronaut to orient

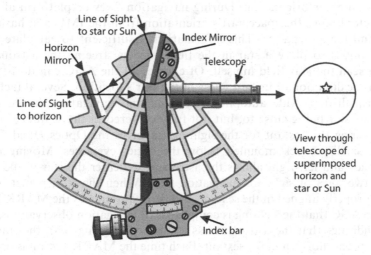

Figure 68: Schematic of a sextant

himself by observing a large area of the sky. The sextant uses a higher magnification, 28X, and so has a far narrower 1.6° field of view in order to provide the desired 10′ arc-second accuracy.

The sextant and telescope view the heavens through windows in the Command Module's hull, near the apex of the spacecraft and on the side opposite from the main hatch. When at their "zeroed" position, the sextant and telescope both point in a direction perpendicular to the conical hull. As with the nautical sextant, the sextant has two lines of sight. The first, aligned along the boresight of the sextant is known as the Landmark Line Of Sight (LLOS) and is always perpendicular to the hull. With the sextant LLOS only able to point in one direction, the spacecraft must maneuver itself to place the celestial body within the sextant's field of view. For cislunar navigation, this body is the Earth or Moon, while platform alignments have one of the two required stars along the LLOS. The second view through the sextant, the Star Line Of Sight (SLOS), is oriented manually or using computer commands. As in the nautical sextant, superimposing both fields of view in a single eyepiece makes it possible to observe two objects at once. Replacing the nautical sextant's simple scale is a set of gears and resolvers which measure the angular difference between the two lines of sight to within 10 arc-seconds.

Several options are available to the crew when pointing the telescope and sextant at a target. First, before taking any measurements, both the sextant and telescope are swung to a known orientation and the shaft and trunnion counters are reset to zero. Known as "zeroing the optics", this procedure ensures that the drive mechanisms are operating correctly, that the optics are in a well-defined orientation, and that the counters do not contain residual data. Any motion of the sextant from this point on is recorded as an offset from its initial position and is sent to the AGC. Only the sextant can be commanded to a new orientation, but the telescope can be "slaved" with the sextant, to follow all of its motions. When not slaved together, the telescope does not move while the sextant's SLOS is pointed to another orientation. During navigation fixes or platform alignments, the computer knows the spacecraft's orientation from the IMU and has a catalog of stars and their locations. This information is sufficient to calculate where to point the optics to place a star in the field of view, freeing the astronaut from having to scan the star field himself. Of course, all the AGC can do is point the sextant in the direction of its best guess of the star's location. Several factors, such as platform drift, a wide autopilot deadband and navigation inaccuracies can result in the star being close to, but not fully centered in the sextant's reticle.

Maneuvering the sextant for the sightings involves the Optics Hand Controller (OHC), a small joystick mounted below the optics eyepieces. Moving the OHC forward-back and left-right allows the astronaut to center the star in the reticle or align the two lines of sight for navigation fixes. When the star, planet or surface feature is properly aligned in the sextant, the astronaut presses the MARK button to inform the AGC that the sighting is completed. In navigation observations, pressing MARK indicates that the astronaut has completed aligning both the star and the Earth or Moon's horizon in the sextant. Each time the MARK button is pressed, the AGC records the shaft and trunnion angles from the CDU registers, and the exact time of the mark. We will see in later sections how this data is processed by the navigation and alignment software.

The sextant is also used to point precisely at non-celestial targets. Tracking a fixed object on the lunar surface as the spacecraft flies overhead allows the crew to observe and photograph features through the sextant eyepiece. Of particular importance is using the sextant to locate the LM's landing site. More than just confirming that the LM has indeed landed in a given area, locator marks made by the CM and relayed to controllers on the ground give mission planners critical information on the position of the LM relative to its geology and science objectives. During the rendezvous sequence, the CMP uses the sextant to locate and track the LM's ascent stage after this has returned to lunar orbit. These sightings update the Command Module's AGC with the LM's position, in order to have the most accurate data in the event that the CSM is obliged to assume the active role in the rendezvous. Combining the sextant's angular data with distance information from the VHF ranging equipment, enables the AGC to calculate rendezvous targeting solutions.

The Alignment Optical Telescope in the LM

The LM's use of a sextant is much more limited, reflecting that vehicle's specialized role as a "space taxi" between lunar orbit and the surface. Development of the LM was a particularly torturous process, with much of the effort dedicated to reducing its weight. What resulted was a machine like no other, with capabilities singularly focused on the task of landing a man on the Moon. A notable example is that

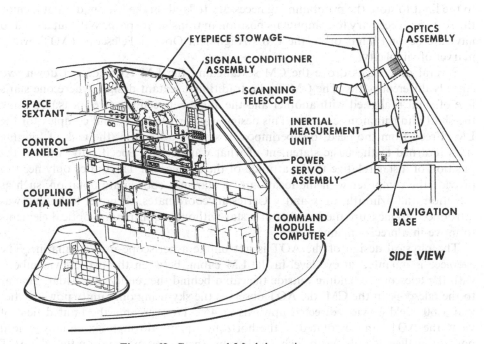

Figure 69: Command Module optics systems

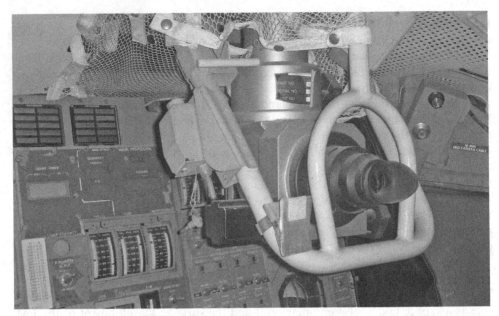

Figure 70: LM Alignment Optical Telescope

navigation duties between the Earth and Moon were handled exclusively by the Command Module AGC and its optics system. The LM computer, already pressed to the limit to store the programming necessary to land on the Moon, did not include the software necessary to compute its position in translunar space. Without a need to navigate beyond lunar orbit, the LM's Alignment Optical Telescope (AOT) was a marvel of simplicity.

Several differences drove the CM sextant and the LM AOT design down two entirely different paths. The CM used a traditional sextant design, where one star's line of sight is aligned with another and the angle between the two measured using the shaft and trunnion resolvers. This design was far too heavy and complex for the LM. Freed from the need to superimpose a star on a horizon, the task of sighting stars returned to the basic statement of what was needed. Specifically, what is the position of a star relative to the attitude of the spacecraft? The optics only need to provide the computer with the location of a star in a coordinate notation, such as elevation and azimuth, or x and y Cartesian coordinates. A novel technique was introduced to measure the location of a star without requiring any optical elements to move in a precise manner.

The physical design of the AOT is not unlike a periscope in a submarine. The eyepiece is mounted at eye level in the LM cabin, between the two crewmembers, with the telescope extending outside the cabin behind the rendezvous radar. Similar to the telescope in the CM, the AOT observes the sky using only unity power optics and a 60° field of view directed upwards at 45°. To overcome the limited field of view, the AOT can be rotated in the horizontal plane through six equally spaced positions called detents. With such a wide view, there is no expectation that the AOT will have the star centered in its field. However, a novel means was devised to determine the position of the star with a great deal of accuracy. Within the AOT is a reticle with two distinctive patterns. The first is a vertical line called the cursor, or Y-line, which forms a part of the crosshair centered on the reticle. The horizontal component of the crosshair is called the X-line. Another line spirals once around the crosshair, beginning at the center and ending at the edge of the reticle. This curve, known as an Archimedean Spiral, has the special property of moving outwards from the center at a constant rate. Rotating along the 360° field of the AOT, the spiral progressively moves outwards from the center so that at the 90° position it has moved one-quarter of the distance to the edge of the reticle, at 270° it is three-quarters of the way, and after a full 360° the spiral has arrived at the outside edge of the reticle.

Aligning the inertial platform using the AOT is possible using two different techniques. The first is used in the unlikely event that the IMU has lost its attitude reference and the Abort Guidance System is incapable of providing an attitude reference. Using Program 51, a detent position is selected and the LM is rotated to bring the selected navigation star near the center of the AOT field of view. The LM is then maneuvered so that the star image crosses the reticle crosshair. When the star image is coincident with the Y-line, the astronaut presses the MARK Y pushbutton; when it is coincident with the X-line, he presses MARK X. When the MARK X or MARK Y pushbutton is pressed, a discrete is sent to the AGC, and the computer records the time and the IMU gimbal angles at the instant of the mark. During this procedure, the LM is not maneuvered to line up the star with the crosshair, but is instead set into a slow rotation that allows the star to "drift" across the AOT field of view, and the crewmember must wait for the moment when the star reaches the crosshair lines. While awkward to perform, this procedure has the advantage of not requiring a previously aligned IMU to hold the spacecraft steady during the observations. The primary limitation of this technique is that only a "coarse align" of the platform is performed, leaving the IMU with only an approximation of the desired orientation. A "fine alignment" will then be necessary to more accurately orient the IMU. This requires the LM to hold its attitude fixed relative to the stars. Once again, the AOT is rotated in the general direction of the star using one of the six fixed detents. The reticle with its scribed lines is not fixed, as the entire reticle can be rotated using a knob mounted on the side of the AOT. As the control knob is turned, a small counter on the side of the AOT displays the angle the reticle has rotated to the tenth of a degree. The location of a star in the AOT's field of view is most easily understood using the polar coordinate system in which the center of the AOT's field is set as the origin of the coordinate system, and the location of the star is defined as the angle from the zero degree mark and the distance from the origin. Both the cursor and spiral lines are etched into the reticle, and rotating the reticle control moves the entire set of lines clockwise or counter-clockwise.

Figure 71 shows how the cursor is rotated so that the cursor line aligns with the star. This angle, also known as the shaft angle, is read off the counter and entered into Noun 79. Next, the reticle is rotated so that the spiral line coincides with the

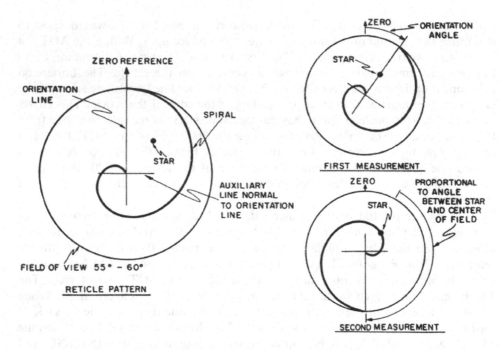

Figure 71: Star sighting through the AOT

star. As the reticle rotates, the spiral moves out from the center at a constant rate, and the distance from the center is proportional to the angle the reticle has rotated. Known as the reticle angle, its value is also read from the counter and is also entered into Noun 79. Finally, the polar coordinates from the cursor and spiral values are translated into the three dimensional vector to enable the alignment software to define the star's location.

As is the case in the Command Module, the LM's IMU has no attitude information when it is first powered up. Already in coasting flight or lunar orbit, there are no gravity references available to gyrocompass the platform in the manner applied to the CM's IMU as part of its pre-launch initialization. However, by taking advantage of the fact that the computer and IMU are identical in the two spacecraft, the crew can transfer the gimbal angles of the CM's platform to the LM. An elaborate AGC to AGC networking protocol to transfer the angles between the two computers might be envisioned for this task, but, in fact, all data transfers between the two computers are done manually, with the crew writing the information on a preprinted form, making some calculations, and keying the data in by hand on the computer. Earlier in the mission, the CMP floated over to the tunnel connecting the CM and LM to determine the accuracy of the docking alignment. Although the two vehicles are held tightly together, the spacecraft are usually not aligned perfectly along the X-axis. This angle is measured using marks inside the tunnel called the Docking Tunnel Index. The discrepancy is small, usually only a fraction of a degree, and the angle is noted on the form. On the day of the lunar landing, the LM's computer and inertial system are powered up together for the first time in the mission. When performing the initial alignment of the LM's IMU, ground controllers first send a copy of the landing site REFSMMAT and a state vector to the LM's AGC. Next, the CMP displays the current CM gimbal angles using Noun 20 and writes these values on the same form that contains the Docking Tunnel Index value. Simple addition and subtraction are all that are necessary to compute the LM's gimbal angles, which are then entered into its computer through Noun 22. This will torque the LM's gimbals to an orientation that is sufficiently accurate for the crew to perform the fine alignment.

THE IMPORTANT QUESTIONS IN GUIDANCE AND NAVIGATION

At this point, a discussion of the IMU, the optics system, and how the computer processes data, must extend beyond the spacecraft, and consider its environment. Here, an abbreviated introduction to celestial navigation is in order. Clearly, this is a topic heavily biased towards mathematical descriptions; after all, it *is* rocket science. Fortunately, there is no reason to delve into the depths of mathematics to obtain a conceptual understanding of the major issues. In the most basic terms, only three things need to be known:

Which way is up?
Where am I?
Which way am I going?

Far from being an attempt at trivializing such an important subject, these three questions neatly sum up the major requirements of a guidance computer. The answers to these questions provide all the elements needed for understanding each of the major tasks in guidance and navigation. Rather than being isolated topics, all three are interrelated and build upon one another.

QUESTION 1: WHICH WAY IS UP?

Schoolchildren are taught that "there is no up or down in space". Perhaps there is a point here, as in coasting flight you are weightless and thus do not have a sense of gravity to help define "which way is up". In reality, flight without a gravity vector forcing your attention towards specific direction creates the greatest freedom of all: "Up" can be any direction you want.

Before dismissing that last statement as overly flippant, consider how to define an orientation in space. On Earth, things fall "down" due to gravity, and we look "up". If a person lives in New York City, up and down, as a practical matter, are in the same directions regardless of the location in the city. A similar set of statements is possible when looking up and down in London or Sidney. It quickly becomes apparent, however, that the trivial concept of up or down becomes complex, as the definition is relative to the person's location on Earth. Taken to its obvious extreme and viewing the Earth as a whole, a person pointing "up" at the North Pole points in the opposite direction to their counterpart at the South Pole. Using the local gravity vector is impractical for determining an orientation in space. For a reference to be useful, it must both be unchanging and independent of whether the observer is on the Earth or Moon.

Once again, the answer begins with the objects that have guided voyagers for years: the stars.

Coordinate systems

An integrated system of the AGC, IMU and optics, contains all the hardware necessary to maintain an orientation in space. The challenge, therefore, is to create a set of standards and references that can define an orientation that is both well-defined and easily usable by the crew. Stating that the platform maintains a steady reference and that rotational motion is measurable to an astounding accuracy, is irrelevant if there is no framework to measure this performance against. In practice, using a single definition for all orientations is impractical, as many references are not easily expressed in terms of a universal coordinate system. Several distinct coordinate systems are needed, each of which describes its orientation and origin in very precise terms, independently from the others. Nevertheless, it is a straightforward matter to move from one coordinate system into another using simple mathematical transformations.

Coordinate systems exist everywhere that we encounter an object which has an orientation as one of its attributes. As an example, I can define my "personal" coordinate system as standing up with one axis perpendicular to the Earth's surface, another facing north, and my outstretched arm pointing east. Note that it is important to remember that once established, this coordinate system is fixed and does not move with me. Any new orientations can be expressed as angular rotations and translations from the origin of my coordinate system. A second coordinate system might exist for my bed, whose long horizontal dimension exists along a northeast/southwest line. The two frames of reference do not depend on each other for their definitions, yet there is sufficient information to determine my new orientation, expressed in my personal coordinate system, when I go to bed. For simplicity, assume I am aligned with my personal coordinate system as I stand and face north. When I lay down to sleep, my orientation should align itself with the bed's coordinate system. How can I determine what my new orientation should be? The transformation has two steps: First, I rotate in the vertical axis 45 degrees to the right, and then I lay down using a 90 degree rotation. These two rotations define the transformation between the two coordinate systems, and, most importantly, it is a commutative operation. I can move from one coordinate system to another by applying the transform, and by reversing the process can return to my original orientation.

Five distinct coordinate systems are used in Apollo, each determining a reference point for celestial bodies, the spacecraft and the inertial platform. Orientation within each coordinate system is known through actual measurements such as sextant observations, or through the calculations used by applying the laws of celestial mechanics. But knowing only the orientation within the current frame of reference is not very useful if that information cannot be translated to other coordinate systems. As seen in the bedtime example, the process begins with knowing the orientation within one coordinate system and the rules that relate one system to another. By applying the appropriate transform operation, the current orientation may be expressed in any of the other coordinate systems.

The Basic Reference Coordinate System

The first and most familiar coordinate system has the mundane name of the Basic Reference Coordinate System. Unlike the other reference systems, this one has the novel attribute of having a single orientation, although it may be centered on either the Earth or Moon's center of mass. Its orientation is defined at a specific moment of time, with the X-axis along the path of the Earth in its orbit at that point in time, the Z-axis going through the North Pole, and the Y-axis completing the right-handed system. The Basic Reference Coordinate System is also unique in that its orientation does not change when its origin shifts from Earth to the Moon. The origin of this coordinate system is the body that has the dominant gravitational sphere of influence over the spacecraft. In the case where the spacecraft is in Earth orbit, the Basic Reference Coordinate System is centered on the Earth. After the spacecraft enters the lunar gravitational field, the coordinate system switches to the Moon-centric orientation. Regardless of the origin, the Basic Reference Coordinate System is not intended as a reference for objects on or close to their surfaces, but is designed to define vectors to celestial objects such as the sun, stars and planets. Additionally, the Basic Reference Coordinate System is used when defining the spacecraft's state vector, expressing the vehicle's position and velocity relative to the Earth or Moon as appropriate.

Maintaining the same orientation in space despite the origin shifting between the Earth and Moon has the important advantage of maintaining entries in the AGC's star catalog. The distance the Basic Reference Coordinate System's origin moves when shifting between the Earth and Moon is only about 225,000 nm (400,000 km), which is an extremely small distance when compared with the scale of the solar

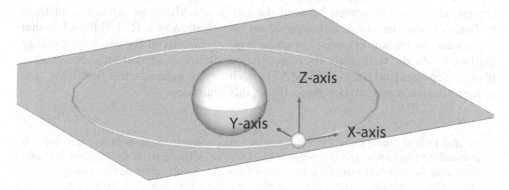

Figure 72: Basic reference coordinate system (Earth centered)

system and even more so with stellar distances. The parallax angles to stars between the Earth and Moon are far smaller than the sextant can resolve, so the same vectors used to locate stars and planets are usable when the reference system shifts its origin.

The Inertial Measurement Unit Coordinate System

The second coordinate system uses the IMU's inertial platform, or stable member, for its reference orientation. Unlike the Basic Reference Coordinate System, the crew can visualize the platform's orientation through the FDAI, the Flight Director[2] Attitude Indicator. Colloquially known as "8-Balls", both the CM and LM contain two FDAIs on their instrument panels. In the CM, the Stable Member Coordinate System has its origin at the platform's center of mass. The LM, on the other hand, uses the Moon's center of mass as its origin during descent, ascent and abort maneuvers. During other mission phases, notably rendezvous, the LM returns to using the center of the platform as the coordinate system origin.

As seen previously, the platform can be aligned to any orientation, providing flexibility for all mission phases. This freedom of orientation is key, as the platform is often aligned to the spacecraft's attitude during major mission events. In aligning the platform to the spacecraft's attitude for Lunar Orbit Insertion, for example, the crew notes that the attitude display indicates zero pitch, roll and yaw when the vehicle is in the proper attitude for the burn. Similarly, the platform alignment for the lunar landing is defined so that all three axes display zero degrees when the LM is on a level surface at the intended landing site. Aligning the platform to other orientations has few technical limitations, but quickly fails when human factors become involved. As demonstrated most dramatically in the last phases of a lunar landing, visibility during the last 50 feet (15 meters) is poor to nil because of blowing dust. It is far easier to interpret an attitude which has pitch, roll and yaw at zero as a "straight and level" flight attitude during this critical phase.

A vector called a REFSMMAT, the Reference Stable Member Matrix, is not a coordinate system, but specifies the orientation of the platform. A vector expressed in the Basic Reference Coordinate System defines the REFSMMAT, making the stars an ideal reference source for orienting the platform. Unlike coordinate systems, REFSMMATs do not imply an explicit origin, so the platform's orientation will not change as the vehicle travels between the Earth and Moon or orbits one of these bodies. A particular constraint imposed on the selection of a REFSMMAT is that any spacecraft maneuvers must not cause large excursions of the platform's middle gimbal, or Z-axis, that might introduce the possibility of gimbal lock. Pre-mission planning ensures that the REFSMMATs are chosen to minimize this possibility, and spacecraft maneuvers are designed around this limitation.

---

[2]   In this context, the phrase "Flight Director" has nothing to do with the lead mission controller in Houston. In aviation, a Flight Director is that part of the attitude indicator that obtains turn and attitude information from the autopilot or navigation system. In effect, the instrument is "directing" the pilot to change his attitude in flight to maintain a proper course.

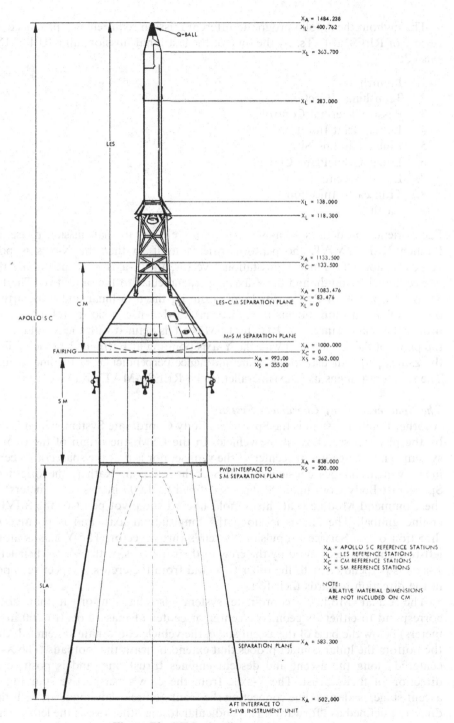

Figure 73: CSM coordinate system

Throughout the Apollo program, different mission requirements produced a wide variety of REFSMMATs. By the time of the last lunar mission, nine REFSMMATs existed:

1. Launch
2. Translunar Injection
3. Passive Thermal Control
4. Lunar Orbit Insertion
5. Lunar Landing Site
6. Lunar Orbit Plane Change
7. Lunar Ascent
8. Transearth Injection
9. Earth Entry.

Each orientation defines an axis that is most relevant to that mission phase. In the Launch REFSMMAT, the platform orients itself so that the X-axis is pointing vertically along the Earth's gravitational vector, the Y-axis is aligned along the 72-degree launch azimuth, and the Z-axis is perpendicular to the other two. The Passive Thermal Control, or PTC, attitude, used during the translunar and transearth coast phases of the mission, orients the spacecraft so that it can slowly roll to maintain a proper thermal balance. The PTC REFSMMAT is defined with the X- and Y-axes in the plane of the ecliptic, and with the Y-axis pointing at Earth at the time of TLI and the Z-axis perpendicular to the plane and hence perpendicular to the line to the Sun. The spacecraft aligns its X-axis parallel to the REFSMMAT's Z-axis.

The Spacecraft Body Coordinate System
Another familiar system is the Spacecraft Body Coordinate System, which is defined by the physical structure of the vehicle. In the CSM, the origin of the coordinate system is not the geometric center of the vehicle, nor is it at the spacecraft's center of mass, which changes considerably during the mission. Rather, the origin of the Spacecraft Body Coordinate System is defined as 1,000 inches (25.4 meters) below the Command Module heat shield mold line, at the pivot point of the S-IVB's J-2 engine gimbal. The X-axis is along the longitudinal axis, and is positive in the direction of the Service Propulsion System's thrust vector. The Y-axis is along the left-to-right axis, as viewed by the crew in their couches, with +Y to their left. The Z-axis is perpendicular to the other two, and from the crew's perspective is positive in the direction towards their feet.

The Lunar Module's coordinate system also has an origin that does not correspond to either its geometric center or center of mass. Located 200 inches (5 meters) below the base of the ascent stage, the vehicle coordinate system is located at the bottom the lunar contact probes that extend beneath the footpads. The X-axis is centered along the ascent and descent engines thrust line, and is positive in the direction of their thrust. The Z-axis, from the crew's perspective standing in the ascent stage, is along a line connecting the front and rear landing gear, with the +Z direction defined as "forward". Perpendicular to the other two is the left-to-right Y-axis, which is positive to the right.

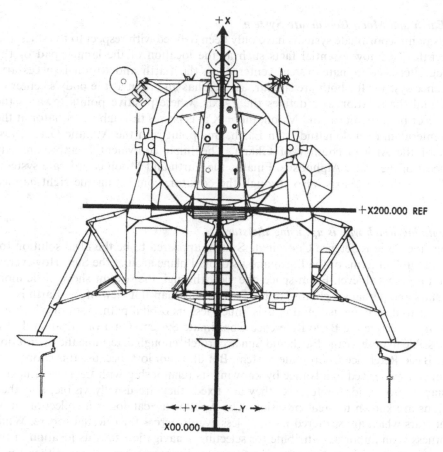

+X

+X200.000 REF

←+Y→ ←−Y→

X00.000

Figure 74: LM coordinate system

The Navigation Base Coordinate System

How the IMU is mounted in the spacecraft is the source for another frame of reference, the Navigation Base Coordinate System. A reasonable first assumption might be that the casing is attached with its axis parallel to the vehicle's body coordinate system. In fact, the navigation base, the structure on which the stable platform is mounted, is aligned to the spacecraft's optics. As noted earlier, the optics in the Command Module are aligned perpendicular to the hull of the vehicle. Along with the sextant and telescope, the IMU is mounted on the CM navigation base, and aligned with the X-Z plane. The navigation base is rotated 32° 31' 29.19" in the Y-axis towards the +Z axis, reflecting the mold line of the spacecraft's exterior. In the Lunar Module the IMU is mounted on the same assembly as the Alignment Optical Telescope, which itself is aligned with the vehicle's X-axis. In the LM's configuration, the axes are indeed parallel, and so there is no need to rotate one coordinate system to another. Still, the origins are different, so a transformation is necessary in some cases.

The Earth and Moon Coordinate System

At this point, coordinate systems have only been defined with respect to the stars and spacecraft. To know essential facts such as the location of the launch pad or the landing site, a coordinate system centered on the Earth and Moon is necessary. Reference systems for both are similar, as each has an origin at the body's center of mass, and the rotation axis defines the Z-axis, where positive points towards the north. For both the Earth and Moon, the +X axis passes through the equator at the prime meridian of 0° longitude. On Earth, this point is in the Atlantic Ocean, just south of the African country of Ghana. The lunar equivalent is approximately northeast of the crater Alphonsus. Finally, the Earth and Moon coordinate systems conclude with the Y-axis, also through the equator, completing the right-handed system.

Determining which way is up, using the stars

Earlier, the Basic Reference Coordinate System appeared to be the ideal solution for defining "up" in terms of the Earth and its orbital plane around the Sun. However, it fails on a practical level. From space, the view through the sextant shows little more than stars and perhaps a glimpse of the Earth. This snapshot view of the Earth is not sufficient to determine its rotation axis, much less its orbital path, both of which are necessary to define the Basic Reference Coordinate System. On a practical level, it is not possible to observe the Earth and Sun accurately enough determine the orientation of the Basic Reference Coordinate System. But all is not lost, because this coordinate system can be derived in advance by knowing its relationship with the stars. Luckily, the stars allow an ideal reference: they are fixed, they are usually visible, and their positions are known to great precision. The Apollo star catalog is a collection of 37 bright stars which are scattered more or less evenly across the celestial sphere. While brightness is an important attribute for selecting a navigation star, its location in the sky is equally important. For all navigation and alignment observations, either a pair of stars, or a star and either the Earth or Moon must be simultaneously visible in the sextant. The Command Module and Lunar Module optical systems have limited fields of view, having at most the ability to view a 60° wide section of the sky. The choice of navigation stars and their distribution in the sky make it possible, in most cases, to place two known objects within the field of view simultaneously.

Using the stars for orienting oneself in space is more than picking a point and assigning the term "up" to it. A proper orientation requires using two stars to define a three-dimensional set of coordinates. Using only the North Star as a reference is a reasonable beginning, but without a second point of reference there is no restriction on rotating around the vector to the star. Adding a second star, ideally 90 degrees away in any direction, is the basis for defining an orientation within the Basic Reference Coordinate System. As Figure 75 shows, two intersecting vectors, each defining a star's location, define a plane in space. A third vector, perpendicular to the other two, is created by taking the cross product of our two vectors. Three mutually perpendicular vectors quickly follow to create an orthogonal, right-handed system. This final, three-dimensional vector is the definition that fixes our reference in space.

Using a set of coordinates fixed with the stars, "up" can be defined precisely and

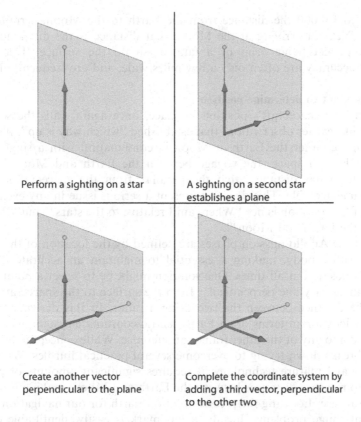

Perform a sighting on a star	A sighting on a second star establishes a plane
Create another vector perpendicular to the plane	Complete the coordinate system by adding a third vector, perpendicular to the other two

Figure 75: Two stars creating a fixed orientation in space

in any arbitrary direction. An important property of vectors is their ability to rotate to other orientations using a mathematical operator called a transformation matrix. Since the locations of the stars used in the navigation sightings are expressed in the Basic Reference Coordinate System, simple matrix operations relate one coordinate system to another. An example of one sequence of transformations begins with the Basic Reference Coordinate System and finishes with the spacecraft's attitude. Guidance, navigation and control tasks often require moving from one coordinate system to another, making transformations a common operation in the software.

QUESTION 2: WHERE AM I?

Now that the essentials of coordinate systems and spacecraft orientations have been established, the next pressing question in spaceflight can be addressed:

Where am I?

During the journey to the Moon, a navigation error of only 60 nm, which is equivalent to 1/4,000 the distance from the Earth to the Moon, certainly sounds impressive. Yet, on arriving at the Moon, that distance is the difference between entering a perfect lunar orbit or a fatal crash on the surface. Tolerances for navigation accuracy are often only a few miles wide, and are frequently less.

Using the sextant to determine position

Determining the spacecraft's position in space, once again, calls the sextant into service. Unlike earlier observations that established "which way is up", the sightings are made using either the Earth or Moon in combination with a single star. For navigating through space, the voyage between the Earth and Moon is so small compared to stellar distances that the parallax from the spacecraft's motion is essentially undetectable. This does not present a serious issue in any event, because the focus of the question is not "Where am I relative to the stars?" but "Where am I relative to the Earth and Moon?".

Many of the Apollo mission phases are defined by the location of the spacecraft relative to either body, making it essential to maintain an accurate idea of the spacecraft's position at all times. One solution might be to select a point on Earth, extend an imaginary line perpendicular from the surface to the spacecraft, and then measure the distance between the two using a radar. In this design, we have the spacecraft's location in terms of an Earth-centric coordinate system, which is easily transformed into any of the other frames of reference. While conceptually sound, the approach breaks down trying to overcome several practical hurdles. While radar is an accurate and mature technology, it requires significant amounts of power and hardware to measure distances between the Earth and Moon, making it impractical for spacecraft use. Locating the precise point on Earth for our navigation landmark also presents huge problems. Ideally, a landmark is easily identifiable even from large distances, is approximately on a line between the spacecraft and the center of the Earth, and is on the sunlit side of the Earth when the measurement is taken. Landmarks with all of these attributes are exceedingly rare in the best of conditions, given the amount of cloud cover and the vast expanses of featureless oceans. Given the challenges involved, this technique is not suitable as a general navigation methodology. A more robust method is required. In fact, while an Apollo mission is in cislunar space, the positions of the Earth and Moon relative to the star fields change in a very predictable manner.

Choosing an appropriate navigation star is the first step, as both the star and the Earth or Moon must be within the range of the sextant's Star Line Of Sight. Using the Earth as the reference body for the rest of this example, the astronaut uses the sextant to superimpose the star on the horizon of the Earth. Within the sextant, the two optical paths combine into a single image for the astronaut, who uses the joystick to drive the Star Line Of Sight optics. Pressing the MARK button, he instructs the AGC to record the values from the sextant's shaft and trunnion resolvers to calculate the angle between the Earth and star. Figure 76 shows the two-dimensional geometry of this sighting, where the two lines of sight intersect at the spacecraft's position.

Given these two lines of sight, and the fixed angle between them, only one point in

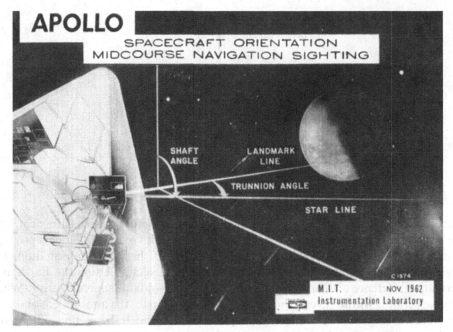

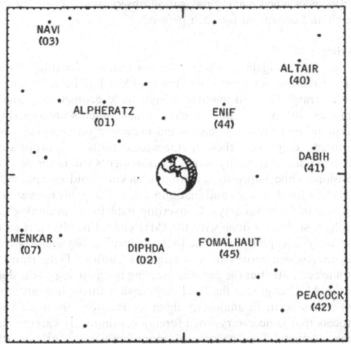

Navigation star locations for Apollo 11 at 6 hours g.e.t.

Figure 76: Cislunar navigation sighting

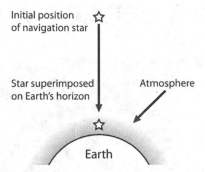

Figure 77: View of horizon and star through the sextant

space is possible. By combining the estimated position before the measurement, and the angular measurements from the sextant, it is possible to calculate a new position vector for the spacecraft relative to the Earth. With this latest information, the navigation process completes with updating the state vector. Using the Earth as a navigation reference also introduces a unique but unavoidable source of error. When the spacecraft is within several thousand miles of the Earth, the atmosphere obscures the horizon (Figure 77). Different people interpret the horizon line differently, leading to differences when making navigation observations. Fortunately, this error is usually small and consistent for each person.

Kalman filtering
In the discussion of navigation, whether during cislunar coasting, Earth or lunar orbit, or during rendezvous operations, the problem has been presented as being dealt with in a straightforward manner though basic geometry, trigonometry and matrix operations. In an ideal world, where measurements are superbly accurate, capable of astounding levels of precision and processed on computers with equally amazing resources, employing these mathematical tools is appropriate. Unfortunately, the hard engineering reality is that measurements will never be as accurate as desired; precision, while impressive, is always lacking; and computing speed and memory are constrained. Errors and inaccuracies, while small in every component, often accumulate in insidious ways. Converting data from an analog device to its digital equivalent, such as is done with the IMU gimbal resolvers, is an example of one source of error. The platform in the IMU is always a physical device, inherently analog in nature, which moves in a continuous motion. Data from the gimbal resolvers is quite accurate, but the act of converting from analog to the digital format required by the AGC degrades the level of precision through quantization errors. Quantization errors occur in analog to digital conversions from the truncation or rounding process that is necessary when forcing continuously varying data into the discrete chunks of digital data. An example more familiar to everyday experiences is the digitization of music for storage on CDs or music files. Converting the complex waveform of sound to a format that is readable by a computer requires sampling it several thousand times per second, and assigning a numeric value to that sample.

Analog sound waves can vary infinitely, yet digitization forces us to choose from a large but finite set of values. Devices such as gimbals are likewise affected. Incapable of generating a continuous range of values, the hardware must inevitably decide between two discrete values.

Computational rounding and truncation is more difficult to manage because of the finite resources of the processor. Unlike sound digitization, with a limited error range that does not propagate past the sample, rounding and truncation errors during extended computations increase after each reprocessing of the data. As we have seen, the AGC's fractional notation is a good but imperfect approximation of the numbers we wish to represent. In the case of irrational numbers, such as 1/3, the true value is truncated to 0.33333. While acceptably close for most situations, this may not be sufficiently precise for critical applications. Using additional words to represent the data helps significantly, but the fundamental issue of our data being an imperfect representation of the real world remains. Additionally, using multiple words for high precision datatypes does not come freely; they demand additional processing and memory resources everywhere they are used. All of these problems transform the much desired, but unattainable "perfect" data into good, but "noisy" data.

Ultimately, the need to navigate prevails despite the errors inherent in even the best hardware. Simply ignoring the errors, hoping they will eventually balance out might be acceptable for a single observation, but they cannot be ignored over time. Errors will accumulate, making calculations diverge from reality. For successful navigation, a method is necessary to determine the spacecraft's position and velocity, and then extrapolate those variables forward in time despite the errors or noise in the data. Our solution to this problem is through a technique known as a Kalman filter. Rather fortuitously, Kalman filters were invented only a few years prior to the development of the AGC. NASA eagerly embraced the technique for the AGC software. Now extensively used in all areas of control systems, Kalman filters provide excellent estimates of the state a system (here, the state vector) in the presence of noisy data. While the actual mathematics are far beyond the scope of this book, the fundamentals are easy to summarize:

> Kalman filters are designed with the assumption that the data is noisy. Knowing the exact amount of error in each measurement is impossible, but it is reasonable to assume that it follows a known statistical distribution. Input to the filter is the system state (i.e. the state vector) at a previous point in time, the noisy data, a statistical estimate of the extent of the error, and the navigation equations of motion. A matrix, called the covariance matrix, contains the estimated statistical error the system has accumulated over time, and is an integral part of the navigation calculations. While the exact error in each set of data is unknown, assuming that it follows a known distribution makes it possible to offset some of the error and compensate for the effects of noise in the system. The Kalman filter, operating recursively over time, uses the inputs to make a prediction of the state vector at a future point in time. In an ideal world, the textbook formulas for motion would be sufficient for extrapolating

to the new state, but this ignores that the input data is has systematic errors. Each time a new state vector is calculated, the equations defining the new state incorporate the values in the covariance matrix. At the same time, the covariance matrix is updated with the latest estimates of the error in the system, which is applied to the next iteration.

This overview of how a Kalman filter works reflects the textbook description of the technique. It should come as no surprise that the implementation in the AGC varies from the ideal in order to reduce processor demands. Most importantly, the AGC uses an alternative implementation of the covariance matrix, called the "error-transition matrix" or "W-matrix". A further modification recognizes that some of the system noise is somewhat predictable. With these important hints, the W-matrix is initialized to preset values for the error terms, rather than starting off at zero and allowing the filter to develop over time. By using this alternative method, the Kalman filter calculations are faster and more accurate.

QUESTION 3: WHICH WAY AM I GOING?

Modern aircraft navigation uses either a network of radio beacons located across the country or GPS satellite information, providing the pilot with the luxury of a continuous display of the vehicle's location. However, in the 1960s this technology was impractical to adapt for spaceflight, or in the case of GPS simply did not exist. Without the means for real-time position updates, the only option available to pilots is the technique of taking a navigation fix and calculating future positions based on time and the velocity of the aircraft. This is the time-honored method of Dead Reckoning navigation. Flying over a distinctive bend in a river, for example, a pilot notes his exact location on a map and the time. The next landmark, an intersection of two highways, is a known distance away and the course to that location is obtained from the map. Factoring in wind speed and direction, our aviator calculates the heading to the waypoint (the highways) and the estimated time he will arrive at that waypoint. If all goes well, and the plane maintains course and speed, it should arrive over the highway intersection exactly on time. During the leg between the two points, continually checking the map is unnecessary, although such checks are useful to verify that the estimates of airspeed and wind are accurate. Choosing the appropriate landmark is necessarily a careful process. First, landmarks are useful only if they are reasonably close together. Winds are never exactly as forecast, it is difficult to hold an aircraft exactly on a given heading, and airspeed always varies slightly. Each factor reduces the accuracy of navigation, which over time may result in venturing far from the desired route. Closely spaced landmarks are chosen so that the aircraft's location is confirmed before the unavoidable errors accumulate to unacceptable levels. Despite its failings and complications, dead reckoning is a perfectly reasonable navigation method. So much so, that it is used in space.

Perhaps surprisingly, onboard navigation with the AGC during coasting flight uses a form of dead reckoning that would be familiar to pilots and mariners alike.

Like aviators and sailors, astronauts must take navigation fixes at regular intervals during a flight. Except for the few minutes during the mission when the engines are accelerating the vehicle, the entire journey is unpowered coasting flight. Unlike the aircraft's course, which is frequently a straight line between waypoints, the path a spacecraft takes in space is invariably curved. Although the equations used to define a vehicle's motion in space are more complex than the time-speed-distance relationships used in aviation, both operate by extrapolating to a future point in time. And space has the tremendous advantage of being a vacuum. Without the atmosphere's disturbing forces, a spacecraft's trajectory very accurately tracks the path described by the equations of motion.

Orbits around a body are traditionally described as ellipses, which strictly speaking is correct. However, ellipses are only one type of motion that is possible in spaceflight. Using a higher level description, all trajectories follow paths described by conic sections, which include circles, ellipses, parabolas and hyperbolas (Figure 78). Routines in the AGC use generalized forms of the equations for conic sections, and the motions of bodies that follow those paths. This eliminates the need to distinguish between the various types of path found in conic sections.

It is not necessary to have a detailed discussion of the specific equations used for calculating the position and velocity of a body, but a general overview is useful. Three distinct methods are used for calculating this state vector, each offering its own operational advantages. First, a highly precise technique called Encke's method of differential accelerations is used for the initial integration of the state vector, as well as the updates during the coasting flight. A second technique called conic integration provides a good approximation of the state vector and has the advantage of generating a solution relatively quickly. Both techniques are used during coasting flight, where the vehicle is in free fall and no external accelerations are expected. Accelerated flight, whether experienced from the thrust of the launch vehicle or the deceleration during entry, demands rapid updates of the state vector. The routine called "Average G" is part of the Servicer, and runs only in programs that are expecting the spacecraft to experience accelerated flight.

In addition to its numerical accuracy, Encke's method includes terms that account for the oblateness of the Earth. The Earth is slightly pear-shaped, with its small end in the northern hemisphere. This is enough to induce gravitational accelerations that are measurable over time. Modeling the pear-shaped Earth is an established technique, but knowledge of the Moon's complex gravity distribution was so limited that it was not possible to develop a similar model for the AGC. The complex equations required in Encke's method demand a large amount of processing time to arrive at a solution. In an ideal world, the AGC would be continuously integrating new state vectors throughout the mission, but this would quickly become prohibitive, limiting the resources available to other important tasks. Fortunately, maintaining the state vector in real time is possible without requiring the continuous integration of a new solution.

Although state vectors are often presented as values used only for guidance and navigation purposes, other housekeeping tasks use knowledge of the vehicle's position. Consider the case of the spacecraft in lunar orbit on the "far side" of the

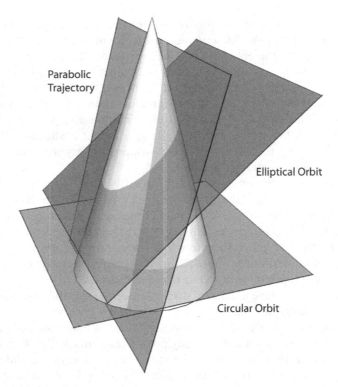

Figure 78: Conic sections defined by planes intersecting a cone

Moon. As the spacecraft travels around the Moon, the communications antenna needs to be pointed towards the Earth as this comes into view. The routine that points the antenna first needs to know the spacecraft's attitude and its position relative to the Earth. Fortunately, the conic integration technique for calculating the state vector arrives at a solution without requiring large amounts of processing time. In contrast to Encke's "precision integration", conic integration yields a result whose accuracy is more than sufficient for this purpose. The speed of its computations are due to its less complicated equations, as well as deleting terms from the calculations that account for the Earth's oblateness. Continuing with the antenna example, assume that the position calculated by conic integration differed from its Encke's counterpart by one kilometer. When in lunar orbit, this difference turns into a minute angular difference when calculating the antenna pointing angles. Extreme accuracy in antenna pointing is unnecessary, as the antenna's beam width is much wider than this small angular error.

Rendezvous navigation also employs conic integration. Fundamentally, rendezvous is an exercise of maneuvering to arrive at a specific point in space at a specified time. During each phase of the rendezvous process, conic routines take an initial position and velocity for each vehicle and then extrapolate forward in time. These calculations are used to create a solution for each major maneuver or midcourse correction. Onboard both the Command Module and Lunar Module, astronauts are taking frequent fixes on the other spacecraft during the rendezvous, ensuring that rendezvous solutions are always using the latest calculations. Given the large number of sightings and the long computation time required for Encke's method, conic integration is the only possible choice for calculating these solutions. This is not a serious problem, as extraordinary accuracy is not necessary during the early phases of rendezvous.[3] Each rendezvous maneuver will correct for the accumulated errors, which become smaller as the two vehicles become closer.

Cislunar navigation sightings and ground supplied updates provide the basic data necessary for state vector integration. Rather fortuitously, integration of the state vector generates much more than the spacecraft's position and velocity. Mathematical coefficients used to define the conic equations of motion are produced as an essential byproduct of the integration process. After each navigation sighting or ground-based state vector update, a very accurate solution for the state vector and the coefficients becomes available to describe future motion. Since continuously making navigation measurements or directly updating the state vector is impractical, another technique is necessary to determine the state vector in the future. Using the coefficients generated from the integration, the previous state vector is extrapolated forward to the current time, to create fresh values for position and velocity. When compared to the integration processing performed immediately after navigation fixes, extrapolation uses far fewer resources and is able to maintain its accuracy for hours at a time. Unfortunately, the extrapolation process gradually and invariably produces errors. Truncation and rounding take their toll, although the incorporation of the W-matrix does minimize this degradation. Eventually, the W-matrix becomes saturated and is no longer able to absorb any more corrections. At this point, the only option is to reinitialize the W-matrix by making a new navigation sighting.

Extrapolating the state vector is not a process that occurs according to a fixed timetable. Rather, the update is scheduled as a function of the distance from the primary gravitational body, either the Earth or Moon. When the spacecraft is in low Earth orbit the extrapolation is scheduled every 15 to 20 minutes, depending on the exact parameters of the orbit. For example, if the vehicle is in a 150 nm (280 km) Earth orbit, the state vector is updated every 14 minutes, 20 seconds. A much higher, 1,000 nm (1,850 km) orbit extends the time between extrapolations to almost an hour. As the distance increases further, so, too, does the time between updates, to a maximum of 4,000 seconds, or about 1 hour and 7 minutes. This schedule is not restricted to a spacecraft in low Earth orbit, as the extrapolation routine performs the same processing whether in a stable orbit or in cislunar space. Such a long time between updates might represent a concern, especially during critical maneuvers such as rendezvous where the knowledge of a spacecraft's position is required in nearly real time. This is not a significant issue, as the conic routines are sufficiently accurate

---

[3] Indeed, if the AGC were to fail completely, a rendezvous could be successfully executed using only pre-printed charts and tables.

for rendezvous and other tracking problems. Also, conic routines only calculate values for use within the mission programs, and do not alter the state vector for the vehicle. This assures that the less accurate conic solutions will not taint the carefully maintained state vector that continues to be updated only through the regularly scheduled program using Encke's method.

Accelerated flight and the Servicer
Coasting flight presents a straightforward case of measuring and extrapolating the state vector according to the equations of motion. Accelerated flight, whether during a thrusting maneuver or during atmospheric entry, imposes severe demands on maintaining an accurate state vector. During the lunar landing and entry, for example, real-time awareness of the spacecraft's position and velocity is not just a good idea, it is essential for mission safety and success. The third means of updating the state vector is a number of routines that fall under the collective name of the Servicer. This operates only under computer controlled thrusting maneuvers or, in the case of the CM, during entry. Fundamentally, its role is to read data from the IMU's accelerometers and update the spacecraft's state vector with the updated position and velocity. Rather than using complicated equations that describe the motions of the spacecraft through space, the Servicer updates the state vector using the current estimate of the state vector, new velocity data, and the local force of gravity.

Except during the launch into Earth orbit, the thrusting or entry programs start the Servicer 30 seconds before the expected start of the accelerated phase of flight, to enable it to initialize and begin collecting data. From this point until the end of the accelerated flight, a Servicer routine called Average G is responsible for updating the state vector. Data from the accelerometers is obviously an important part of the calculation, but so is the gravitational influence of whatever body the spacecraft is near. The value of G, the acceleration due to gravity of the Earth or Moon, is averaged over the 2-second sampling interval and integrated into the state vector calculations. Extrapolating the state vector using Encke's method is suspended during the time the Servicer routine is running, as state vector updates are done exclusively by the Servicer.

The three mutually perpendicular Pulsed Integrating Pendulous Accelerometers, or PIPAs, are mounted on the IMU's stable member. Integrating acceleration over time, the PIPAs compute the change in velocity for each axis of the vehicle. The data, sent as counter pulses to the AGC, appear in the PIPA counter registers in locations 00037_8 to 00041_8. When called by the thrusting or entry program, the Servicer sequence starts by scheduling the READACCS routine in the waitlist. READACCS has the essential task of reading and saving velocity information from the PIPAs. After reading the data, the PIPA counters are zeroed, preparing them for the next pass through the Servicer. Immediately afterwards, the AGC clock is saved to place a timestamp on the data. Next, the spacecraft's attitude is captured by reading the CDU counters for each of the three IMU axes. As a waitlist task, READACCS executes with interrupts inhibited and must complete quickly. With only a limited amount of time to complete its work, READACCS performs two final acts before exiting. First, it reschedules itself in the waitlist to run two seconds in the future, ensuring that data continues to be read and processed throughout the period of accelerated flight. The final task READACCS performs is to schedule its eponymous job, the Servicer, to perform the actual state vector updates.

In completing the discussion of READACCS, there may be the impression that the task it performs is not particularly complex. This point has some merit, but it ignores the fact that capturing data in a regularly scheduled task, for processing by a subsequent job, invites many different failure scenarios – and designing a robust recovery environment is difficult. Unlike other routines, especially those that can compute their data again without timing considerations, a number of timing windows exist where a restart during READACCS processing will result in lost or corrupted accelerometer data. A failure in reading PIPA data results in state vector errors since it does not include the latest velocity data, but this problem is correctable with ground updates. A far more significant error is during a thrusting maneuver or entry, where the loss of accelerometer data means that the most recent velocity change is not recognized. Without the latest accelerometer data included in the state vector, a longer engine burn or errors during entry result.

The hard work of calculating the new state vector during accelerated flight is left to the Servicer job. Running as Routine 11 in the CM software, and Routine 12 in the LM, the Servicer is scheduled every two seconds by the READACCS task. The Servicer begins with processing the PIPA values and the timestamp captured by the READACCS task. Using the velocity data from the accelerometers and the timestamp, it is a straightforward matter to calculate the change in position and velocity of the spacecraft. However, additional steps remain before the state vector update is complete. Our calculations thus far have been using accelerometer data directly from the IMU. With the data expressed using the IMU's coordinate system, we first must rotate the data into the state vector's coordinate system. Accelerometer data also does not account for the local gravity field. While in orbit, for example, the body beneath the spacecraft still is exerting a gravity vector that must be included in the state vector terms. This gravity term is a function of distance, and must be included in the final calculations. Finally, the Servicer updates the state vector and terminates until it is rescheduled by READACCS.

The Servicer in the Lunar Module

In the LM, the Servicer does more than simply update the state vector. It is an integral part of other key software routines: specifically the Digital Autopilot, powered flight programs, and landing radar management. Coordinating the operation of the landing radar is a significant part of the LM's Servicer routine. During the latter part of the descent, the accelerometers are not the sole source of data for position and velocity information. The landing radar, mounted on the bottom of the LM's descent stage, collects data on altitude and velocity in all three axes. Data from the radar is superior to the state vector computed using data from the PIPAs, which may include errors of hundreds of feet. In orbit, this error is of little consequence, but the final phase of landing demands measurements with an accuracy of less than one foot (0.3 meter).

The altitude at which the landing radar first acquires the surface is dependent on a number of factors, but by approximately 35,000 feet (10.7 km) a solid lock-on is possible. Both the crew and ground controllers evaluate the validity of the data, and if it is valid, the crew is allowed to accept it using extended Verb 57. By accepting the landing radar data, the Servicer incorporates the altitude and velocity data into the state vector calculations. Six seconds before pitchover, when Program 63's braking phase is about to transition to the visibility phase of Program 64, the Servicer stops accepting landing radar data. The landing radar, which is only able to point in one of two different directions, will lose lock on the surface during pitchover. To prevent unnecessary radar alarms and questionable state vector updates, the Servicer inhibits landing radar data until the LM has completed its maneuver to a more upright attitude. During this pitch maneuver, the Servicer commands the landing radar to its second position, creating a more favorable geometry for the radar to scan the surface. As with the state vector updates, where the accelerometer data must be transformed from the IMU's coordinate system to that of the state vector, the landing radar has its own unique orientation that forms a "coordinate system" of its own. The radar beams are not aligned with any previously defined coordinate system, but it can be defined in terms of the LM's body axes. Additionally, the LM's attitude during the landing results in the landing radar not pointed directly at the surface, and requires additional calculations to account for this oblique line of sight. Once the radar locks on to the surface again, reasonableness checks are done on both the altitude and velocity data, and if valid, radar data will again be used to update the state vector.

During the LM's descent to the Moon, data is flowing in from many sources and must be coordinated. Accelerometer, gimbal and radar data is being brought into the computer, each with its own routine to process it. Real-time data for analog displays must not only be current, it must also be synchronized with each other and with the DSKY display. Consider the problem for the pilot when he manually tries to slow the descent rate. The rate of descent on the DSKY and the panel tapemeters must display the same information; if some data was skewed in time, where one source displays the correct value and the other is delayed a second or two, the pilot is presented with a conflict. Routines controlling the analog instrumentation and the landing radar data must be properly synchronized with the READACCS routine. Part of the Servicer's responsibility is to schedule these jobs so that as data is flows in, it is processed and displayed in a consistent manner for the crew.

Accelerated flight outside the control of the Servicer
Not all accelerated flight is under the watchful eye of the Servicer. Translational maneuvers, such as the CSM separating from the S-IVB stage, turning around and docking with the LM, are performed without the Servicer running. As such, these maneuvers are performed without any of the position and velocity changes being recorded in the state vector. Other subtle sources of translational maneuvering can occur during the mission. Using a single RCS jet for rotational maneuvers introduces small but measurable translational components, as do waste water dumps. Residual oxygen in the tunnel between the CM and the LM will push the vehicles apart when

they undock. A quite real source arises with the exhaust from the LM sublimator, used to cool the vehicle's electronics. Although producing the faintest amount of thrust, it also operates for hours at a time and can build up a measurable error – as evident in Apollo 11's descent trajectory.[4] Without the Servicer running, these translational motions remain unrecorded. Ignorant of the disturbance being introduced, the state vector is extrapolated with gradually increasing error, until corrected by a navigation fix or ground-based update. During maneuvers where translational motions are expected, but not directly controlled by the AGC, the crew does have an option to capture the velocity changes. Program 47, Thrust Monitor, will both start the Servicer and display Noun 83, which allows the crew to monitor velocity changes in all three axes.

[4] An excellent discussion of this issue is found in Bill Tindall's whimsically titled memo, "Vent Bent, Descent Lament".

4

Mission programs and operations

INTRODUCTION

Up to this point, the discussion of the AGC has necessarily taken the approach of an anatomy lesson. Each organ is placed in full view, described and dissected, and its interactions with other body parts are described. Now that the major components of the AGC are introduced, it is possible to take the next step and consider how they perform real mission functions. Of particular interest is to extend the scope outside the narrowly defined world of the AGC, its platform and optics, and the DSKY. A proper explanation enlarges the view to a more holistic one where the computer and its related components do not operate in isolation, but as an integral part of the spacecraft. In general, the major mission events will be described in the sequence in which they occur over time, showing how each spacecraft system works to solve its particular mission task. It is not possible to focus exclusively on the operations and procedures of a specific mission. Much of the success of Apollo derived from the fact that its systems and infrastructure grew and matured over the life of the program. This evolution allowed mission planners to develop increasingly sophisticated missions, climaxing with Apollo 15, 16 and 17. These descriptions are an aggregation of all missions, with a bias towards the later flights. Specific event timelines, so critical to a given mission, will be replaced with approximate but realistic times. There are also cases where two missions might use entirely different techniques due to the unique requirements of each flight. Many of these procedural variants will be described, if only to highlight the capabilities that were available at that time.

LAUNCH FROM EARTH

A flight to the Moon begins weeks before launch, when the Saturn V rocket and the Apollo spacecraft rolls out of the Vehicle Assembly Building to the pad, where it is carefully lowered onto the launch platform's stanchions. All the systems are tested and retested to ensure that they are ready for launch. Hours before liftoff, the backup flight crew prepares the spacecraft, slowly bringing it to life by powering up its systems and positioning hundreds of switches, dials and circuit breakers. By the time the prime crew arrives at the pad, approximately 90 minutes before launch, much of the activation process has been completed. Initializing the AGC for launch begins with a number of basic checks.

First, a quick lamp test using Verb 35 confirms that all the lights and displays will illuminate. This is followed by two separate checks of the memory and hardware to insure that it is ready for flight. The self-check routine is started by entering V21N27, and placing the self-check option into Register 1. A number of self-check options are available, but many map to the same operation. Entering a positive or negative 4 into Register 1, for example, starts the ERASCHK erasable storage checks. Designed to ensure both a '0' and a '1' are readable and writable into every bit of erasable memory, ERASCHK reads, complements, stores and retrieves each erasable word. A value of 5 for the self-check option tells CNTLCHK to test the operation of the special and control registers. After placing a known value into each register and reading it back out, the result is compared against the expected output. For example, if 01234_8 is placed into location 00021_8, which shifts data to the right, CNTLCHK expects to see 00516_8 in the register. A second check is started using Verb 91, which begins executing the ROPECHK routine. ROPECHK verifies that each word in fixed memory is readable, and that its content is correct. The test sums together all the words in each fixed memory bank, plus a value unique to each bank known as the "bugger word". The bugger word is used as a checksum to validate the memory contents, and its value, plus the summed total of the banks words, should equal the fixed bank number that is under test. Summing words together to validate the memory contents is imperfect however, because multiple errors can cancel one another out, giving the incorrect impression that the memory contents are correct. Nevertheless, it does provide a high degree of confidence that the core ropes are ready to fly.

Throughout the time the AGC is operating, it regularly downloads essential data from erasable storage to ground controllers for analysis, providing an exceptionally detailed view of its operation. Controllers will check this data for soundness, as a verification that the hardware and software are operating as expected. At important milestones during the prelaunch preparations, the entire contents of erasable memory are dumped so that engineers can view the complete programming environment running in the AGC.

The backup crew initiates the process of preparing the IMU by powering up the platform and running Program 01, Prelaunch Initialization. When the IMU first powers up, it has no reference to orient itself to, and is not positioned properly to begin the alignment process. The counter registers in the AGC, used to record the angles and velocities calculated by the IMU, contain unusable data. Program 01 begins by orienting the stable member to its coarse alignment orientation, with the Z-axis pointing down along the gravity vector, and the X-axis along the launch azimuth. The PIPA counters are cleared and their output flagwords are set, and input bits detailing the IMU status are checked. When all the status bits report a properly initialized platform, Program 01 automatically transfers to Program 02, Prelaunch Gyrocompassing.

Using the optics to align the platform is not possible before launch, if for no other reason than the Command Module Boost Protective Cover and the optics dust cover prevent any use of the sextant. Without external references, it is not possible to align the platform using any of the techniques developed for use during the flight. Aligning the IMU now becomes an exercise in abandoning the late 20th century technology to turn to a technique rooted in a 19th century principle that will be familiar to science museum visitors, namely a Foucault Pendulum which slowly swings back and forth with its weight suspended at the end of a wire extending several floors above. The weight at the end of the pendulum is large enough that the inertia of the swinging bob overwhelms smaller forces such as air resistance. True to the laws of Isaac Newton, this property of inertia assures that the path of the pendulum remains fixed in space. With the arc of the pendulum fixed, and the Earth spinning on its axis, the arc of the pendulum appears to rotate over time. If the pendulum were placed at the north or south pole, an observer would see the pendulum's arc rotate once per day. Moving the pendulum away from the pole, the motion becomes more complex and the time to complete one revolution increases. Upon reaching the equator, the rotational motion halts, and the pendulum swings in a constant arc.

Although using a pendulum to align the IMU might create an interesting but impractical engineering problem, it is possible exploit two important concepts from Foucault's Pendulum: rigidity in space, and the rotation of the Earth. Once powered up during the prelaunch preparations, the gyros in the IMU ensure that the stable member holds a fixed orientation in space. The Earth, however, is not so cooperative, as it continuously rotates around its axis. At first glance, aligning the IMU appears to be an intractable problem; there is no option for aligning it to an external reference, and the Earth will not hold still for us to take a measurement. In actuality, all the necessary ingredients exist for aligning the platform. Remembering from the discussion on coordinate systems relating to the need to define "which way is up", only two vectors to known references are necessary to derive a three-dimensional reference in space. The first vector comes easily. Gravity provides the perfect force vector; operating in a single direction, it is reliable, immune to interference, and does not require special hardware other than an accelerometer to sense it. Identifying one vector solves half of the problem, but a second vector is still necessary. The only other influence that the IMU is experiencing at this time is the spacecraft slowly rotating around the stable member, exactly as a Foucault's Pendulum rotates in a museum gallery. At the Kennedy Space Center, located 28.5° north of the equator, the spacecraft rotation around the stable member is slow and complex but follows well-understood laws of physics. Because the motion is so well defined, it is used as a vector for the second attitude reference. Known as gyrocompassing, this process is initiated early in the launch preparations and runs as AGC Program 02. For the initial alignment of the stable member, P02 performs gyrocompassing for 640 seconds to accurately align the platform. A launch REFSMMAT is defined at the location of the launch pad, and is only valid for the expected time of liftoff. Operating continuously during the countdown, P02 ends only after detecting the liftoff discrete, at which point it automatically switches to Program 11.

A 72-degree launch azimuth is used for all missions that liftoff at the scheduled time. If the launch is held to a later time, the azimuth shifts to account for the Earth's rotation during the delay, but this does not alter the flight path over the surface of the Earth. Rather, the change in azimuth reflects the change in trajectory with respect to the Earth-centric coordinate system. Apollo 17, with its unscheduled 2 hour, 40 minute hold, launched on an azimuth of 91° 30', but flew the same ground track as an on-time launch. In the event of such a launch delay, Program 02 allows the crew to update the launch azimuth using Verb 78. A flashing V06N29 appears on the DSKY, prompting the crew to enter new azimuth. Once entered, the new data is not usable immediately, because P02 requires another 320 seconds to reorient the platform to its new azimuth. After this reorientation is complete, P02 continues the normal gyrocompassing process.

Gyrocompassing is not a passive operation that allows ground controllers a time to relax. While P02 is running, controllers are monitoring for any unexpected drift in the platform, as well as any spurious readings from the accelerometers. Despite the extraordinary precision used in its manufacture, no mechanical device is perfect or free of friction losses and some gyro drift is inevitable. Each IMU is extensively tested to determine its drifting characteristics, and terms to compensate for the expected drift are derived from this data. On average, the platform might drift about 1 meru or 0° 21' 36" per day.[1] The drift compensation values are loaded into the AGC, and are applied evenly over time. During the T4RUPT processing cycle, the need to compensate for accumulated drift is checked, and if an adjustment is necessary the gimbal counters are incremented or decremented one bit at a time.

The final minutes before launch
During the last several minutes of the countdown, the AGC's primary task is to continue gyrocompassing, waiting for the liftoff discrete which signals that the mission is underway. All the updates for the boost and insertion phases are loaded, and the mood inside the CM is quiet and businesslike. All the systems are powered up and ready, and only a few communications checks and a final alignment of the stabilization and control system remain. About three minutes before the Saturn booster ignites, the crew makes a final verification that the AGC is continuing its gyrocompassing in P02. Satisfied that the computer is ready, the Commander sets up extended Verb 75 on the DSKY, but most importantly, will not press Enter to begin processing the Verb. Used as a backup in the event that the AGC does not receive the liftoff discrete, Verb 75 will switch the Major Mode to Program 11, the Boost Monitor program. Ignition begins at T-8.9 seconds, when the five monstrous engines at the base of the S-IC stage receive their start commands. Only after the engines have built up sufficient thrust is it committed to launch, and the hold-down arms release the vehicle from the launch platform. Umbilical cables, used to provide power, propellant, telemetry and command signals, pull out from the stages and send

---

[1]   One meru, or Milli-Earth Rotational Unit is 1/1,000 of the Earth's rotational rate.

the all-important liftoff discrete to the AGC. Upon receiving the liftoff indication, Program 02 immediately switches the Major Mode to Program 11. This starts the Servicer and begins the state vector updates. The mission clock also starts counting up immediately after receiving its own liftoff discrete. Of all the instruments monitored by the crew during the first few seconds of flight, the Commander keeps a particularly keen eye on the DSKY to verify that P11 starts as expected. If for any reason the AGC fails to process the liftoff discrete or does not automatically switch to P11, the Commander presses the Enter key. Verb 75, which was loaded on the DSKY during the prelaunch activities, manually ends the gyrocompassing program and switches the AGC to P11. If all goes well, P11 begins as expected and clears the Verb 75 entry and replaces it with its own display.

For the first ten seconds of the mission, the Saturn V moves slowly, struggling to travel just the height of its launch tower. In those ten seconds, the journey covers only 450 feet (137 meters), yet already has burned 140 tons (127 metric tons) of propellant, a full 4 percent of the vehicle's liftoff weight! Despite its size, or perhaps because of it, those few seconds are when the booster is in danger of drifting into the launch tower from an ill-timed gust of wind. Almost imperceptibly, the vehicle yaws 1.25 degrees away from the tower, holding that angle for the moments it takes to clear the obstruction. By the time the Saturn has cleared the tower, it has returned to vertical. It then initiates a programmed roll maneuver to align itself with the launch azimuth. Once the roll completes, the stack slowly begins to pitch over, placing the crew in a heads-down attitude as it begins its flight downrange over the Atlantic.

With Program 11 running, the AGC is monitoring the trajectory of the launch vehicle. The Servicer is running the Average G routine, integrating the outputs of the IMU's accelerometers and applying the results to the state vector. When P11 started, it switched to the Verb 06, Noun 62 display, to provide velocity, altitude rate and altitude on the three DSKY registers. This display is continuously updated during the boost phase of the mission, and the crew compares its output against the data on a pre-computed cue card Velcro'd to the instrument console. The AGC is also driving the error and rate needles on the attitude indicators, or FDAIs, showing the relative difference between the expected trajectory and the one being flown. Combined with the cue card on the console, the crew has multiple sources of information to assure them that the flight is following the correct trajectory.

While it might seem natural that the AGC should control all of these events and steer the vehicle, it is normally just a passive observer of the entire event. The AGC is completely unaware of, and certainly does not control, the hundreds of events involved with igniting the engines, managing the tank pressures and the like. Sequencers and controllers onboard each stage control these functions, all under the direction of the Launch Vehicle Digital Computer, or LVDC. Using a completely different architecture and programming philosophy, the LVDC initiates the major events in the launch vehicle and performs all the guidance calculations necessary during the powered flight. The LVDC is located in the Instrument Unit at the top of the S-IVB stage and provides the same type of guidance and control service for the

SATURN BOOST			3/8/72 APR 16
DET θ	VI	Ḧ	H
00:00 90	1341	0	-.0
:30 85	1400	299	.6
1 68	1883	817	3.3
1:30 49	3044	1504	9.0
2 34	5087	2234	18.2
a 2:18 28	6783	2726	25.4
2:30 25	7902	2961	31.1
b 2:41 22	8976	3174	36.3
3 22	9164	2777	46.0
3:30 24	9702	2290	58.4
4 21	10341	1861	68.7
4:30 19	11079	1464	76.9
5 17	11914	1102	83.2
5:30 14	12852	778	87.8
6 12	13899	497	90.9
6:30 9	15067	263	92.8
7 6	16374	86	93.6
7:30 3	17842	-23	93.7
8 6	19262	-58	93.5
8:30 0	20618	-28	93.3
9 357	22003	21	93.2
c 9:18 355	22869	86	93.4
9:30 353	22998	36	93.5
10 350	23535	-40	93.5
10:30 347	24100	-66	93.3
11 344	24690	-69	92.9
11:30 342	25306	-36	92.6
d 11:44 342	25599	-1	92.6

Figure 79: Launch monitor cue card

Saturn booster that the AGC, IMU and optics perform for the Apollo spacecraft. In addition to the LVDC, the three foot (one meter) tall Instrument Unit also contains its own stable member to measure accelerations and attitudes, and a very specialized optics system used to align the IU's stable member to a known orientation. No human interface exists except for telemetry between the computer and ground controllers, and there is no direct connection between the LVDC and the AGC. Operating independently, the AGC performs the same guidance calculations as the LVDC, and the attitude errors displayed on the FDAIs are not dependent on LVDC input.

The AGC does not perform any guidance calculations on its own until after 11.85 seconds, when the yaw maneuver has completed and the vehicle is clear of the tower. Guidance from the moment of launch is "open loop", issuing only predetermined pitch and roll commands that are designed to direct the vehicle through the thickest part of the atmosphere while limiting stresses on the airframe. During this open loop, or "polynomial", guidance, there is no maneuvering to bring the vehicle to any specific course, because the LVDC is blindly steering the vehicle. A fixed set of equations issue steering commands for roll, pitch and yaw, with little consideration as to where the trajectory will end up. Although this sounds rather imprecise, in actuality the equations incorporate expected vehicle performance and data on the winds aloft in the upper atmosphere to produce a rather predictable trajectory. Open loop guidance continues until just a few seconds before the end of flight on the first stage. At this point, all maneuvering of the vehicle stops and the attitude is held fixed until the engines shut down and the stage is cast off. By freezing the stack in one fixed attitude, the two stages will separate cleanly without risk of recontacting. Continuing to maintain a constant attitude after the first stage separation, the S-II stage ignites and begins its six and one half minute burn. Even though staging is completed, the vehicle's attitude must remain fixed for the next 30 seconds. The 4.9 meter high interstage ring at the base of the S-II fires small solid rockets to move the upper stages clear of the discarded first stage, and to settle the propellants. With its job completed, the 11,000 pound (5,000 kg) interstage ring must also be cast off, but the clearances are very tight between the structure and the second stage engines. Continuing to maintain a steady attitude allows the ring to part company with room to spare. Shortly after the interstage separates, a small separation rocket fires on the Launch Escape Tower, pulling it and the Boost Protective Cover away from the Command Module.

As the Saturn V accelerates through the thick layers of the atmosphere during the first two minutes of flight, forces on the vehicle build up rapidly. The gargantuan size of the Saturn V and its dumbbell shaped mass distribution make it a rather flexible structure, causing the entire stack to bend and vibrate in flight. During this early phase, a major objective of polynomial guidance is to minimize aerodynamic and bending loads. About 80 seconds into the flight aerodynamic pressure on the vehicle reaches a peak of 700 pounds/square foot (33.5 KPa), and shortly thereafter the vehicle slips through Mach 1. After three minutes, it has left much of the sensible atmosphere and dynamic pressure reduces to almost zero. Once past the S-IC/S-II staging event, the Saturn V has lost almost 40 percent of its length and 80 percent of its mass, and the stresses from atmospheric buffeting and bending are greatly diminished. While the role of the first stage is to push the vehicle above the densest part of the atmosphere, the S-II stage is also a workhorse, providing much of the velocity and altitude necessary to achieve orbit.

The LVDC ends the Polynomial Guidance Mode at 3 minutes, 22 seconds into the flight, and begins the Iterative Guidance Mode. Given that the Polynomial Guidance flies the vehicle in a preprogrammed path, it is not surprising that it is somewhat off course by the time this terminates. The LVDC, having computed the corrections necessary to bring the vehicle back on course, steers the vehicle onto the proper trajectory with the goal of an accurate insertion into orbit. For the rest of the ascent, the Iterative Guidance software computes divergence from the desired path and gimbals the engines to maneuver the vehicle back onto the correct course. At the same time as the LVDC is switching its guidance mode, the AGC does so as well in order to track the progress of the LVDC. Both ground controllers and the crew monitor this process, watching for when the vehicle is solidly on its intended course. When it has eliminated the errors from the early Polynomial Guidance phase, guidance is said to have "converged".

With the tower and the interstage jettisoned, and the guidance converging, the crew settles into a more relaxed routine. Acceleration forces are low, never exceeding two g's, and the physical assault from the first stage is behind them. Fewer tasks demand attention by the crew, and everyone onboard is focused on monitoring systems, looking for any indication of trouble. As with the S-IC stage, the center engine of the S-II is shut down early, in this case about 1 minute, 30 seconds before staging. The S-II is a marvel of engineering efficiency, using the lightest construction possible and the highest performance propellants available in spaceflight. Of course, much of this effort is wasted if propellants remain in the tank at the end of the burn. Small differences in engine operation may result in burning more hydrogen or oxygen, and the mixture ratio is varied to even out propellant usage. The event, called Propellant Utilization Shift, or PU Shift, occurs about 60 seconds before staging. Originally running at a 5.5:1 ratio of oxidizer to fuel, the engines switch to a more fuel rich 4.8:1. With the PU Shift comes a reduction in thrust and g forces, but the effect is so small that it is not always noticeable by the crew.

About 8 minutes, 30 seconds into the flight, preparations begin for S-II staging. To prevent a premature shutdown of the engines, the process of arming the sensors that detect fuel depletion is delayed until shortly before the expected shutdown time. As the burn progresses, the fuel uncovers a sensor located at a level in the tank where less than 45 seconds worth of propellant remains. Electronics in the S-II stage use this event to activate, or arm, the fuel depletion sensors, and begin readying the vehicle for staging. Once again, the vehicle freezes its attitude to assure a clean separation of the S-II and the S-IVB. When the fuel uncovers two of the four sensors at the bottom of the tank, the staging sequence begins. At 9 minutes, 12 seconds the four outboard engines shut down and an explosive cord around the circumference of the stage fires to cut the empty second stage from the rest of the vehicle. The third stage does not ignite immediately, as it takes a few seconds for the stages to separate, and rockets must fire to settle the propellants in their tanks. This done, the third stage engine ignites to add the remaining 2,700 ft/sec (820 m/sec) necessary to achieve orbit.

Steadily accelerating for the 2 minute, 30 second burn, the ride is surprisingly gentle. G forces on the crew are less than 0.75 g, which is a stark contrast to the nearly 4 g's at the end of the first stage burn. There are no major tasks for the crew during these last few minutes of the ascent aside from systems monitoring and go/no-go status reports every minute. About 11 minutes, 45 seconds into the flight, the vehicle is traveling at about 25,600 ft/sec (7,800 m/sec) and is in parking orbit. At shutdown, the crew's first task is to record the final display on the DSKY. Values from Program 11, velocity, altitude rate and altitude are written into the flight plan and reported verbally to the ground. Routine 30, the Orbit Parameter Display, has been running during the ascent and has illuminated the KEY REL light indicating that it has data to display. Pressing the KEY REL button on the DSKY brings up the data as Noun 44. This shows the vehicle's apogee, perigee and time to complete one orbit. For the later Apollo missions, the orbit is quite low, only about 90 nm (167 km) high. Atmospheric drag at this altitude limits the orbital lifetime to a few days at most, but this is of little concern since the Translunar Injection burn is only about two hours away.

Each of the major systems goes through a comprehensive check to confirm it is functioning properly. Temperatures, pressures and voltages are verified, but so are the numerous small indicators called "talkbacks" that show the state of a specific system component. A crewmember might flip a switch to open a valve, but did the valve actually open? A talkback connected to the valve reports whether it is indeed open or closed, providing the crew with a direct indication of its status. Not only is the onboard crew busy with system checks, so too are the ground controllers in the back rooms in Houston. Despite the fact that hundreds of switches and displays cover the interior of the Command Module, only a fraction of the data available on each system is presented to the crew. The rest of this data provides a detailed engineering perspective on the components of each system in the spacecraft. During the brief passes over tracking stations, this data is radioed to the ground where specialists for each system review it in detail. Together with the onboard readings taken by the crew, controllers and engineers build a picture of the spacecraft's health to determine whether it is ready for a trip to the Moon.

Translunar injection
Preparations for Translunar Injection, or TLI, are started shortly after the S-IVB achieves parking orbit. None of the TLI parameters – ignition time, burn duration or vehicle attitude – are computed onboard by either the AGC or the LVDC. Ground-based tracking stations make precise measurements of the vehicle's position and velocity and feed that data to the Manned Spacecraft Center in Houston, where the computers in the Real-Time Computer Complex (RTCC) generate a solution to the complex equations necessary for a flight to the Moon. The parameters are uplinked directly to the LVDC, and read verbally to the crew in a "Pre Advisory Data", or PAD, exchange.

One of the first and most important preparations for TLI begins shortly after entering the first period of orbital darkness. Having been subjected to the accelerations and vibrations of the launch the inertial platform will have drifted somewhat, and must be brought back into alignment. Glare from the Sun and the bright surface of the Earth interferes with sextant observations, making the optics usable only when the vehicle is on the night side of the planet. As the spacecraft passes into darkness, the Command Module Pilot moves down to the lower equipment bay to begin the platform alignment. After powering up and zeroing the optics, he jettisons the protective cover over the optics window, allowing the sextant and telescope their first view of the heavens. Maneuvering the combined S-IVB and CSM/LM stack to the optimal attitude for an alignment is impractical, and limits the choice of stars visible through the sextant. Fortunately, this is not a serious issue as the navigation stars are more or less equally distributed across the sky.

In the "orbital rate" attitude mode, the stack rotates at a leisurely 4°/minute, keeping itself aligned with its path, or velocity vector, around the Earth. In this attitude, the stack's longitudinal axis is always parallel to the Earth's surface, much like an airplane. Frictional heating from passing through the tenuous atmosphere at an altitude of 90 to 100 nm (167 to 185 km) is significant enough to require measures to limit its effect. The solution is to keep the CSM pointed into the direction of flight to minimize the heating on the skin of the S-IVB, whose insulation is marginal at best. Even with this precaution, there are serious limits to the time the vehicle can remain in orbit before committing itself to the TLI burn. Liquid hydrogen, the fuel in the third stage, exists at a temperature of -252° C. As the hydrogen is heated, it begins to boil like a kettle of water on the stove, and must be vented to avoid overpressurizing the tank. Released through the engine, the hydrogen gas imparts only a slight acceleration, but enough to have the beneficial effect of nudging the propellants to the bottom of the tank. Unfortunately, venting is just another word for dumping valuable fuel overboard, and if allowed to continue there will not be enough fuel for the burn to put the spacecraft on a trajectory to the Moon. This loss of hydrogen is the primary constraint on how long the stack can remain in orbit before either performing or abandoning the TLI maneuver. Given the need for thoroughly checking the spacecraft systems before committing to a trip to the Moon, the burn cannot happen immediately after arriving in Earth orbit. The flight plan nominally places the TLI burn on the second orbit, about two and a half hours after launch, when the vehicle is over the Pacific Ocean. There is another chance approximately one orbit later. If this second opportunity is missed, the rate of hydrogen venting may rule out a third attempt and force the abandonment of the lunar mission.

With the platform aligned and the spacecraft systems configured and verified, the crew begins their final preparations for the TLI maneuver about 25 minutes before scheduled ignition. Loaded with the burn parameters calculated by the ground-based computers, the LVDC controls the entire maneuver without further assistance from the ground. Software in the LVDC that is used to control the third stage reignition is named Time Base 6,[2] and begins 9 minutes, 38 seconds before the engine ignites. During this time, several restart activities occur, from controlling pressurization to firing the maneuvering engines to settle the propellants in their tanks.

Manual backup for the LVDC
Despite the highly reliable design of the LVDC, the crew employs two systems to add additional layers of redundancy. During launch and the ascent to orbit, the Commander has the option of terminating a failing LVDC's control of the vehicle and using the AGC for guidance and control. The AGC is also able to take control during the TLI burn. Here again, the Digital Autopilot is configured to control the booster, but is not enabled unless the LAUNCH VEHICLE GUIDANCE switch is flipped to CMC from IU.[3] Left in the IU position, the AGC passively monitors the burn, displaying attitude information on the FDAIs, and velocity data on the DSKY. If the LVDC and other booster hardware are performing normally, there is no reason for the AGC to assume guidance and control responsibilities. Setting the

[2]   Time Bases are the major program phases in the LVDC, and are comparable to the Major Modes in the AGC.
[3]   CMC is the name of the Apollo Guidance Computer when it is installed in the Command Module, with LGC used when the computer is in the Lunar Module.

switch to CMC directs steering commands generated by the AGC to control electronics in the third stage, which in turn gimbals the engine to maneuver the entire stack. Manually steering the vehicle by using the hand controller is also possible, with the Commander using his FDAI as an attitude reference.

The next system configured in the Command Module for TLI is the Entry Monitor System, or EMS. While its name reflects its role in assisting the crew during atmospheric entry at the end of the mission, some of its capabilities are useful during any accelerated flight. Inside the EMS is an accelerometer that is sensitive to velocity changes along the X-axis. The accelerometer is connected to a numerical display on the EMS, and indicates the amount of velocity remaining in the maneuver. Set by the crew using data from the ground or other sources, the display continuously updates when sensing acceleration over time. A positive acceleration resulting from an engine firing or from atmospheric drag is noted on the display as a decreasing velocity value, and reaches zero when the spacecraft has achieved the expected velocity change. Although the EMS cannot control the startup or shutdown of the third stage, the velocity display provides a useful visual backup to the crew during TLI.

Program 15 and TLI
Program 15, TLI Initiation/Cutoff, monitors the Translunar Injection burn in the same manner that Program 11 watched over the launch into orbit. Using data calculated from ground-based computers, the crew starts Program 15 and enters the key parameters into the DSKY. When started, P15 flashes V06N33, requesting that the crew enter the time when Time Base 6 begins. Next, V06N14 flashes, asking the change in velocity required for the TLI maneuver. After entering these values and pressing PRO each time to continue, P15 displays Noun 95, which counts down the time to ignition, the velocity to be gained, and the velocity gained thus far. The AGC configuration is now complete, and Noun 95 will remain on the display throughout the maneuver.

After configuring the DAP, EMS and P15, there is little to do except monitor the systems until Time Base 6 begins. Lacking dedicated displays in the Command Module, the LVDC reuses indicators to signal that it has started Time Base 6. At 9 minutes, 38 seconds prior to ignition, both the UPLNK ACTY light on the DSKY and the S-II SEP light on the instrument panel illuminate. Time Base 6 is now underway, and the lights remain on for ten seconds. Over the next eight minutes the LVDC prepares the J-2 engine for ignition, but details on its progress are visible only to the controllers on the ground. The crew can only monitor tank pressures, which must remain within a narrow range of values. At 100 seconds before ignition, the DSKY blanks to indicate the beginning of the Average G routine, and 30 seconds later the S-II SEP light illuminates once again, this time notifying the crew that the ullage maneuver is starting. Ullage, originally a brewers' term for the gas in the top of a barrel above the liquid, describes the state the propellant tanks need to be in before the burn: all of the liquid propellants need to be at the bottom of the tank, with the pressurizing gas above it. After one minute, the S-II SEP light extinguishes. The ignition of the S-IVB's engine follows 18 seconds later.

Acceleration is light but noticeable, starting at a modest 0.5 g. Noun 95 on the

DSKY is now counting down the time remaining in the burn, with the velocity-to-be-gained register decreasing and the total velocity increasing. The crew is continually monitoring the FDAIs to ensure that the vehicle does not stray from its attitude limits, and the DSKY for assurance that the burn is performing according to plan. Although there is no capability to directly monitor engine performance, a cue card is Velcro'd to the panel that shows the expected velocities for every 30-second increment of the burn. Like the cue card used during launch, entries on the card are compared against the DSKY display and the FDAI. During the 5 minute, 45 second burn, the diminishing propellant enables the rate of acceleration to grow, increasing the load to about 1.5 g. When shutdown occurs, the third stage has accelerated the spacecraft by approximately 10,400 ft/sec (3,170 m/sec) to nearly 35,600 ft/sec (10,850 m/sec), enough to break free from the Earth and begin the voyage to the Moon.

Aborts: a bad day at work

Even before the moment of ignition on the launch pad, crew safety has the absolute highest priority. A launch escape tower containing a powerful solid-fuel rocket is attached to the apex of the Command Module. In the event of an emergency, such as a fire or loss of control of the booster, the launch escape system will violently pull the spacecraft and its occupants to safety.

Many abort situations demand immediate action, and in some cases they are completely automatic. Especially during the early phases of the ascent, there is very

NOMINAL SIVB TLI 2

LAUNCH APR 16   2/10/72

DET	θ	ψ	VI	$\dot{H}$	H
0:00	112	359.6	25601	17	97
:30	107	1.9	26218	13	97
1	105	1.4	26916	51	97
1:30	104	0.7	27649	154	98
2	103	0.0	28417	330	99
2:30	102	359.2	29224	585	101
3	101	358.5	30074	929	105
3:30	98	357.7	30971	1369	111
4	98	357.0	31922	1912	119
4:30	97	356.3	32936	2566	130
5	94	355.7	34021	3331	144
5:30	91	355.1	35193	4185	163
5:39	91	355.2	35579	4480	169

Figure 80: TLI cue card

little time for troubleshooting and options are often limited to firing the escape rocket. Of all the problems the Commander must worry about during ascent, control of the launch vehicle is certainly most important. The lack of time to react during an emergency makes the task of alerting the crew to a problem especially difficult. Only the most basic indictors are used to alert the crew to failures in the Saturn guidance, consisting of two lights on the instrument console. The LV GUID light illuminates if there is a failure in the Instrument Unit's inertial platform, leaving the LVDC without attitude and acceleration data. This is a serious condition, as the LVDC can no longer steer the vehicle. Even more urgent is where the vehicle begins rotating rapidly, stressing the airframe and threatening its breakup. During the burn of the first stage of the Saturn V, the entire structure is surprisingly flexible and sensitive to bending loads. Pitch and yaw rates of only 4°/sec, or a roll rate of 20°/sec, are beyond what the vehicle can safely withstand and will illuminate the LV RATE light and trigger an automatic abort. After S-IC staging, the pitch and yaw limits are relaxed to 9°/sec, and the crew has the option of taking corrective action or initiating an abort.

During the early minutes of flight, three consecutive abort regions, or "abort modes", are defined, each of which is tailored to a given range of altitudes. The first abort mode, 1A or 1-Alpha, is in effect while the vehicle is still on the pad through to T+42 seconds. 1-Alpha aborts begin with the nearly simultaneous severing of the Command Module from the Saturn stack followed by the firing of the escape tower motor. A smaller rocket motor near the top of the tower fires perpendicularly to the flight path, pitching the CM and its attached tower over and away from the failing booster. Now safely away, the CM must place its blunt heat shield towards the direction of flight to position itself properly for deploying its parachutes. The RCS jets on the CM do not have sufficient authority in the thick atmosphere, so flap-like canards deploy from the top of the escape rocket. Canards are small aerosurfaces that flip the spacecraft around without the need for thrusters. Finally, the tower is jettisoned, its lifesaving work done, and the parachute deployment sequence begins. From this point, the abort sequence takes on all the same events as a normal entry, and the CM settles into the ocean.

As the Saturn V rises further and faster, the normal sequences used for a Mode 1-Alpha abort are no longer appropriate. By T+42 seconds, the vehicle is over the ocean, well into its pitchover maneuver and passing through 30,000 feet (10 km), and entering the next abort region, Mode 1B, or 1-Bravo, which continues until the vehicle reaches 100,000 feet (30.5 km). An abort during this period pulls the CM free in the same manner as a Mode 1-Alpha abort, but it is too high for the parachutes to deploy immediately. Using both the escape tower canards and the RCS system to orient itself with the heat shield forward, the CM follows a ballistic trajectory until falling to 24,000 feet (7.3 km), jettisoning the tower along the way. At this point, two small drogue parachutes deploy, stabilizing the spacecraft for the first part of the descent. The drogue parachutes are jettisoned on reaching 10,000 feet (3 km), and the three main parachutes carry the CM to the ocean.

After passing 100,000 feet, the vehicle enters the abort regime named Mode 1C, or 1-Charlie. Here, the atmosphere is too thin for the canards to be effective, and the RCS jets alone maneuver the CM to the correct entry attitude. Maneuvering quickly to the entry attitude while still high above the Earth is especially prudent. If the combined CM and tower were to enter the denser air facing forward, the entire assembly would be whipped around when the canards became effective, certainly injuring the crew. Upon reaching 24,000 feet, parachute deployment begins with the two drogues.

Two minutes after launch, the Commander disables the automatic abort circuitry, transferring the job of initiating an abort to the crew. Given the level of violence that accompanies S-IC/S-II staging, there is the possibility that the emergency detection system might misinterpret the event and abort unnecessarily. Abort Mode 1-Charlie ends after the spent S-IC separates. With the most dynamic part of the launch completed, there is less of a chance of aerodynamic stresses causing the vehicle to break up. The launch escape tower, no longer needed to pull the Command Module away from an exploding booster, is jettisoned to free the vehicle of its weight. Without the launch escape tower, Abort Mode II will use the Service Module's Service Propulsion System to separate the CSM from the rest of the stack. Once clear, the CM jettisons the SM and rolls heads-up to a fully lifting entry attitude. An abort during Mode II results in a steep entry, which generates very high g loadings on the crew. A lifting entry extends the flight path and lowers the forces to a more tolerable level. Mode II continues until a safe orbit is achievable using either the S-IVB or the SPS engine.

Three different abort options are available after Mode II, and the times when they apply effectively overlap each other at different points. Both Abort Mode III, a Contingency Orbit Insertion option using the third stage, and Abort Mode IV, an SPS burn to orbit, are available to place the CSM in a stable orbit to perform an alternate mission. The abort option chosen is determined by the nature of the failure forcing the abort and the velocity of the vehicle. A failure late in the second stage burn might result in an early staging and using the S-IVB to achieve orbit. However, this extended burn may leave the third stage with insufficient fuel to perform the Translunar Injection burn.

All of these abort modes either terminate the mission or result in an orbit where few of the original mission objectives are possible. However, a failure of the Saturn Instrument Unit, with its independent guidance platform and computer, does not always mean the mission is lost. At liftoff, the AGC begins running Program 11 to monitor the progress of the ascent. If the Commander sees the LV GUID light illuminate, or on the advice of ground controllers, he can switch the source of the booster's guidance commands to the AGC from the Instrument Unit. Program 11 now is no longer passively monitoring the ascent, but is actively steering the entire stack into orbit. Staging, engine ignition and other booster functions continue as dedicated controllers within each stage manage these operations, and not the LVDC itself. Unfortunately, a worst case situation exists, where the AGC is providing accurate guidance but cannot positively control the vehicle. Perhaps an engine has shut down unexpectedly or a gimbal is stuck, making the normal steering commands ineffective. Incredible as it may sound, the crew can enter Verb 46 to activate the Digital Autopilot, using the previously loaded parameters for a manual takeover of the Saturn. Rather than sending commands to fire the spacecraft's RCS jets, the hand controller now uses the AGC to send commands to the launch vehicle's engine gimbal controllers. Using only the attitude indicator and a small chart on the panel as a reference, the Commander uses his Rotational Hand Controller and manually guides the stack into orbit.

THE LUNAR LANDING

If we can land on the Moon, how come we can't .... Countless times since 1969, people have looked at some unsolved problem and wondered aloud that if we could conquer the incredible task of landing a man on the Moon, then why can't we find a cure for cancer, world peace, or where the car keys are.

This then begs the question: How difficult was it to land on the Moon? What were the procedures used for the landing? What piloting skills were necessary to descend towards the surface, find a predetermined landing spot, and accurately put the LM down? And what happens if despite the best efforts, it just isn't meant to be and an abort is necessary? What follows is a composite description of the procedures the crews of the LM used to land on the Moon. Each mission was different, as mission profiles and experience evolved, so many of the values are approximations. This section presents a detailed look at the procedures the crews used to land. First, an overview of the descent is useful.

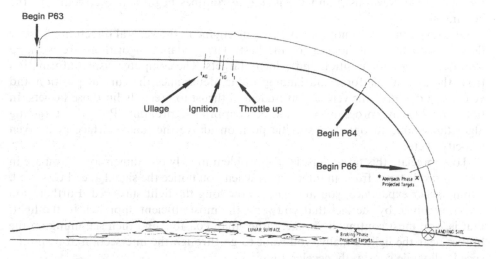

Figure 81: Descent profile with braking, approach and landing phases

Lunar landing overview
The landing process is broken down into three phases: braking, approach and landing. These are defined by a compromise between fuel efficiency and the crew's ability to view the landing site. Each phase is controlled by a separate computer

program. The LM is in a 9 × 60 nautical mile descent orbit, having been lowered from the CSM's circular 60 mile orbit by its descent engine (Apollo 11 and 12) or the CSM's Service Propulsion System (Apollo 14 through 17). Powered Descent Initiation (PDI) begins about 240 miles uprange of the landing site, when the LM reaches the 50,000 foot (15 km) perilune of its orbit. During the braking phase, the objective is to reduce the 5,500 ft/sec (1,670 m/sec) orbital speed to about 700 ft/sec (210 m/sec) at an altitude of 8,000 feet (2.4 km). At this point, the approach phase begins, and the LM pitches to give the crew their first clear look at the landing site. Once an acceptable spot has been found, the Commander manually guides the vehicle down through the final few hundred feet to the surface.

Final preparations for descent are begun in the last orbit before landing. The guidance platform has been aligned, Houston has sent up the latest updates of their position and velocity, and all systems are configured for landing. About 15 minutes before ignition, the crew instructs the computer to run Program 63 – Powered Descent, Braking Phase. At 5 seconds before ignition, the registers in the DSKY blank and the display flashes V99N62 to ask the crew if it has permission to continue for engine ignition. The crew must press the PRO key within those five seconds, instructing the AGC to issue the command at zero seconds to ignite the engine. At ignition, the engine is running at only 10 percent thrust (1,500 pounds force, 6,680 Newtons). While the crew checks the engine instruments, the computer is using the gimbal motors to align the engine's thrust vector through the LM's center of mass. 26 seconds after ignition, the engine is brought up to full throttle of 9,870 pounds of thrust (43,900 Newtons),[4] and the guidance routines begin active targeting to the landing site.

Of course, the LM is not merely firing its engine in the general direction of "slow down", with the crew hoping for the best. The guidance algorithms are working towards two specific position and velocity targets. The computer, using information from the inertial platform and landing radar, determines the current position and velocity and adjusts the vehicle's attitude and thrust to precisely hit those targets. In both the braking program P63 and the approach program P64, the targeting algorithms hope to obtain a specific position above the lunar surface at a given velocity vector.

To appreciate this task, consider the problem in only two dimensions. You are in a car, a far distance from an intersection when you notice the signal ahead turns red. From earlier experience, you also know how long the light stays red. Further, you are constrained by the fact that you want the most efficient approach to the light; that is, you must continually brake, rather than coasting and then slamming on the brakes near the light. Finally, you want to go through the intersection at a specific speed, all while constantly decelerating.

So that's the task: pass through the intersection at a precise speed exactly when the light turns green (too early, and you'll get a ticket when you pass through a red

---

[4]   Full thrust on the LM descent engine is at 94 percent of its rated 10,500 pounds force. Thrust is limited to this level because of excessive combustion chamber and nozzle erosion.

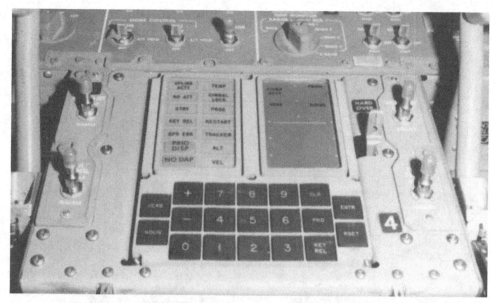

Figure 82: Lunar Module DSKY

light – too late and it's inefficient). The problem becomes an exercise in knowing the exact amount of braking, at a constant rate if possible, that is just enough that you cross the intersection at the speed you want, entering the intersection at the moment the light turns green. For pilots, extending the problem into three dimensions is a familiar test of timing and coordination. A circling approach to land or a Cuban-8 maneuver both have the goal of arriving at a specific position, altitude, airspeed at a given time. The constant change of position, altitude and velocity places demands on pilots that are often difficult to achieve precisely, and require a lots of practice to master. Luckily for the LM crews, the computer handles these tasks exceptionally well. Variables such as position, velocity and acceleration are obvious components of this equation, but equally as important is the time when the endpoint goal must be achieved.

For the next several minutes, the descent engine is maintained at full thrust, to gradually slow the LM from orbital velocity. For the crew, a large part of the time is spent monitoring the progress of the descent. As they pass about 40,000 feet, approximately three minutes from the time of ignition, the landing radar locks onto the surface, providing accurate altitude and altitude-rate data to the computer. In all missions except Apollo 11, tracking from the ground provided data to refine the targeting to the landing site. Using doppler radar data from Earth, combined with navigation data from the LM's guidance system, ground controllers were able to identify targeting errors and calculate updates for the computer to insure an accurate landing. These updates were relayed verbally, and the crew entered the data manually into the computer. In what may seem to be reverse logic, the computer is not told that it is off course, but rather that the landing site has moved! While the targeting could be altered in all three axes (up- or down-range, or north or south of

Figure 83: Engine gauges and tape instruments

track, and accommodate differences in altitude), it was only used for up- and down-range corrections.

Around 6 minutes and 20 seconds into the descent,[5] the engine is throttled down to approximately 57 percent thrust (5,900 pounds of thrust, 26,300 Newtons). This sets the stage for the next phase of the landing: the approach phase. The engine has throttled down as needed to satisfy the guidance parameters, all the while taking into account the decreasing mass of the LM. The crew is monitoring the DSKY, which is counting down the time to starting the approach phase.

Program 64 is automatically called from Program 63 at about 8,000 feet and 700 feet/second velocity, and maneuvers the LM from a nearly face-up, feet-forward attitude to approximately 30 degrees from vertical. Here, the crew gets the first clear view of the landing area. If the braking phase is characterized by monitoring, the approach phase is where they get busy with finding a suitable landing site. A key function of Program 64 is to display altitude and velocity, and the angle known as the Landing Point Designator (LPD). One of the tasks of the Lunar Module Pilot is to relay the LPD to the Commander, who uses a scale on his window to see where the LM is headed.

As the LM continues through the approach phase, the Commander focuses his attention outside the window, and spends little of his time reviewing the displays in the cockpit. The LMP is reading off the altitude, descent rate and the LPD, acting as a human "heads-up" display in order to allow the Commander to concentrate on the landing.

One instrument is important to the LMP at this time. The cross-pointer (or "X-Pointer") displays the horizontal velocity of the LM (forward/backward and left/right). This is essential information, especially during the last moments of the descent when the surface is obscured by dust raised by the engine plume. The horizontal line is forward and backward velocity, and the vertical line is for lateral motions. A scale is on the left and bottom to give the actual velocity components. When all horizontal velocities are nulled, the cross-pointer is centered. By the time the spacecraft has descended to within a few hundred feet above the surface, the Commander takes manual control of the LM. Using the rotational hand controller in his right hand, he maneuvers the LM in the horizontal plane. A toggle switch by his left hand adjusts the rate of descent by 1 ft/sec (0.3 meter) increments. At this point, his attention is focused exclusively outside, and the LMP is monitoring the instruments and providing verbal readouts of the altitude, vertical rate and angle of descent. In some missions, the amount of dust kicked up by the engine completely obscured the surface, making a visual descent impossible. When this happened, the Commander had to "come inside" and rely on his instruments and the LMP's altitude and velocity callouts. Eventually, the tip of one of the five-foot-long probes extending from three of the footpads touches the surface, and a blue light illuminates in the

---

[5]   Throttle down occurred at approximately 6:20 into the powered descent on Apollo 11, 12 and 14. Due to their larger mass, the final three missions, Apollo 15 through 17 required another minute to slow the LM before the engine could be throttled back.

Figure 84: Landing point designator in commander's window

cockpit to indicate lunar contact. The crew quickly shuts down the engine, safes the descent systems, and prepares for a possible emergency return. Once the review of the systems confirms that all is well, the LM is powered down, and preparations for the lunar stay are started.

This brief overview should not give the impression that the crew is sitting back, punching a few buttons and casually waiting for their moment in history. The next section follows the checklists and details in a step-by-step manner what is occurring onboard the LM and behind the scenes in Houston, using Apollo 14 as a case study. The undocking sequence is scheduled to begin at about 104:10 MET (104 hours, 10 minutes Mission Elapsed Time), less than five hours before landing. A note about the times of the various events: in some parts of the flight, the moment of performing a specific task was somewhat flexible. Crews were usually anxious to "get ahead" in

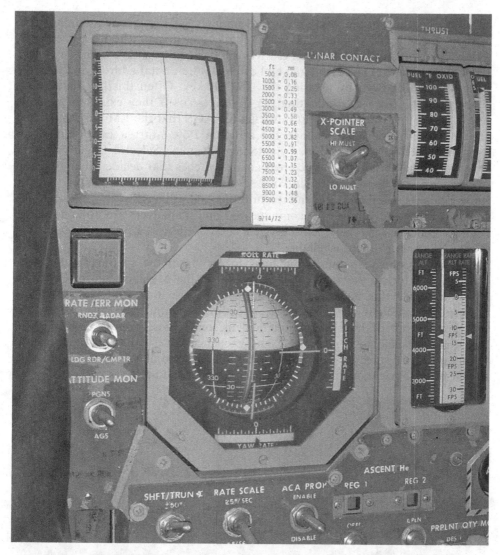

Figure 85: Commander's cross-pointer and attitude indicator

the timeline where they could, if only to leave themselves with a cushion in case there were problems. Other items, however, especially engine burns, must be performed exactly on time.

By their nature, checklists are necessarily short and somewhat cryptic, and the key to following the cockpit procedures is to understand this shorthand. In general, there are three categories of checklist items: switch positioning, DSKY entries and AGS entries. Operating the Abort Guidance System computer (the AGS) is quite different than the DSKY, and reflects its frightfully limited memory (only 4K words!) and processing power. Clever mnemonics such as Verb and

Noun do not exist in the AGS; rather, all interactions are through manipulating memory directly. To display data, the 3-digit octal memory address is entered, followed by pressing a button marked READOUT, and the data comes up in the single data register. Data is entered by keying a 3-digit octal address and then the 5-digit value, followed by pressing the ENTER key. Often this key sequence is abbreviated to ### + x, eliminating the trailing zeros if they exist. An example would be 413 + 1, the command to instruct the AGS that the LM has just landed on the Moon. Finally, to change the program flow, start a new function or perform a specific action is done by entering values in "Selector Logic" memory locations. Data in Selector Logic areas act as switches that decide which function to perform, such as which rendezvous solution needs to be calculated. All the while guidance commands are entered into the AGS, this is shadowing the AGC, not controlling the vehicle. All computations and maneuvers are done using the Primary Guidance and Navigation System (PGNS), unless a problem means that the PGNS cannot function reliably.

When discussing communication links to and from the spacecraft, the usual convention is to say the crew communicates with the Mission Control Center in Houston. While this is accurate, checklists often use the name "MSFN", the Manned Space Flight Network, to refer to this communication link. Through a complex network of tracking stations and in some cases, aircraft, controllers in Houston were able to communicate and exchange data with Apollo spacecraft. Communications satellites were still in their infancy, and all of the voice and data flowed through telephone lines using terrestrial links and undersea cables.

Figure 86: AGS data entry and display assembly

The cockpit is of the LM Mission Simulator on display at the Cradle of Aviation Museum in Garden City, NY. Far from the "fresh-from-the-factory" cockpit seen in movies, the condition of the LM Simulator reflects the wear and tear of years of nonstop training.

You will notice that the paint and lettering is worn, small bits of corrosion are visible, and a switch or two is non-standard. That's OK. Its legacy is that every LM crew climbed into this simulator and sweated, cursed and mastered one of the greatest machines ever built.

A bit of missing paint is a small price to pay for history.

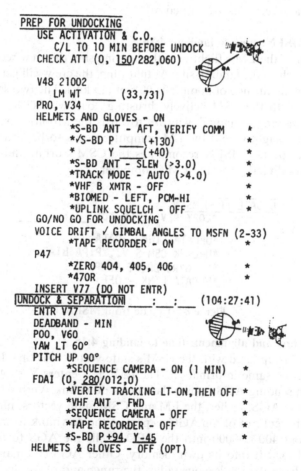

```
PREP FOR UNDOCKING
    USE ACTIVATION & C.O.
        C/L TO 10 MIN BEFORE UNDOCK
    CHECK ATT (0, 150/282,060)

    V48 21002
        LM WT ____  (33,731)
    PRO, V34
    HELMETS AND GLOVES - ON
            *S-BD ANT - AFT, VERIFY COMM    *
            *√S-BD P ____ (+130)            *
            *      Y ____ (+40)             *
            *S-BD ANT - SLEW (>3.0)         *
            *TRACK MODE - AUTO (>4.0)       *
            *VHF B XMTR - OFF               *
            *BIOMED - LEFT, PCM-HI          *
            *UPLINK SQUELCH - OFF           *
    GO/NO GO FOR UNDOCKING
    VOICE DRIFT √ GIMBAL ANGLES TO MSFN (2-33)
            *TAPE RECORDER - ON             *
    P47
            *ZERO 404, 405, 406             *
            *470R                           *
    INSERT V77 (DO NOT ENTR)
UNDOCK & SEPARATION ___:___:___ (104:27:41)
    ENTR V77
    DEADBAND - MIN
    P00, V60
    YAW LT 60°
    PITCH UP 90°
            *SEQUENCE CAMERA - ON (1 MIN)   *
    FDAI (0, 280/012,0)
            *VERIFY TRACKING LT-ON,THEN OFF *
            *VHF ANT - FWD                  *
            *SEQUENCE CAMERA - OFF          *
            *TAPE RECORDER - OFF            *
            *S-BD P +94, Y -45              *
    HELMETS & GLOVES - OFF (OPT)
```

Figure 87: Prep for undocking

Prep for undocking: time to landing: 4:45
All of the LM systems have been prepped by this point. The final communications configurations have been established and the signal strengths verified. Just prior to undocking, Program 47, Thrust Monitor, is started to monitor the LM's velocity changes during the undocking and separations sequence, and the AGS velocity monitors are zeroed. During the undocking, and for the first moments of the separation maneuver, the LM is essentially in "free drift", with no active attitude control enabled. After separating from the CSM and the possibility of any recontact is past, V77, Enable Rate Command and Attitude Hold, is entered and the computer assumes the task of actively maintaining the spacecraft in a fixed attitude. Once the two spacecraft are at a safe distance, the LM maneuvers to "face" the CSM and the rendezvous tracking light is tested. Radios are reconfigured for flight, and the cameras and tape recorder are turned off.

Update from MSFN: time to landing: 4:20
In anticipation of the LM's flight over the landing site, the crew receives the Time of Closest Approach to the landing site. At that time, the crew will get an excellent look at the site from an altitude of about 50,000 feet (15 km). The two spacecraft are now moving apart, with the CSM actively thrusting "backwards" to ensure separation (initially, it is moving radially downward towards the Moon). This maneuver alters the CSM's orbit slightly, and the LM computer needs to be aware of this change. Radar tracking by the MSFN recomputes the CSM's orbit, and flight controllers uplink this data directly into the LM's computer.

```
UPDATE FROM MSFN
            *COPY REV 12 LS TCA ___:___:___

      *UPDATA LINK - DATA
      *UPLINK CSM S.V., PIPA BIAS,
      *    GYRO DRIFT COMP
      *UPDATA LINK - OFF
```

Figure 88: Update from MSFN 1

AGS initialization and alignment: time to landing 4:15
After the AGC is updated with the CSM's state vector, it is time for the AGS to be provided with the same information. The LM crew enters Verb 47, which tells the AGC to begin sending AGS update data in its downlink. Verb 47 does not interact directly with the AGS; rather, the LM and CSM state vectors, plus the LM's IMU orientation, are sent out of the AGC as part of the downlink to the Earth. Entering 414 + 10000 and 400 + 30000 into the DEDA tells the AGS to monitor the AGC downlink and load it into its own memory. Other AGS commands are entered to display input from the rendezvous radar, the range and range rate to the CSM, and also direct the LM to maintain its front ($+Z$ axis) facing towards the CSM.

```
AGS INITIALIZE AND ALIGN
      *V47, 414+1, 400+3
  V83, SET ORDEAL ON LMP FDAI
      *317R,440R, 277R
      *400+2, 507+0
      *CAMERA SETTINGS
      *LM3/DAC/10/CEX-ULC (f2.8,250,∞)
      *  1 FPS, .05 MAG, (5 MIN)
      *LM_/DC/60/HCEX-(f2.8,500,∞)5
```

Figure 89: AGS initialization and alignment

Figure 90: Data Entry and Display Assembly

Two cameras are configured for photographing the CSM. The Maurer 16mm sequence camera, positioned to look through the LMP's window, is set to run at 1 frame per second for 5 minutes. The other camera is a 70mm Hasselblad, set at f/2.8, 1/500th second and focus at infinity, using Ektachrome transparency film.

DPS throttle check: time to landing: 4:05
Checking the descent engine throttle does not check the engine response per se, as the engine is not burning during the test. What is actually evaluated is the "command override" function of the Thrust/Translation Hand Controller. A TTHC is at the left hand of each LM crewmember, and performs double duty: first as a translational controller (complementing the rotational controller at the right hand of each

crewmember), and as an override to computer commanded engine throttle settings. When configured as the thrust controller, three important ranges need to be checked: the minimum throttle setting (10 percent of the rated 10,500 pounds of thrust), the "soft stop", which defines the lower end of the allowed throttle range (about 57 percent), and full thrust at 94 percent. To confirm that both controllers are operating correctly, the crew checks the physical Thrust Controller settings against the commanded thrust meter on Panel 1. Although the actual engine thrust remains at zero, the commanded thrust meter should reflect the position of the TTHC.

```
DPS THROTTLE CHECK
    THROT CONT - MAN/CDR
    TTCA (BOTH) - THROTTLE (MIN)
        (SET FRICTION)
            *VERIFY MSFN CONTACT
    ENG STOP - PUSH
    ENG ARM - DES (DES REG LT - ON)
    TTCA MIN (6.6% - 13.4%)
        THEN SOFT STOP (46.2% - 59.2%)
        THEN MAX (93.6% - 100+%)
        THEN MIN
    ADJUST FRICTION
    MAN THROT- LMP
            *REPEAT TEST FOR LMP TTCA
    ENG ARM - OFF
    CYCLE CWEA (DES REG LT - OFF)
    ENG STOP - RESET
    THROT CONT - AUTO/CDR
    TTCA (BOTH) - JETS
```

Figure 91: DPS throttle check

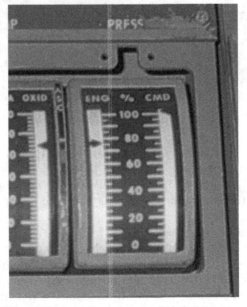

Figure 92: Actual vs commanded thrust meter

Figure 93: CDR thrust/translational hand controller

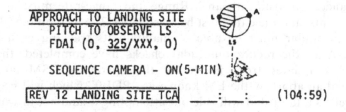

Figure 94: Approach to landing site

Approach to landing site: time to landing: 4:00
This is the first time the LM crew will be viewing their landing site from 50,000 feet. They spend several minutes observing and photographing the site and familiarizing themselves with the approach. If all goes well, they will be on that very surface in only in four hours.

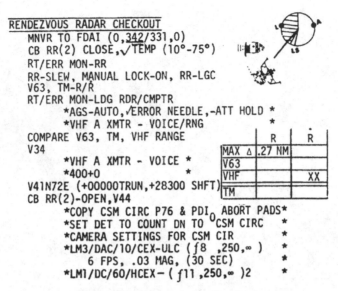

```
RENDEZVOUS RADAR CHECKOUT
    MNVR TO FDAI (0,342/331,0)
    CB RR(2) CLOSE,√TEMP (10°-75°)
    RT/ERR MON-RR
    RR-SLEW, MANUAL LOCK-ON, RR-LGC
    V63, TM-R/Ṙ
    RT/ERR MON-LDG RDR/CMPTR
            *AGS-AUTO,√ERROR NEEDLE,-ATT HOLD *
            *VHF A XMTR - VOICE/RNG            *
    COMPARE V63, TM, VHF RANGE
    V34
            *VHF A XMTR - VOICE *
            *400+0              *
    V41N72E (+00000TRUN,+28300 SHFT)
    CB RR(2)-OPEN,V44
            *COPY CSM CIRC P76 & PDI₀ ABORT PADS*
            *SET DET TO COUNT DN TO CSM CIRC    *
            *CAMERA SETTINGS FOR CSM CIR        *
            *LM3/DAC/10/CEX-ULC ( ƒ8 ,250,∞ )   *
                 6 FPS, .03 MAG, (30 SEC)       *
            *LM1/DC/60/HCEX- ( ƒ11 ,250,∞ )2    *
```

	R	Ṙ
MAX Δ .27 NM		
V63		
VHF		XX
TM		

Figure 95: Rendezvous radar checkout

Rendezvous radar checkout: time to landing: 3:45

Before the CSM and LM separate further, the operation of the rendezvous radar is checked. The LM is maneuvered so that the radar is pointing towards the CSM, the radar is powered up and its temperature checked, and the computer is told to start accepting its data. Data from the radar is displayed on the cross-pointers (located at eye level on both the main instrument panels) and the tapemeters on the Commander's instrument panel. Range and range-rate must agree between these instruments, and the data must be reasonable. The lock on the CSM must be solid, and the radar must be able to maintain its tracking even as the LM maneuvers. After the rendezvous radar checks have completed, the LM crew receives PAD updates from Earth for the upcoming CSM circularization maneuver and an abort by the LM known as "PDI 0" which will be made if the vehicle develops a fault that will rule out attempting a landing, in which case the LM will use its descent engine to initiate the rendezvous.

NO PDI + 12 ABORT PAD																
+	0	0					+	0	0					HRS (104)		N33
+	0	0	0				+	0	0	0				MIN (40)		TIG
+	0			•			+	0			•			SEC (24.70)		**E**
			•							•				ΔVX (+113.9)		N81
			•							•				ΔVY (+0.0)		LV
			•							•				ΔVZ (-46.8)		**F**
+			•				+			•				HA (+146.6)		N42
														HP (+8.8)		
+			•				+			•				ΔVR (+123.1)		
X	X	X	•				X	X	X	•				BT (0:40)		
X	X	X					X	X	X					R (000)		FDAI
X	X	X					X	X	X					P (271)		INER
+							+				•			TIG (280.4)		373
			•							•				ΔVX (+114.3)		N86
			•							•				ΔVY (+0.0)		AGS
			•							•				ΔVZ (-45.9)		
+	0	0					+	0	0					HRS (107)		N11
+	0	0	0				+	0	0	0				MIN (35)		CSI
+	0			•			+	0			•			SEC (38.90)		**G**
+	0	0					+	0	0					HRS (109)		N37
+	0	0	0				+	0	0	0				MIN (18)		TPI
+	0			•			+	0			•			SEC (38.30)		**H**

RESIDUALS						
PGNS				AGS		
	•	ΔVX	N85		•	ΔVX	500
	•	ΔVY			•	ΔVY	501
	•	ΔVZ			•	ΔVZ	502

Figure 96: No PDI + 12 abort data card

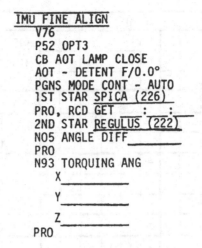

```
IMU FINE ALIGN
    V76
    P52 OPT3
    CB AOT LAMP CLOSE
    AOT - DETENT F/0.0°
    PGNS MODE CONT - AUTO
    1ST STAR SPICA (226)
    PRO, RCD GET    :   :
    2ND STAR REGULUS (222)
    NO5 ANGLE DIFF_____
    PRO
    N93 TORQUING ANG
        X_____
        Y_____
        Z_____
    PRO
```

Figure 97: IMU fine alignment 1

IMU fine alignment: time to landing: 3:34

A well understood property of the IMU is that it slowly drifts out of alignment, and periodic realignments are necessary. The computer knows the exact position of the stars, but the crew must use the optics to take sightings so that the AGC can align the IMU to a predetermined orientation. The two stars that are used for this alignment are Spica in the constellation Virgo, and Regulus in Leo. Program 52 uses the Alignment Optical Telescope, the AOT, to perform several marks on the pair of stars. When the star sightings are completed, the computer rotates the platform to the correct alignment in all three axes, usually only a few fractions of a degree. These Gyro Torquing Angles are displayed on the DSKY, and are relayed to Houston.

```
CONFIGURE COMM FOR LOS
*MATCH INDICATED ANGLES
*TRACK MODE - SLEW
*S-BD ANT - AFT
*SET P____ (+14)
*    Y____ (-14)
*VHF B XMTR - DATA
*BIOMED - OFF, PCM - LO
*UPLINK SQUELCH - ENABLE
*S-BD ANT-FWD(AFTER LOS)
```

Figure 98: Configure communications for LOS

Configure communications for LOS: time to landing: 3:45

The LM is now close to the point where its orbit carries it around the far side of the Moon, and out of communication with the Earth. In anticipation of reacquiring the Earth, the High Gain S-Band antenna is configured so that it will be pointing towards the Earth when it reappears about 40 minutes after loss of signal. The S-Band antenna is gimbaled in two axes (pitch and yaw) and the LMP can adjust the antenna in each axis. Two meters show the actual position of the antenna, and a signal strength meter assists the crew in ensuring that the antenna is pointing in the optimal direction.

Figure 99: LMP's communications panel

COAS and LPD calibration: time to landing 3:25
The Crewman Optical Alignment Sight, or COAS, is used to provide an attitude reference for the crew. Attached to its mounting bracket above the Commander's window, it projects a known line of sight onto its combining glass. It is used during docking and as a backup to the Alignment Optical Telescope and Landing Point Designator. As the COAS can be removed from its mount and repositioned or stowed, a quick calibration is necessary to capture the small alignment errors that occur when it is reinstalled. The COAS calibration is performed as an option of the P52 IMU Realignment program, which automatically maneuvers the LM such that the COAS is pointing at a reference star. The Commander notes the differences between the center of the COAS reticle, which is assumed to be pointing directly at the star, but does not enter these values into the computer.

The Landing Point Designator is simply a series on calibrated lines painted on the Commander's window. During the approach phase of the descent, the computer

```
LPD CALIBRATION
  PRO, ENTR
  N70, ENTR 013 (CAPELLA), PRO BIAS AZ____
  N87, (+35580,+32040) PRO,PRO       EL____
          *DETENT CL
          *CB AOT LAMP - OPEN
  V34, P00
  PGNS MODE CONT - AUTO
          *400+3
```

Figure 100: LPD calibration

calculates the current angle of descent and displays it in the DSKY. Taking advantage of the two panes used in the LM windows construction, the Commander visually lines up the angle indexes of the LPD. From this vantage point, he can visualize the current descent path and the computer's projected landing location.

To calibrate the LPD markings on the Commander's window, the computer is asked again to locate a star (in this case Capella in Auriga), this time using the Landing Point Designator as its reticle. Noun 87 tells Program 52 the orientation of the "backup optics", in this case, the angle of the 40 degree mark on the LPD. If the star is not centered in the 40 degree mark, the deviation is noted. Since the computer knows the location of the star and the angles defined by the LPD markings, this test determines how much error, or bias, exists between the LPD markings and the computer's understanding of where the LPD marks are pointing. Usually this is a small number, but the test adds to the confidence that the computer and platform are operating correctly.

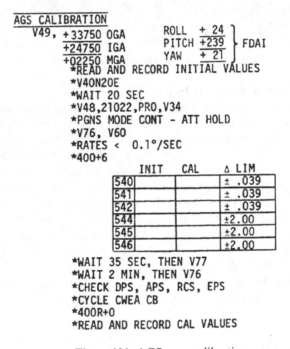

Figure 101: AGS gyro calibration

AGS calibration: time from landing 3:25
Similar to the Primary Guidance and Navigation System, the Abort Guidance System has a set of gyroscopes that supply attitude information to its computer. These gyros also drift and lose their accuracy over time, and understanding the extent of this drift is necessary for accurate measurements. Since this drift usually occurs at a constant rate, compensation for the error is a straightforward process of measuring their drift and incorporating this value in the computer software. The LM

is placed in a known attitude, which it maintains within 0.1°. The calibration routine is started, and the drift of the gyros is measured for several minutes. While the calibration is proceeding, the crew has a few moments to visually check other systems, such as temperatures and pressures of the propellants in the descent and ascent stages, and the cabin environmental systems. Only visual checks are possible, as during the calibration period the LM must not introduce any motion that might skew the calibration data. Once the calibration period has ended, the values are recorded in the checklist, to be read back to ground controllers when the LM reappears on the near side of the Moon.

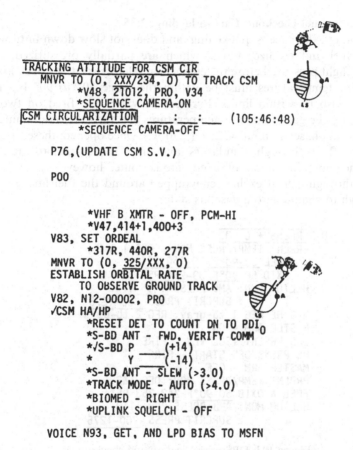

```
TRACKING ATTITUDE FOR CSM CIR
   MNVR TO (0, XXX/234, 0) TO TRACK CSM
        *V48, 21012, PRO, V34
          *SEQUENCE CAMERA-ON
CSM CIRCULARIZATION        :   :___   (105:46:48)
          *SEQUENCE CAMERA-OFF

    P76,(UPDATE CSM S.V.)

    P00


          *VHF B XMTR - OFF, PCM-HI
          *V47,414+1,400+3
    V83, SET ORDEAL
          *317R, 440R, 277R
    MNVR TO (0, 325/XXX, 0)
    ESTABLISH ORBITAL RATE
        TO OBSERVE GROUND TRACK
    V82, N12-00002, PRO
    √CSM HA/HP
          *RESET DET TO COUNT DN TO PDI
          *S-BD ANT - FWD, VERIFY COMM
          *√S-BD P ___(+14)
          *      Y ___(-14)
          *S-BD ANT - SLEW (>3.0)
          *TRACK MODE - AUTO (>4.0)
          *BIOMED - RIGHT
          *UPLINK SQUELCH - OFF

    VOICE N93, GET, AND LPD BIAS TO MSFN
```

Figure 102: Tracking attitude for CSM CIRC burn

Tracking attitude for CSM CIRC burn: time to landing: 3:10
Starting with Apollo 14, the CSM's engine performed the Descent Orbit Insertion, or DOI, burn, placing the combined CSM and LM into an orbit with an apolune of 60 nm (110 km) and a perilune of 50,000 feet (15 km). This maneuver saves valuable LM fuel, but places the CSM in an orbit where its rendezvous options are limited. By

circularizing its orbit, the CSM is able to rendezvous with the LM in both the nominal and abort cases.

The LM crew orients the spacecraft to observe and photograph the burn using the 16mm Maurer sequence camera. The vehicles are still on the far side of the Moon, and tracking and updates by the MSFN are not possible. Once communication is reestablished, reports from the CSM and tracking data provide the essential data to recalculate its state vector. Controllers in Houston then uplink the new CSM state vector data to the LM computer. As with any time the AGC receives an update, that data must also be passed along to the AGS.

DPS pressurization and checkout: time to landing: 2:35

From this point on, the pace is quickening, and does not slow down until well after landing. The fuel and oxidizer tanks, which are partially pressurized, must be brought up to flight pressure before the engine can reliably start. A check is made of the tanks. Their temperatures must be within the range 50° to 90° F, and their pressures must also be within limits. Pyrotechnic valves are fired on two helium tanks, one storing the gas at ambient temperatures and the other containing helium in the supercritical phase at about -452° F (-270° C). Once opened, these valves allow the helium to flow through regulators and into the propellant tanks. The supercritical helium takes a bit of a circuitous route, however. From its tank, helium flows through a heat exchanger wrapped around the fuel line so that it is warmed enough to change into a gaseous state.

```
DPS PRESS + C. O.
    PRPLNT TEMP/PRESS MON - DES 1 & 2
        FUEL 50-90°F 70-160 PSI
        OXID 50-90°F 39-254 PSI
    HELIUM MON: AMB PRESS 1495-1750
            : SUPCRIT PRESS 700-1275
    DES HE REG 1 tb-gray, REG 2 tb-bp
    MASTER ARM - ON
    DES PRPLNT ISOL VLV - FIRE
    HE PRESS/DES START - FIRE
    MASTER ARM - OFF
    PRPLNT TEMP/PRESS MON: DES 1&2
    FUEL & OXID 50-90°F 242-253 PSI
    HELIUM MON: AMB PRESS 200-1110
            : SUPCRIT PRESS 700-1275
```

Figure 103: DPS pressurization and checkout

Figure 104: Explosive devices panel

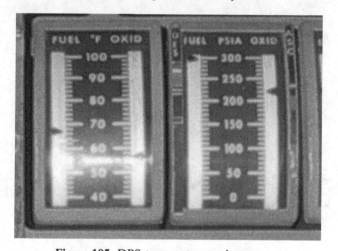

Figure 105: DPS temperature and pressure

LANDING RADAR CHECKOUT
CB LR CLOSE, CK TEMP (60° - 95°)
RATE ERR MON-LDG RDR/CMPTR

X-PNTRS-HI MULT, TM SW-H/H
LDG ANT-AUTO, MODE SEL-LR
RDR TEST - LDG
TEST MON-ALT/VEL XMTR (2.1 - 5.0), AGC

TM (8000 ± 100)/H (-480 ±2)
V63, N12 OPT 2, PRO
N66 8286 ± 10, ANT POS 1 (00001), PRO
N67 V_x (-00495 ±2), V_y (+01862 ±2)
   V_z (+01331 ±2)
V34, RDR TEST OFF
CB LR - OPEN

Figure 106: Landing radar checkout

Landing radar checkout: time to landing 2:30
The operation of the landing radar, which provides altitude and velocity data to the AGC during the descent, is checked. After letting the radar electronics warm up for a few minutes, test signals are exchanged between the radar and computer. The computer accepts these signals as valid data, and the test "altitudes" and "velocities" are displayed on the DSKY. Tape meters on Panel 1 displaying altitude and altitude rate, are also driven by the test signal to confirm that these instruments are in working order.

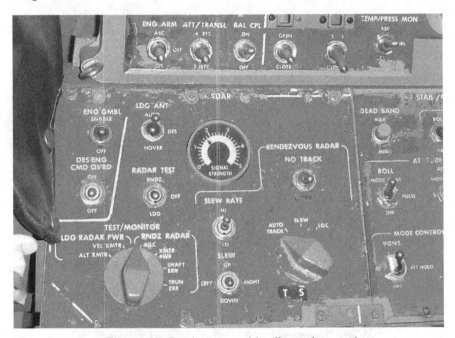

Figure 107: Rendezvous and landing radar panel

Update from MSFN: time to landing: 2:25
In tracking the CSM and LM over the last ten to fifteen minutes, ground-based radar and tracking systems derive accurate position and velocity information. This is uplinked to the AGC in the form of the state vector, and will eventually be used to update the AGS. Next, the tedious work of copying abort times and options begins. These updates, or PADs, are verbally given to the LMP, who copies them down on standardized forms. Each update addresses a particular abort scenario; for example, the "NO PDI + 12 Abort" is used if the landing is called off, and is executed at 12 minutes past the intended descent initiation time. An onboard document, LM Data Cards, contains the preprinted PAD forms.

The ability of the AGS to track the CSM with the new state vector is checked using the rendezvous radar. The CSM's range and range-rate as computed by the AGS are compared against the AGC estimates. The ORDEAL (Orbital Rate Display – Earth and Lunar) is set up, which automatically maintains the spacecraft

```
UPDATE FROM MSFN
          UPDATA LINK - DATA
          UPLINK CSM/LM S. V.. PIPA BIAS,
            DESCENT TARGETING, LPD BIAS
            (IF Δ>2° IN AZ OR 1° IN EL)
          COPY PADS FOR
              NO PDI + 12 ABORT,
              PDI,
              PDI EARLY ABORT,
              PDI LATE ABORT,
              T2 ABORT
              T3 TIG
          UPDATA LINK - OFF
          V47, 414+1, 400+3
     V83, SET ORDEAL
          *317R, 440R, 277R
     VERIFY NO PDI₀ ABORT WITH MSFN
```

Figure 108: Update from MSFN 2

in the same attitude relative to the surface. Without such a system, the spacecraft would maintain its attitude with respect to the stars, and would appear to rotate when viewed from the surface.

Figure 109: Orbital Rate Display – Earth and Lunar (ORDEAL)

```
LPD ALT CHECK
     MNVR TO AND MAINTAIN FDAI (0,295/XXX,0)


     BEGIN LPD ALT MARKS (IF DESIRED)
     PDI LMK LPD ALT CHECK

PDI 0 ___:___:___  (106:48:19)

     PITCH TO OBSERVE LS


     START PITCH TO P52 ATT (0,XXX/325, 0)



     *CAMERA SETTINGS (PDI)
     *LM3/DAC/10/CEX-
     *(f2.8, 500, ∞) 12 FPS,
     *  0.75 MAG, (6 MIN)
     *LM3/DC/60/HCEX-(f5.6,250, ∞)10
     *RELOCATE CAMERA
```

Figure 110: LPD altitude determination check

LPD altitude check: time to landing: 2:20

An optional visual check of the LM's approximate altitude can be made as the spacecraft approaches a known landmark on the surface. The LM is maintaining an "orbital rate" attitude, whereby it slowly rotates with respect to the stars, to maintain a fixed attitude with respect to the surface. Measuring the LM's altitude is a simple application of Kepler's law of orbital motion. In lower orbits, the speed of the vehicle (relative to the surface) is higher than at higher orbits. By measuring the time required to fly over a fixed distance, the velocity can be determined. The relationship between velocity and altitude is well known, and the altitude is easily determined. Maneuvering the LM, the Commander sights the landmark on the zero degree Landing Point Designator mark at the upper part of his window, and notes the time. As the LM moves along its ground track, the landmark "moves" lower in the window to the 50 degree mark, where the time is measured again. Using a graph in the LM Data Card manual, the vehicle's altitude is found. For example, if the landmark traverses the LPD in 10 seconds, its altitude is about 9.7 miles.

During the pass, the LMP is operating the 16mm camera, recording 6 minutes of flight over the lunar surface in the area of the landing site. Several still pictures are also taken using the Hasselblad cameras. Next, the Commander pitches to the attitude favorable for performing the next guidance platform alignment. An important milestone occurs at this point. An abort option, known as "PDI 0", is available to the crew if their checks of the LM's systems reveal a fault that will rule out attempting a landing. If an abort is necessary, the LM can fire its descent engine to begin the process of returning to the CSM.

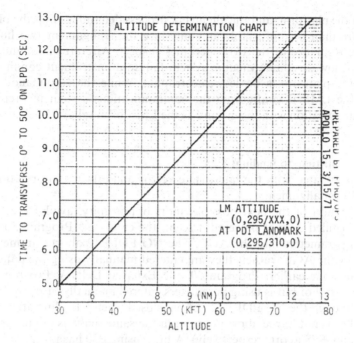

Figure 111: LPD altitude determination chart

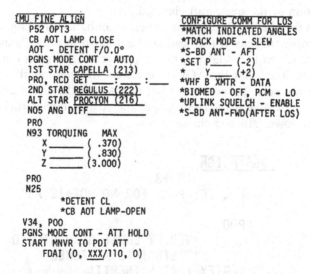

Figure 112: IMU fine alignment 2, and configuring for LOS

IMU fine alignment and configuring for LOS: time to landing: 1:35
A final tune-up of the platform's alignment is performed, for more than the obvious reason of having the most precise alignment before descent. It is known that the platform drifts, and that rate of drift is known with some precision. If there is a

problem in the platform, and the alignment process determines that the platform has drifted more than a comfortable amount since the realignment two hours earlier, then the performance of the IMU becomes suspect. Aircraft pilots understand the danger of a slow, insidious vacuum system failure in instrument conditions; a slow failure of the platform is no less catastrophic. While the Commander aligns the platform, the LMP is configuring the communications system in preparation for losing communication with the ground as the LM's orbit takes it back around the far side of the Moon.

AGS alignment: time to landing: 1:25
The AGS is updated with the latest state vector and attitude information from the earlier platform alignment.

In anticipation of the next possible abort opportunity, available at 12 minutes past the scheduled time of powered descent, the crew stars Program 30 to load the abort burn parameters into the AGC. The NO PDI + 12 abort parameters are not calculated by the AGC; rather, they are entered manually using the data radioed to the crew an hour earlier. Program 30, the External Delta-V Program, facilitates computer control of burns using parameters entered manually or by uplink from the ground. Although the NO PDI + 12 option is used when a landing attempt has been abandoned, it is not a true abort program in the same sense as are programs P70 or P71 (DPS and APS aborts respectively). A burn using P30 has no special meaning to the AGC, and does not attempt to place the LM into a predetermined orbit. Nor is the configuration of the spacecraft changed.

For the last time before the descent, the crew configures the Digital Autopilot. To maintain a proper trajectory, the deadband is limited to only 1° and the maximum rate for any maneuver to a relatively fast 2° per second. This has the effect of rapidly assuming any attitude that the AGC or the crew will command, as well as quickly reacting to forces, such as fuel slosh, that might alter the LM's attitude. However, this relatively aggressive maneuvering is at the cost of higher fuel consumption in comparison to slower attitude changes.

```
AGS ALIGN
        *400+3
    P30 TGT PGNS FOR NO PDI+12 ABORT

    P00
         *VERIFY LOOSE GEAR STOWED
         *RESTRAINTS ATTACHED
    VERIFY FDAI'S INERTIAL
    V48,21112,PRO
    V34
```

Figure 113: AGS alignment

```
P63 IGNITION ALGORITHM TEST
    P63
            *RESET DET TO COUNT DN TO PDI
    ENTR-BYPASS ALIGN, PGNS MODE CONT - AUTO
    N18 R, P, Y (0, 110, 0) PRO
    POO, PGNS MODE CONT - ATT HOLD,V77
    COAS TO OVERHEAD WINDOW
```

Figure 114: P63 ignition algorithm check

P63 ignition algorithm check: time before landing 1:10

A critical test of the computer is made by starting the descent program, Program 63. It is too early to begin the descent, and there is no intention of starting the engine, but this tests P63 to see if the descent algorithms are calculated correctly, as well as the proper time of ignition and the attitude necessary to begin descent. Once these variables are confirmed, P63 is terminated and the computer is told to maintain the current attitude.

Pre-PDI ECS configuration: time to landing: 1:00

The crew don their helmets and gloves, then configure the environmental control system to flow oxygen directly to the suits. As there is plenty of oxygen in the cabin, the suits are not stiffly pressurized against a vacuum, a condition often called a "hard suit". The suit and cabin pressure are approximately the same, which allows a degree of comfort and mobility. Placing the oxygen control valves to the EGRESS position sends oxygen only to the suit, and not into the cabin. Egress, in this sense, is normally the occasion when the crew will be depressurizing the cabin in preparation for an EVA.

```
PRE-PDI ECS CHECKOUT
            *HELMETS AND GLOVES ON
            *CABIN REPRESS-CLOSE
            *SUIT GAS DIVERTER - EGRESS
            *CABIN GAS RETURN - EGRESS
            *PRESS REGS A&B -   EGRESS
```

Figure 115: Pre-PDI environmental control system checkout

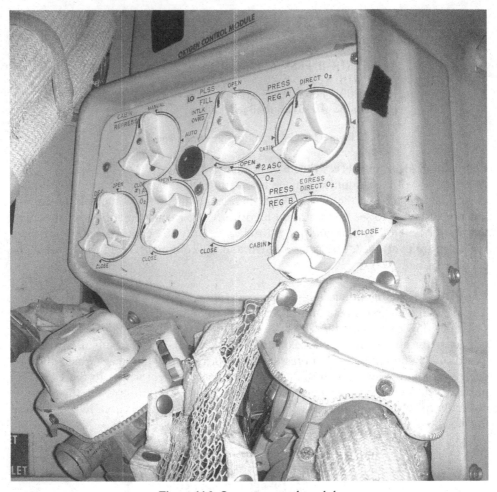

Figure 116: Oxygen control module

Pre-PDI switch setting check: time to landing 0:50

Just before the LM reappears from the far side, the crew performs a final check
and configuration of its systems. The ascent stage is readied for an immediate
abort, with its batteries brought online, and circuit breakers necessary to stage the
vehicle and arm the ascent engine controller (AELD) are closed. When
communications are reestablished with the Earth, the S-Band antenna is
configured and updates are relayed between the ground and the crew. Both
Thrust/Translational Hand Controllers are configured for descent, and the LMP's
attitude hand controller is enabled. In the very unlikely event that the
Commander's attitude controller fails, the LMP can take over maneuvering the
vehicle. Temperatures, pressures, voltages and quantities are verified for all
critical systems, and the descent and ascent engine systems, as well as the power
and environmental systems, are checked closely.

```
PRE-PDI SWITCH SETTING CHECK
           *VHF ANT - FWD
    CB INV 1 - CLOSED
             *SELECT INV 1
    CB AELD (2) - CLOSE
    CB ABORT STAGE (2) - CLOSE
             *CYCLE CWEA CB
             *BATS 5&6 NORM FEED - ON
             *RECORD GET ____:____:____
    RESET ENG STOP PB
    SET WINDOW BARS /
             *S-BD ANT - FWD, VERIFY COMM
             *√S-BD P _____  (-2)
             *      Y _____  (+2)
             *S-BD ANT - SLEW (>3.0)
             *TRACK MODE - AUTO (>4.0)
             *VHF B XMTR - OFF
             *VHF A XMTR - VOICE/RNG
             *BIOMED - LEFT, PCM - HI
             *UPLINK SQUELCH - OFF
    VOICE N93, GET, AND ASC BATT
       ON TIME TO MSFN
    THROT CONT - AUTO
    CDR TTCA - THROTTLE - MIN
             *LMP TTCA - THROTTLE - SOFT STOP
             *ACA PROP (LMP) - ENABLE
             *ACA/4JET (LMP) - ENABLE
             *TTCA/TRANSL (LMP) - ENABLE
             *CHECK DPS, APS, RCS, ECS, EPS
```

Figure 117: Pre-PDI switch setting check

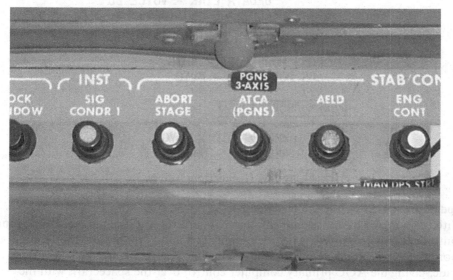

Figure 118: Abort Stage and AELD circuit breakers

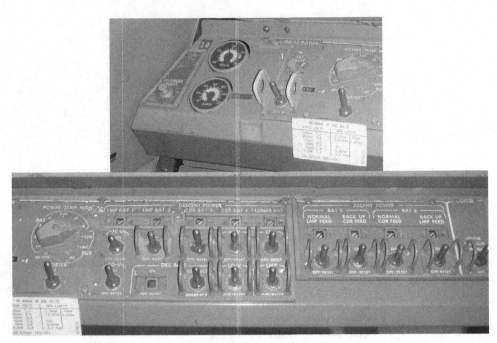

Figure 119: Electrical power system battery monitoring and selection

UPDATE FROM MSFN
　　　　　　　*UPDATA LINK - DATA
　　　　　　　*UPLINK LM S.V., RLS,
　　　　　　　*　MSFN GYRO DRIFT COMP
　　　　　　　*UPDATA LINK - VOICE BU
　　　　　　　*COPY AGS RLS (231)
　　　　PRPLNT QTY MON - DES 1 /
　　　　MODE SEL - PGNS /
　　　　PGNS MODE CONT - AUTO
　　　　AGS MODE CONT - AUTO
　　　　V77
BURN ABORT RULES

　　*AUDIO MODE (BOTH) - VOX
　　*TAPE RECORDER - ON

Figure 120: Update from MSFN 3

Update from MSFN: time to landing: 0:30
With the powered descent imminent, the crew receives the final updates from ground controllers. An improved state vector and the latest update on the landing site coordinates are uplinked to the computer. The final guidance switch settings are made, giving the AGC primary computer control over the spacecraft, with the AGS ready to take over if needed.

The crew makes a quick review of the abort rules, as specified by the Cue Cards manual. These are the most fundamental rules of the mission: if certain problems arise, the crew has no option but to abort. Problems such as electrical bus failures, vehicle gyrations, and environmental system failures can place the crew in serious danger and all require a mandatory abort. Surprisingly, some failures are tolerable until late in the descent. For example, in the last few minutes of the descent, and even to the point of touchdown, a failure such as a loss of cooling water from the descent storage tanks does not require an immediate abort; nor does an AC bus failure. These types of failures may not prevent the crew from landing, but an extended stay on the Moon is impossible. More importantly, it is sometimes more reasonable to land, assess the situation and set up for a coordinated return to the CSM.

PDI RULES/ALTITUDE CHECK

```
                        PDI RULES
 1.  NO AUTO ULLAGE - BACKUP VIA +X OVERRIDE
     (+NO AUTO IGNITION - PDI NO-GO)
 2.  NO IGN (WITH AUTO ULL) DELAY 2 SEC, THEN START
     PB-PUSH; THEN DES OVRD - ON AT 5 SEC
 3.  T/W >1.6 AND DSKY CHANGES >18 fps/2 SEC
 4.  ATT/RATE <5°/SEC
 5.  ΔH IN LIMITS >10 SEC, NOT OUT OF LIMITS >60 SEC
 6.  DATA GOOD AT > 6,000 ft
 7.  IF NO THROTTLE DOWN BY P64 + 15 SEC - ABORT
 8.  BINGO FUEL 1 MIN 31 SEC AFTER LOW LEVEL OR WHEN
       FUEL QTY <2% UNLESS LANDING IMMINENT
NOTE:  FOR FLASHING LR ALT OR VEL LIGHTS PRECEDED BY
   STEADY LR LT, CYCLE RADAR TEST SWITCH
```

Figure 121: PDI rules/altitude check cue card

```
         ABORT RULES
1   ATT & RATE LIMITS
      DPS >5° SEC
      APS >10° SEC
2   DPS SHUTDOWN
      <30 fps - STAGE & RCS
      >30 fps - ABORT STAGE
3   APS UNDER BURN
           PGNS              AGS
      <400 NULL              AUTO,
           RESIDUALS         A/H 15FPS
      >400 A/H BURN
           Ha,Hp,HDOT
4   INSERTION
    WITH VOICE-GROUND RECOMMENDS TRIM SOURCE AT TGO ~ 1 MIN
    TRIM TO <2fps (AGS X AXIS ONLY) AND STANDBY FOR TWEAK
    AT INSERTION +2 MINUTES(10° OHW OR 257° FDAI)
5   NO VOICE
      AGS & PGNS <10fps TRIM ACTIVE SYSTEM
      AGS & PGNS >10fps TRIM SYSTEM THAT AGREES WITH RR
```

Figure 122: Abort rules cue card

MISSION RULES NO-GO'S

PRE PDI		PDI TO PDI +6+10	PDI +6+10 TO HI GATE	HI GATE TO TD
EPS	ONE DC BUS	ABORT	ABORT	ABORT
	ONE DESCENT FEEDER SHORTED	ABORT	ABORT	ABORT
	ONE ASCENT FEEDER SHORTED	ABORT	ABORT	ABORT
	4 DESCENT BATS	ABORT	ABORT	GO
	ONE ASCENT BAT	ABORT	ABORT	GO
	BOTH INVERTERS	ABORT	ABORT	GO
	AC BUS A AND B	ABORT	ABORT	GO
ED	ONE PYRO SYSTEM ARMED	ABORT	ABORT	ABORT
	ONE PYRO SYSTEM DEARMED	ABORT	GO	GO
	ONE STAGING RELAY CLOSED	ABORT	ABORT	ABORT
	ONE PYRO SYSTEM BATTERY	ABORT	GO	GO
ECS	CABIN PRESS <4.4	ABORT	ABORT	GO
	SUIT LEAK	ABORT	ABORT	ABORT
	BOTH SUIT FANS	ABORT	ABORT	GO
	BOTH DEMAND REGS	ABORT	ABORT	GO
	BOTH H20 SEPS	ABORT	ABORT	GO
	BOTH DESCENT 02 TANKS	ABORT	ABORT	GO
	BOTH ASC 02 TANKS	ABORT	ABORT	GO
	PRI OR SEC COOLANT LOOP	ABORT	ABORT	GO
	PRI OR SEC H20 FEED	ABORT	ABORT	GO
	BOTH DESCENT H20 TANKS	ABORT	ABORT	GO
	BOTH ASC H20 TANKS	ABORT	ABORT	GO
G&C	PGNS GUID STEER	ABORT	ABORT	GO
	3 AXIS ATT CONT			
	PGNS RATE CMD & PGNS AUTO	ABORT	ABORT	OPTION
	AGS RATE CMD	ABORT	ABORT	OPTION
	2 ACA	ABORT	ABORT	OPTION
	AUTO +X & AUTO DPS IGNITION	GO	GO	GO
	2 FDAI-ATT/RATE/ERR	OPTION	OPTION	OPTION
	LR	ABORT	ABORT	GO
	REDNT APS ON	ABORT	GO	GO
	P & R GDA TRIM (IMPING CONST VIOL)	ABORT	ABORT	ABORT
	MANUAL THROTTLE (2 TTCA) & AUTO THROT	ABORT	ABORT	ABORT
DPS	PROP LEAK (ΔQ FU/OX>10%)	ABORT	ABORT	ABORT
	FU OR OX INLET/ULLAGE<160	ABORT	ABORT	ABORT
	BINGO/2%			ABORT
APS	PROP LEAK	ABORT	ABORT	ABORT
	FU/OX INLET PRESS<62,>220	ABORT	ABORT	ABORT
	APS HE 1 OR 2 DECREASING	ABORT	ABORT	ABORT
RCS	HE/PROP LEAK			
	PROP LEAK (DOWNSTREAM OF MAIN)	ABORT	ABORT	GO
	FU/OX MNFLD A OR B PRESS<100	ABORT	ABORT	GO

Figure 123: Mission rules cue card

AGS initialize: time to landing 0:30

The last AGS update is more than a simple update of the CSM and LM state vectors. The AGS clock reference and the coordinates of the landing site are updated, and velocity vector checks are made. Note that the commands to enter the data are not particularly intuitive. The Verb-Noun syntax doesn't exist with the AGS, which only uses the address of a memory location and the data that is to be saved there.

```
AGS INITIALIZE
      *V47, 414+1
      *V83, 317R, 440R, 277R
      *240 + (231 RLS  PAD  )
      *254+08313
      *261-00013
      *262-00151
      *404-12345
```

Figure 124: AGS alignment

```
POWERED DESCENT INITIATION
        CB LR - CLOSE
        √ALT XMTR
        P63
        √DPS CONFIG CARD
              *RESET DET
        ENTR-BYPASS ALIGN
        N18 R, P, Y (0, 110, 0)
        VERIFY FDAI
              *V40N20E, 400+3, 410+0
              *400+1, 433R VI
        PRO-FINAL TRIM
        ENTR, √DET
        GO/NO-GO FOR PDI
        COMM CHECK WITH CSM
        RESET WATCH
 -1:00  MASTER ARM - ON
 -0:30  ENG ARM - DES
 -0:07.5 ULLAGE
 -0:05  PRO
  0:00 PDI   :   :     (108:42:01)
 +0:02  (NO IGN) - START PB - PUSH
 +0:05  DES ENG CMD OVRD - ON
        MASTER ARM - OFF
```

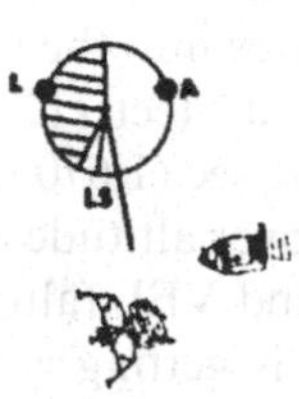

Figure 125: Powered Descent Initiation

Powered descent initiation: time to landing: 0:25

Braking phase: Program 63
This is it! Program 63 was entered about ten to fifteen minutes before ignition, to give the computer time to calculate the ignition time and ensure that the LM is at the correct attitude for descent. On the DSKY, Verb 37, Enter, 63, Enter is keyed. P63 replies with verification of several important pieces of information, starting with a flashing V06N61. The time to go in the Braking Phase, the time from ignition and the cross-range distance are displayed on the DSKY. If these are reasonable, the crew continues to the next piece of data, entering Noun 33 to display the projected time of ignition. While the crew is reviewing this data, the computer illuminates the KEY REL light, as it needs to begin maneuvering the LM to the proper attitude for the descent burn. Pressing the KEY REL button, the AGC flashes V50N18 to request that the

spacecraft be maneuvered to the final PDI attitude. By responding with PRO, the Digital Autopilot adjusts the LM's attitude to that necessary for the beginning of the powered descent. Once at the correct attitude, the DSKY displays V06N62. Of the values on the three registers, the most important is Register 2, the time to ignition. At 35 seconds before ignition, the DSKY blanks. Five seconds later, the display returns, but is now flashing urgently, indicating the last few seconds before the beginning of powered descent. Engine Master Arm is switched to Descent, enabling commands to be sent to the engine. At 7.5 seconds before ignition, the + X (upward) thrusters fire to settle the propellants in their tanks, ensuring a smooth ignition. Five seconds before ignition, a flashing Verb 99 appears on the DSKY, asking permission to continue with ignition and the Commander presses PRO in response.

Ignition!

At 10 percent thrust, the acceleration is only slightly noticeable. The computer slowly gimbals the descent engine to align it though the LM's center of mass. After 26 seconds, the engine throttles up to full thrust. Shortly thereafter, the crew receives an update from Houston as to their landing site location. Using Noun 69 to redesignate the landing point, the adjustment is given in thousands of feet. While the computer is able to accept an update for up/down range, left/right of track and variations on altitude, only the up/downrange correction is generally used. Entering the correction as V21N69, the crew tells the computer that the landing site has "moved" and its targeting needs to be updated accordingly. In fact, of course, the target is the same and this update is to correct for navigational errors of which the AGC is ignorant.

A few minutes into the powered descent, the spacecraft's altitude has lowered to a little over 40,000 feet (12 km), and the velocity has slowed from 5,500 ft/sec to around 4,000 ft/sec (1,670 m/sec down to 1,220 m/sec). Now it is time for the landing radar to take over altitude and velocity determination. On the DSKY, the crew waits for the ALT and VEL (altitude and velocity) lights to extinguish, indicating that the landing radar is getting good data. Entering V06N63 displays the difference in the altitude calculated by the AGC based on the trajectory and the actual measurements by the landing radar. A 1,000 foot (300 meter) discrepancy between the computer's estimate and the measured altitude from the radar is not surprising. In the final moments of the descent, an error of hundreds of feet is the difference between a crash, a soft landing and an abort for lack of fuel. To reconcile the difference between the AGC and the radar, Verb 57 is keyed to accept the landing radar data. This allows the guidance algorithms to begin to incorporate the radar's data in their calculations. A primitive terrain model composed of several segments approximating the average slope of the surface is also incorporated into the targeting. Through V06N63, the difference between the landing radar and computer estimate of altitude is seen on the DSKY. Eventually, the computed altitude and the actual altitude as measured by the landing radar converge to the same value.

V06N63 also displays the computed altitude and altitude rate (H-dot). During the descent, these values are compared against pre-planned estimates printed on a card Velcro'd on the main panel between the two astronauts. Nominal velocity, altitude, altitude rate and descent fuel values are listed on the card. These are cross-checked

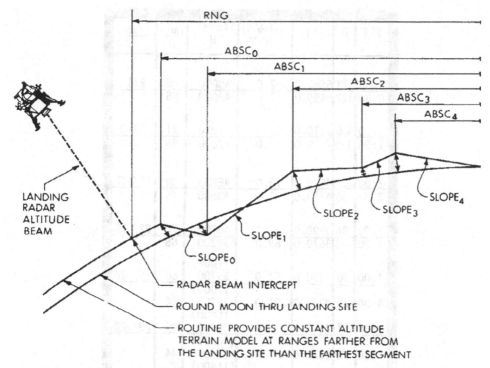

RNG

ABSC$_0$

ABSC$_1$

ABSC$_2$

ABSC$_3$

ABSC$_4$

SLOPE$_4$

SLOPE$_2$ SLOPE$_3$

LANDING
RADAR
ALTITUDE
BEAM

SLOPE$_1$

SLOPE$_0$

— RADAR BEAM INTERCEPT

— ROUND MOON THRU LANDING SITE

— ROUTINE PROVIDES CONSTANT ALTITUDE
TERRAIN MODEL AT RANGES FARTHER FROM
THE LANDING SITE THAN THE FARTHEST SEGMENT

Figure 126: Lunar terrain model

against the actual instrument readings. The Program 63 data is listed at 30-second intervals, giving the crew sufficient time to verify their instrument readings and attend to other duties in the cockpit.

About eight minutes into the descent, the Commander performs a quick check of the manual controls. Changing the setting of the PGNS Mode Control switch to Attitude Hold from Auto, he uses the attitude controller to evaluate the maneuverability of the LM in all three axes. Equally important, this is his first opportunity to get a feel for the handling characteristics of a much lighter LM. Once satisfied with the feel and controllability of the LM, the PGNS Mode Control is set back to Auto and the computer resumes attitude control of the descent.

Although the descent is performed using the AGC, the AGS must be ready to assume control and return the LM back to orbit in an abort. A key task of the LMP is to constantly cross-check the altitude and altitude rate computed within the AGS with information displayed on the DSKY. The AGS, which is not updated by the landing radar, doesn't have as accurate an idea of altitude as the AGC, and it needs to be updated with the height above the surface. Rather than using the complex AGC-to-AGS alignment (which is best done only during non-powered flight), the LMP monitors the altitude readout on the DSKY. As the descent progresses, he queues up an entry on the DEDA to instruct the AGS that it is now at 12,000 feet. As soon as Noun 63 on the DSKY indicates 12,000 feet, he presses Enter on the

TFI	θ	VI	(-ḢMAX) -HDOT	(ΔHMAX) H	DPS	SBD
0:00	113	5560.0	2.0	50000	95	2/1
0:30	112	5490.0	7.0	49900	95	
1:00	106	5210.0	37.0	49300	91	7/-3
1:30	100	4910.0	59.0	47800	86	
2:00	95	4610.0	73.0	45800	80	15/-11
2:30	90	4310.0	82.0	43500	75	
3:00	86	3990.0	87.0	40900	70	22/-16
3:30	83	3670.0	89.0	38300	65	
4:00	80	3330.0	91.0	(17000) 35700	60	26/-20
4:30	78	2990.0	91.0	(17000) 32700	54	
5:00	77	2640.0	93.0	(15800) 30500	49	29/-22
5:30	74	2270.0	92.0	(12800) 26400	44	
6:00	73	1890.0	86.0	(11400) 24700	39	32/-25
6:30	70	1490.0	(432.0) 69.0	(9200) 21800	33	
7:00	66	1230.0	(401.0) 95.0	(8200) 18900	30	39/-29
7:30	65	980.0	(367.0) 119.0	(6900) 16100	27	
8:00	65	730.0	(323.0) 139.0	(5600) 12800	23	40/-29

Figure 127: P63 descent monitoring cue card

DEDA keyboard to register that altitude in the AGS computer. Although this technique is not particularly sophisticated, it is satisfactory for abort functions.

At anytime during P63, the crew can select V16N68. Noun 68 displays velocity, the range to the landing site, and time until visibility phase pitchover, which marks the start of Program 64. The pitchover maneuver is not performed at a fixed time into the descent. Several variables could alter this time by several seconds, and Noun 68 informs the crew when the computer estimates that it will automatically switch programs.

The approach phase: Program 64
At about 7,000 feet (2,100 meters), the computer automatically enters Program 64 to start the Approach Phase. This involves pitching the spacecraft to give the

DEDA ADDRESSES

Vi	433		V16N78 R/RDOT
HDOT	367(360 UPDATE)		V16N92 %THROT/HDOT/H
H	337(223 UPDATE)		V21N69 ΔRLS
Ha	315		
Hp	403	AUTO	MAN
Y	211(-)(100Ft)	417+1	
YDOT	270,263	411+1	411+0
R	317	621 R	415+1
RDOT	440	606+R NEXT	316 E R
θ	277	-RDOT NEXT	503 E RDOT
		411+0 STOP	

Figure 128: DEDA addresses cue card

astronauts their first clear view of the landing site. While not abrupt, the change in pitch angle is significant, from around 70 to about 30 degrees from vertical. The only crew interaction with the computer is to press PRO when the program change occurs. Immediately, V06N64 comes up on the DSKY. This displays the Landing Point Designator angle, as well as the altitude and altitude rate. The LMP verbally relays the LPD to the Commander, who sights that angle through the index lines on his window to see where the computer expects to land. The elegance of the LPD is in its simplicity. Without using elaborate electronics or heavy and complex displays, the crew has a relatively accurate tool for determining the path of the LM. Inevitably, the most desirable landing spot is not where the LM is heading. To alter the path of the LM, the Commander uses his right hand to manipulate the Rotational Hand Controller. Rotating it forward and backward instructs the LM to land long or short; and left or right adjustments move the flight path accordingly. Each forward or backward motion of the hand controller adjusts the flight path up- or down-range by one degree; left or right moves the flight path left or right of track by two degrees. Commanders and their LMPs had their personal information flow protocols finely honed with respect the altitude, altitude rate, and LPD. Some preferred a steady stream of information, while others desired a sparse amount of data. Of course, the perfect amount of information was the one agreed between the two crewmembers, generally reflecting their personal tastes, developed over long hours in the simulator.

Now the real job of landing begins. The Commander is flying "outside the cockpit" by assessing the situation visually, and is dependent on the information called out by his colleague. Margins for altitude, altitude rate and fuel are especially small at this point, and so the LMP is constantly checking a cue card with data for Program 64. Unlike the P63 cue card, which has data for every 30-second interval, the P64 card is divided into 1,000 foot (300 meter) increments. At the beginning of P64, a descent of 1,000 feet takes less than 10 seconds, but as the LM nears the surface, the rate of descent slows to where the same distance takes 30 seconds or more. The LMP rarely gets any time to look out of his window during the visual phase of the lunar landing. His job is more that of a flight engineer, concentrating his attention on the DSKY, fuel status, and horizontal velocity.

H	(-ḢMAX) -HDOT	DPS
	(228.0)	
7000	151.0	19
	(208.0)	
6000	134.0	19
	(187.0)	
5000	113.0	18
	(163.0)	
4000	93.0	17
	(136.3)	
3000	71.0	16
	(105.0)	
2000	48.0	15
	(64.0)	
1000	27.0	13
	(36.0)	
500	17.0	11
	(29.0)	
400	14.0	11
	(21.0)	
300	12.0	11
	(12.0)	
200	9.0	10

Figure 129: P64 approach phase cue card

Finally, the Commander locates the best spot to land, one that is free of boulders, craters or other hazards to a safe landing. It is now time to take manual control of the spacecraft.

Final descent: Program 66

Two things must happen for the Commander to assume manual control during the final descent.[6] First, the PGNS Mode Control is switched from Auto to Attitude Hold. Then, in the final and irrevocable step, the Rate of Descent, or ROD, switch is

---

[6]   This phrase is very misleading. During the last moments of the lunar landing, it is common to state that the Commander has assumed "manual control", strongly implying that the AGC is no longer involved in controlling the vehicle. This is certainly not the case. The guidance mode is in Attitude Hold and the DAP continues to maintain control of the vehicle. However, the DAP now takes its attitude commands from the hand controllers rather than the targeting program.

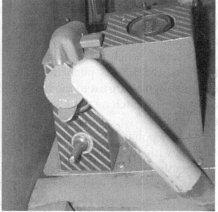

Figure 130: Commander's rate-of-descent switch

toggled. Located next to the Commander's left index finger, it is placed so that he does not have to move his hand away from the Thrust/Translational Hand Controller. Flipping this toggle switch up or down changes the computer program to P66, the Landing Phase Program. Each "click" of the switch changes the rate of descent by 1 foot/sec. Clicking the switch up reduces the descent rate and clicking downwards increases the descent rate. Although the descent can be performed using only the manual thrust controller, the ROD switch is far more intuitive, and is integrated with the computer to precisely adjust vertical velocity. Furthermore, since the vertical velocity calls from the LMP are in feet per second, a switch calibrated in foot-per-second increments simplifies adjustments.

```
PGNS MODE CONT-
     ATT HOLD
      P66


X-PNTR-LO MULT

      BINGO FUEL
DES QTY LT+1+34
TOUCHDOWN
ENG STOP - PUSH
PRO
MODE CONTROL (BOTH) - AUTO
DES ENG CMD OVRD - OFF
ENG ARM - OFF
413 + 1
```

Figure 131: P66 and lunar contact

Flying the LM is certainly a two-handed job, and in some ways it is similar to a helicopter. The Commander's right hand adjusts the spacecraft's attitude (and by extension, the direction of flight). This can be compared to the cyclic stick. The Rate of Descent switch is analogous to the helicopter's collective control. To move forward, the LM is pitched forward, allowing the horizontal component of the descent engine's thrust to move the LM forward. Slowing forward velocity is just the opposite, by pitching backwards. Maneuvering left and right is done in a similar manner.

At this point, the LM is only a few hundred feet above the surface (100 to 200 meters) with a low velocity, and the only objective is to null the horizontal velocities and gently lower the LM to the surface. Sooner or later, the descent engine exhaust begins to kick up dust. Depending on the amount of dust, the surface may be completely obscured. If so, the Commander's attention must be brought "inside the cockpit". Often he slowly eases the spacecraft down using simply the FDAI, cross-pointers and the altitude readouts provided by his LMP. Located just above the FDAI on the main instrument panels, the cross-pointers are a set of needles that

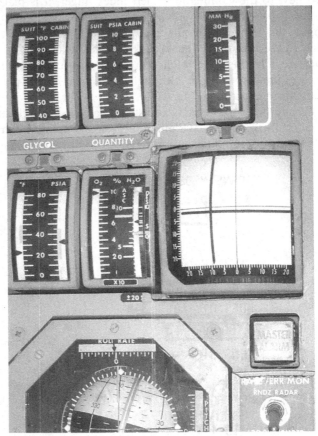

Figure 132: Cross-pointer display for the LMP

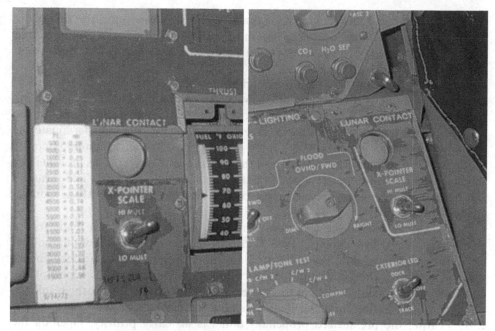

Figure 133: CDR's and LMP's lunar contact lights

move horizontally and vertically. One set depicts forward/backward velocity and another the lateral rates, with their velocity scales along the outside edge. Aircraft pilots would be comfortable with this display, as it is modeled after the localizer/ glideslope needles used during an instrument landing. During P66, the goal is to null any horizontal rates and hold the cross-pointers centered as the LM makes its final vertical descent to the surface.

Then, Contact! One of the three probes extending below the LM footpads touches the surface and sends a signal to indicate that the LM is five feet above the ground. A blue light illuminates at the Commander's eye level, as does one on the lower panel in the LMP's field of view. Quickly, the shutdown checklist is started. The Engine Stop button is pressed, and the LM settles, perhaps not quite softly, to the surface at only a few feet per second.

Post-landing operations
Both guidance systems are returned to automatic mode, meaning they will control all aspects of any emergency ascent that might occur. The LM is not expected to land on a perfectly level surface, and its at-rest attitude is likely tilted by several degrees. Sensing that the LM is not in the attitude that it last commanded, the Digital Autopilot begins firing the thrusters in hopes of "righting" the vehicle. Firmly on the surface, the LM will not move, so the Commander moves the hand controller, or ACA, out of its centered position for just a moment. Because the Digital Autopilot is in the attitude-hold mode, the autopilot perceives the LM's orientation as being the result of the attitude controller input, and no longer tries to change its attitude. Of

course, the LM is no longer moving or changing its attitude, so thruster firings are no longer necessary. Engine arming is disabled, and the LMP enters a command that tells the AGS that the LM has landed – amazingly, the AGS has no idea that it is on the surface of the Moon!.

After the engine has been shut down and the other systems have been placed in their post-landing configuration, the LM crew initiates Program 68, Landing Confirmation. P68 displays the landing coordinates, ensures that the descent engine has shut down, and terminates the Average G routine. P68 does not terminate the Digital Autopilot, but reconfigures it by placing the DAP in Minimum Impulse Mode with the attitude errors zeroed and the widest deadband set. This prevents any thruster activity while the crew performs the post-landing checklist. The thrusters still have the potential to fire, and must be disabled by setting the Mode Control PGNS switch to OFF.

After shaking hands, the LM crew turn their attention to getting ready for an emergency ascent. Two epochs, called T-1 and T-2, are the critical times in the first few minutes after landing. Each point is an abort option, precisely timed for an optimal return to the CSM. In Houston, the Flight Director takes a poll of the controllers specializing in each of the spacecraft systems. If all is well, the result is a "Stay for T-1". Safing the descent stage is now the primary business onboard the LM. Venting the pressurized propellant tanks is an early priority. Opening the vents on the propellant tanks reduces the pressure to low levels, eliminating any chance of the tanks or other plumbing bursting. Program 12, the LM Ascent program, is started just in case an emergency ascent is needed. About ten to fifteen minutes after landing, the Flight Director in Houston takes another poll, this time for the "Stay for T-2". If the controllers determine that all the systems are in acceptable condition, the Flight Director gives permission to remain on the surface for about two hours. This is hardly an arbitrary amount of time, as the end of the T-2 stay corresponds with the next passage of the CSM over the landing site. If problems with the crew or LM systems warrant leaving the surface, returning to the CSM at the end of the T-2 stay uses a standard set of ascent and rendezvous procedures, unlike an earlier abort which would require several complex maneuvers.

Assuming the LM crew is allowed to stay, the time is occupied with systems checks, photography, and commentaries about the landing site. The crew is finally able to get a bit of relief by removing their helmets and gloves, and start powering down the spacecraft to conserve battery life. A realignment of the guidance platform is performed, this time using Program 57, which uses a star and the Moon's gravity as references. Like the initial alignment of the CSM's IMU, a gravity alignment uses the dependable force of the Moon's pull as one of the two vectors necessary for alignment. For the final three missions to the Moon, the platform was powered off for most of the time on the surface, and was successfully powered back up as part of pre-ascent procedures.

LUNAR ORBIT RENDEZVOUS

Arguably, the most demanding and complex maneuvering during an Apollo mission was the rendezvous and docking of the LM ascent stage with the CSM. To be sure, the details of each maneuver are complex and the calculations exacting, but the basic concepts are surprisingly straightforward. The methods for Apollo were developed in large part from the experience with Gemini rendezvous techniques, which, in turn, shaped the capabilities of the Lunar Module. Two plans evolved from the Gemini experience: the coelliptic and direct rendezvous methods, and they used similar underlying assumptions. Common to both methods is that there is one "active" vehicle that performs all of the maneuvering and is launched into a lower orbit to catch up with the target vehicle. At the proper time, the active vehicle fires its engine to raise its orbit and intercept its passive target.

Much of the complexity of rendezvous comes from attempting to optimize a wide variety of constraints. The LM assumes the role of the active vehicle, performing all the maneuvers up to the point where the CSM is ready to begin docking maneuvers. Already seriously constrained by weight limits, this requires that the LM must carry a radar and sufficient fuel for all orbital maneuvers and aborts. Efficient and precise maneuvers are required to maximize these limited resources, as recovering from a poorly executed maneuver may deplete the fuel reserves or other consumables. In the face of this requirement, the engineering realities of engine or guidance performance are that a level of error is unavoidable, as hardware never operates with 100 percent precision. Much of the training and hardware testing for rendezvous focuses on minimizing these errors, collectively known as dispersions, by adjustments incorporated into either an upcoming maneuver or midcourse correction. The flexibility in the techniques makes the rendezvous tolerant of a variety of errors that might be encountered. Finally, the rendezvous maneuvers are designed to create a standardized terminal phase. Here, the approach path, velocity changes and lighting are the same regardless of the maneuvers that preceded the final phase, thereby greatly simplifying procedures, propellant budgets and training.

Orbital mechanics

Orbital mechanics, it is often said, is real rocket science, and many of the concepts are not necessarily intuitive. While the subtleties of orbital mechanics require entire textbooks to detail, the basics of rendezvous requires presenting only four main concepts. First, all orbital movements result from the motion of the object and the influence of the force of the central body's gravitational attraction. Once an object is set in motion at a high enough velocity, it can orbit the Earth forever without applying any additional force. Movement in a straight line is not possible, as it requires a constant expenditure of energy exceeding that available to any existing spacecraft. Isaac Newton provided the familiar example of a cannon on a high mountain top and above the atmosphere. A cannon ball fired at a velocity of 100 meters/sec would travel for some fixed distance before inevitably hitting the ground. As the cannon fires the ball faster and further, a curious thing occurs. Because the Earth is a sphere, the ball travels below the horizon. This comes as little surprise, and

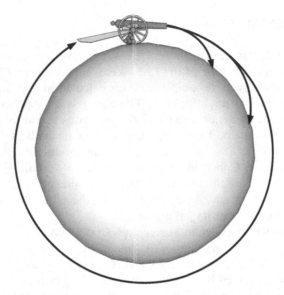

Figure 134: Isaac Newton's cannon ball experiment

Figure 134 shows the geometry quite clearly. As the ball travels outward, gravity pulls it towards the center of the Earth (remembering that gravity acts radially inward). At the same time, the Earth's surface curves away from underneath, effectively "getting out of the way" of the cannon ball. Firing balls faster propels them further around the Earth, until finally, at about 17,500 miles/hour (28,200 km/hour), the ball will travel completely around the Earth. With no atmosphere to slow the ball, it will continue to "fall" forever. In fact, most orbits are not perfect circles, but ellipses. Varying the angle and velocity of the ball when fired from the cannon determines the characteristics of the orbit, which can range from nearly circular to highly elliptical. At a sufficiently high velocity, the ball will escape entirely, never to return.

A second point is that the altitude of the spacecraft determines the time required for one orbit. The higher the orbit, the longer it takes to complete one trip. The differences can be dramatic. A spacecraft in a low Earth orbit of 200 miles altitude travels at 17,270 miles/hour (27,790 km/hour), but when that same spacecraft is in geosynchronous orbit at 22,300 miles (35,800 km), its velocity is reduced to about 6,870 miles/hour (11,050 km/hour). This notion is used when a spacecraft needs to catch up with another: by lowering its orbit, a pursuer can reduce the angular distance between itself and its objective. At the same time, when a spacecraft is in an orbit ahead of its target, maneuvering into a higher orbit will slow the active vehicle, allowing the target to catch up with it. Although the vehicle in the lower orbit travels a shorter distance around the Earth, it travels further than the vehicle in the higher orbit primarily because of its faster velocity.

Regardless of the operational details, the objective of rendezvous is to bring together two spacecraft originating in different orbits. The relative positions of the

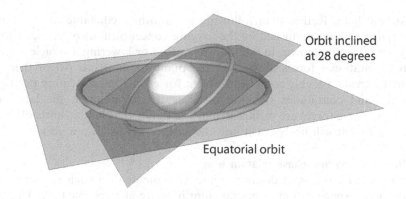

Figure 135: Orbital planes

two vehicles dictate the maneuvers necessary to bring them together, and more than adjustment to the orbital path may be necessary. The need to change one's orbit leads us to the third important concept. The simplest and most efficient technique used for raising or lowering the orbital altitude is called a Hohmann transfer.[7] To raise the apogee of an orbit, thrust is applied along the direction of flight at a point 180° away from the desired new apogee. Reducing the altitude of the orbit is done in a similar fashion. Thrusting against the velocity vector of the spacecraft will reduce its speed, and lower its altitude at a point 180° away. As the spacecraft completes its orbit around the body, it returns to the same altitude the burn was performed at. Two burns are necessary to raise or lower a vehicle's altitude at all points. Raising a spacecraft's circular orbit to 300 miles above the Earth from 200 miles first requires firing the engine to create a new, 300 mile apogee. After coasting 180° to the new apogee, the engine is fired again, this time to raise the 200 mile perigee to create a circular 300 mile orbit.

The fourth and final concept deals with another type of orbital change, when a spacecraft in one orbital inclination, or plane, is required to change to another. The most efficient way to move a spacecraft from one orbital plane to another is to time the burn at the point where the planes of the two orbits intersect. But plane changes are very expensive in terms of the amount of energy that they require. Indeed, the energy needed to move a communications satellite launched from Cape Canaveral into low Earth orbit with an inclination of 28.5° to a geosynchronous orbit over the equator is nearly the amount required to send it to the Moon! Clearly, significant plane changes are not practical in fuel critical vehicles such as the LM ascent stage.

Orbits of the Moon are particularly complex. For a perfectly spherical body with its mass distributed evenly, objects in orbit follow the path defined by an ellipse. In reality, nature is never so flawless. The Earth is slightly pear-shaped, and the small variations in the gravitational field affect spacecraft orbits over time. The Moon is a more complex body whose mass is not distributed evenly, creating an irregular

---

[7] Named for the German scientist Walter Hohmann, who developed the technique.

gravitational field. Rather than following a smooth, predictable orbit around the Moon, spacecraft are perturbed by the varying concentrations of mass, or mascons as they are formally known. In addition to raising or lowering a vehicle's orbit, the uneven gravitational forces also affect the plane of its orbit. All of these forces combine to create a highly complex problem for estimating the future position of a spacecraft, and continuous tracking is necessary to maintain the latest orbital parameters. These forces are sufficiently strong that a vehicle that starts off in a low, 60 nm lunar orbit will likely crash into the surface in less than a year.

Targeting, phasing and phase relationships
Rendezvous is a process of discrete steps, with the goal of each segment being to place the active spacecraft at a specific point in space at a specific time. This process is not arbitrary; rather, it is defined by the location in time and space of the passive vehicle. To say that the CSM is in a 60 nm circular orbit around the Moon is only one of the necessary pieces of information. Since the CSM is continually moving around the Moon, its location over a specific point is defined at a particular instant in time. Both of these pieces of information are known from the CSM's state vector and extrapolating to the desired time. The process of calculating the next maneuver so that the active vehicle arrives at its designated point in space at the same time that the passive vehicle arrives is called targeting. Rendezvous targeting is not unlike using a shotgun to hit a clay pigeon, where the challenge is to understand the path the pigeon is taking, and to shoot ahead, or "lead", so that the target and the birdshot hit. Because orbital maneuvers follow long curving paths, calculating the proper "lead" is the key to successful targeting.

Rendezvous techniques must be able to tolerate the small errors that inevitably creep in, whether due to variations in engine performance, guidance system misalignment or the variations in lunar gravity. The limits to the errors that the rendezvous method can compensate for create a "box" in space and time in the orbit. Targeting focuses on the goal of coordinating the maneuvering of one vehicle such that it arrives in its targeting "box" at the same time the other vehicle arrives in its targeting box. "Phasing" is a term that defines the relationship of one vehicle to another, and whether they will arrive at their targeted positions at the expected time. If the active vehicle will arrive at its target point earlier than the passive spacecraft, the active vehicle is considered to be "ahead" in its phasing, and if arriving later than desired then it is "behind".

Hardware dedicated to rendezvous
While the LM's radar and computer are the primary systems used in rendezvous, it is entirely possible to perform the maneuvers with either system operating marginally or indeed not at all. A defense-in-depth strategy exists to accomplish a successful rendezvous despite single-point hardware failures. An impressive array of systems and their backups are the means to ensure the robustness of the rendezvous procedures. The primary systems found on both spacecraft consist of a computer tightly integrated with tracking and ranging hardware. In the LM, the rendezvous radar and the large tracking light complement the CSM's radar transponder and sextant; and both vehicles carry equipment necessary for the CSM to determine

distances through the VHF ranging system. Additionally, the LM's Abort Guidance System is programmed to handle rendezvous guidance and navigation independently of the AGC. Onboard both spacecraft, a comprehensive set of charts and tables allow the crews to manually plot the rendezvous solutions without requiring a functional computer or radar. Finally, ground controllers use the Real-Time Computer Complex and tracking data to yield yet another source of data. While this impressive list of resources provides highly accurate navigation solutions, the rendezvous design is tolerant of errors and can accept solutions that are less than the highest level of precision. A more constraining issue is the limited amount of propellant and other consumables available on the LM. This drives the need to optimize every maneuver. A poorly executed maneuver might require a lot of fuel to correct, leaving little to address any additional problems.

Providing distance information and angles relative to the CSM is the job of the LM's rendezvous radar. Highly accurate, the radar measures distances to the CSM within 0.1 nautical miles (0.18 km) and angular position to within 0.01°. Using data from the rendezvous radar and its knowledge of both the CSM's and LM's state vector, the AGC computes the parameters necessary for each maneuver. Working together, the radar and computer determine the location of the CSM and its relative position to the LM. The radar continuously takes range and range-rate marks on the CSM, and records the radar antenna's shaft and trunnion angles to establish the relative direction to the CSM. Located on the top of the LM, the radar broadcasts a signal at 9832.8 MHz and is received by the Rendezvous Radar Transponder on the CSM. The transponder replies to the interrogation by transmitting its own signal at 9792.0 MHz. On receiving this response, the radar's internal electronics calculate the distance and relative velocity between the two spacecraft.

Onboard the LM, Program 20 requests radar range and range-rate data by setting a bit code in Output Channel 13, and the data is returned in memory location 00046_8, named RNRAD. This distance, rate and relative angle data is polled by the computer every minute, and P20 uses this data to update its estimate of the CSM's state vector. Using ranging data to provide the distance between the two spacecraft, the radar's shaft and trunnion angles, and the known attitude of the LM, the AGC determines the position the CSM in its orbit. Repeating this process throughout the rendezvous provides a set of values of the CSM's position over time, from which its velocity is calculated. These two sets of data are combined to create an accurate value for the CSM's state vector. For the LM's computer to perform a successful rendezvous, maintaining an accurate copy of the CSM's state vector is essential.

Onboard the CSM, the Command Module Pilot is also taking distance and relative position marks of his own. For measuring the distance between the two spacecraft, the CSM broadcasts a VHF signal at 259.7 MHz, which is received by the LM, "turned around", and then rebroadcast at 296.8 MHz. The Digital Ranging Generator onboard the CSM measures the total time every minute, calculates the CSM-LM distance to within 0.01 nm, and passes this range data to the Command Module's computer. Using the sextant, the CMP observes the LM's tracking strobe in order to determine the angular relationship between the two spacecraft. This light, which is located on the face of the LM, flashes with 9,000 candlepower once per second and is visible up to 140 miles (260 km) with the naked eye or up to 300 (550 km) miles through the sextant. Using the same techniques as for star sightings, the CMP centers the strobe light in the sextant, which then sends its relative angles to the computer. With the distance and position of the LM now known, the computer updates its copy of the LM's state vector. Although the CSM is the passive vehicle, an accurate copy of the LM's state vector allows the CMP to cross-check targeting solutions and assume the role of the active vehicle if necessary.

Backing up the Primary Guidance and Navigation System on the LM is a second complete guidance and navigation system. The Abort Guidance System is composed of the small AGS computer and an independent set of gyros and accelerometers mounted to the body of the LM collectively known as the Abort Sensor Assembly, and it has the capability to navigate the spacecraft through all of the maneuvers necessary for a successful rendezvous. The LMP enters and displays data on the AGS using a keypad and numeric displays in front of him. Unlike the sophisticated DSKY, the AGS Data Entry and Display Assembly (DEDA) more closely resembles a calculator. Far more limited in computational capability than the AGC, and able to display only a single line of numeric data at a time, it is tedious to enter and retrieve data. Unlike the AGC, the AGS is not a complete navigation solution. There is no interface with the LM's optics system, and the AGS needs its initial attitude and state vector data information to be supplied by the AGC. To calculate solutions for the rendezvous burns, the AGS receives the same radar data as the AGC, and by virtue of using similar algorithms, the accuracy of the two computers is very similar.

The complexities in rendezvous targeting make it reasonable to assume that only a computer has enough power and flexibility to arrive at a solution. However, a failure of both of the LM's onboard computers does not mean that the LM must abandon its role as the active vehicle in the rendezvous. Specialized charts and tables were developed to determine targeting solutions using only ranging data and the relative angles between the two spacecraft, with copies kept in both the CSM and LM. By plotting the data onto these charts, a crewmember can determine the parameters of an upcoming maneuver, such as the time of ignition and velocity required. Unlike the computer programs, the manual procedures are usable only for a fixed range of solutions and are not reliable if either spacecraft's orbit lies outside a predetermined set of values. Charts for a normal rendezvous and several standardized abort scenarios were developed, including those where the CSM assumes the active role. Even when all systems are operating normally, crews plot the data and compare the results against those from the computer. A manually derived solution, while less accurate, still is useful as it provides a sanity check against the computed results.

Finally, the data on upcoming maneuvers that is calculated by the RTCC is radioed to crews on both spacecraft, along with state vector updates. Calculated from data obtained by tracking stations, it is sufficient for a successful rendezvous, but because tracking from Earth is possible only when a spacecraft is on the near side of the Moon, its solution is less accurate than measurements made onboard either spacecraft.

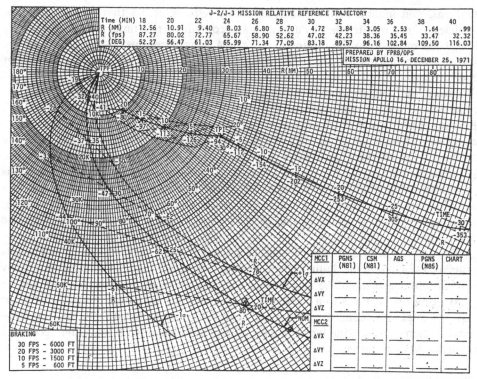

Figure 136: Polar plot from LM rendezvous procedures manual

Rendezvous techniques

During development of the Gemini rendezvous plan, standardizing the terminal phase of the rendezvous (closing, braking and docking) was identified as the key to success. Such standardization mandated specific vehicle-relative positions, closure rates and lighting conditions that simplified the more critical maneuvers of the rendezvous and docking. Mathematical models, simulations, and Gemini flight experiences, determined that the optimal altitude difference between the two spacecraft would be 15 nm (with the active vehicle below), using a transfer angle (the angular distance the active vehicle would traverse during its rise to the higher orbit of the passive vehicle) of about 130 degrees, and the lighting conditions should be such that the Sun would be behind the active vehicle during the braking phase. Mission planners then worked backwards to design ascent trajectories and orbital parameters to set up for the terminal phase.

To rendezvous with the Command Module, the Lunar Module crew has to perform several maneuvers, with each adjusting the altitude, phasing or closing rate of the trajectory. Rather than running the risk of attempting to make one burn of the ascent engine directly into a rendezvous orbit with the CSM, this ballet is composed of several steps, each of which progressively adjusts the various orbital parameters of the LM. The design of the rendezvous procedures reflects the tight LM propulsion margins, and minimizes the effects of small velocity errors that might cause the vehicles to miss one another by a large distance. During this time, the CMP is hardly idle, he is monitoring the progress of the LM, computing his own solutions to the navigation problems for use in the event that the LM's crew require assistance or even rescue.

Two rendezvous methods were developed and used in Apollo. The first, coelliptic rendezvous, applied the experience gained in Gemini to Apollo 9 through 12. Beginning with Apollo 14, and continuing through the end of the program, sufficient expertise had been gained to use a variant known as direct rendezvous. The coelliptic method reflected the conservative nature of early mission design, and required a sequence of burns over 3½ hours. From these missions, the experience gained plus the fuller understanding of spacecraft performance, allowed eliminating two workload-intensive maneuvers. In direct rendezvous,[8] the terminal phase began shortly after ascent from the surface, with docking being accomplished less than one orbit later.

The AGC provided the flexibility necessary to perform the rendezvous maneuvers. Unlike the launch from Earth, where to miss an orbital target by several kilometers might be acceptable, a properly executed rendezvous maneuver must bring two vehicles to a relative halt at a separation of one hundred meters or less. Rendezvous consists of several carefully coordinated sets of targeting measurements and engine burns, each performed independently and managed by a separate AGC program.

The coelliptic method
The goal of the coelliptic method is to create a conservative mission design, featuring a standardized terminal phase that reduces the effects of variations in performance in engine, guidance and other systems. Collectively known as dispersions, the accumulation of these small sources of error can cause significant problems during rendezvous, and must be eliminated or compensated for. Equally important, the coelliptic method provides a number of abort and rescue options in the event of the LM being unable to complete its maneuvers. The coelliptic method also provides a crew with ample time to monitor their progress, perform a number of radar and optical fixes on each other, obtain updates from Earth, and verify systems performance.

While usually referred to as the passive vehicle, the workload onboard the CSM is anything but. Because of the need to be available to rescue a stranded LM, the Command Module Pilot is continually using the CM's optics system and ranging equipment to update his own computer with the position and velocity of the LM.

---

[8] This technique was originally called "early rendezvous" or "short rendezvous", as another method had been assigned the name "direct rendezvous". The first iteration of direct rendezvous had the LM launching directly into an intercept trajectory with the CSM. This technique was soon rejected due to its inability to correct for trajectory errors, and the large amount of fuel required for the final braking phase.

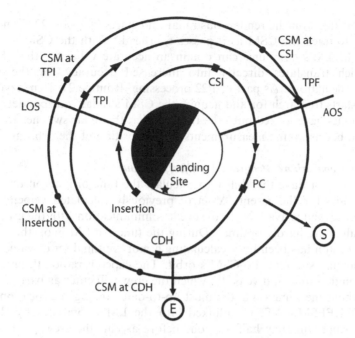

Figure 137: Coelliptic rendezvous

By calculating his own solutions, the CMP provides a sanity check of the solutions computed both on the ground and onboard the LM. Maintaining a 60 nm (110 km) circular orbit during the rendezvous simplifies the procedures, and places the CSM in an optimal position for any possible rescue maneuvers.

Coelliptic rendezvous was used on Apollo 11 and 12, and was an option on all the remaining missions. What follows is a composite version that is true to overall mission design, but does not reflect a specific mission. Figure 137 shows the overall design of the rendezvous using the LM as the active vehicle.

P57 IMU alignment – liftoff -2 hours, 15 minutes

Approximately 2 hours and 15 minutes before liftoff, the LM crew performs an IMU alignment to ensure that the platform has the most accurate orientation possible. The crew selects Program 57, Lunar Surface Alignment. While the P52 program that is used in flight orients the inertial platform by taking sightings on two stars, P57 uses the Moon's gravity as one reference, requiring only a single star sighting to set the IMU's orientation.

Optional P22 CSM tracking with the rendezvous radar: liftoff -2 hours

Two hours before lunar liftoff, the CSM is passing over the landing site for the last time. If there is any uncertainty about the LM's position, using the rendezvous radar to track the CSM will provide the LM's computer with enough data to update its position. The LM's computer can also track the CSM to update its estimate of the CSM's orbit, in the event that an update from ground controllers is not available.

The LM crew turns on the rendezvous radar and starts Program 22, Lunar Surface Navigation, to track the CSM as it passes overhead.[9] With the CSM in such close proximity, the LM's computer can obtain an accurate value for the CSM's state vector, which translates directly into increased accuracy for the upcoming rendezvous calculations. As part of P22 processing, Routine 21, Rendezvous Radar Designate, starts to search for and acquire the CSM's radar transponder. When the radar electronics have locked onto the transponder, the radar switches to automatic tracking and begins taking measurements of the position of the orbiting spacecraft.

Final launch preparations, Program 12: liftoff -1 hour
The AGC does not have the ability to calculate the time of ignition or the desired orbital velocities for the ascent. A set of previously calculated velocity targets is saved in storage, and is available in the event communication with the ground is lost, but these values are hardly optimal. During the time the LM is on the surface, the RTCC in Houston has been busy calculating a more optimal set of values based on the LM's landing site and the CSM's orbit. The capcom radios the most current values for ignition time and velocity, which the crew will enter as part of the ascent program. About the same time, the most up-to-date landing site location data (the lunar liftoff REFSMMAT) is uplinked into the LM's computer by the ground controllers. Approximately half an hour before ascent, the crew performs a final alignment of the guidance platform, again using the Moon's gravity as a vertical reference and a star as its second reference. The Abort Guidance System is initialized and is loaded with the launch parameters. Next, Program 12, Powered Ascent, is started from the DSKY. Program 12 asks the crew to confirm or update the two critical sets of data. First displaying Noun 33, the crew will accept or update the liftoff time. Next the velocity targets that were precomputed are displayed in Noun 76, and are replaced with the ground provided values. After a final check of the guidance system's switch settings is completed, there is little for the crew to do except monitor the spacecraft systems.

To serve its necessary backup role, the Abort Guidance System is also loaded with the details of the launch parameters. Without an interface to the LM optics, the AGS needs to orient itself before it can assume the task of guidance and navigation. The most convenient source of such an orientation is the inertial platform of the AGC. Entering Verb 47 starts the job to send the value of the clock, the CSM and LM state vectors and the current LM attitude to the AGC downlink. At the same time, the LMP enters 400 + 30000 into the DEDA, telling the AGS to align itself using the PGNS downlink as a data source. After the AGS initialization is complete, the LMP enters 410 + 00000, which instructs the AGS to begin processing its orbital insertion routine. Unlike the AGC, there is no formal dialog with the AGS to enter or verify parameters, placing the burden of entering all the data in its correct sequence solely on the crew. Using the data radioed up from Earth, the LMP enters the ignition time, velocity components and steering options, each into its dedicated memory

---

[9]   While scheduled for the early Apollo missions, only Apollo 11 performed the P22 tracking.

location. Managing data for the AGS was slow, tedious and error prone, and was certainly the most manually intensive task during the ascent preparations.

About 70 seconds prior to the launch, the CSM passes over the LM's landing site. All the while, the DSKY in the LM has been counting down. It blanks at T-35 seconds to indicate the Average G routine has started, with the display returning five seconds later. The Servicer routine is now computing velocity changes, although the LM has yet to begin its ascent. The Commander presses Abort Stage ten seconds before ignition, detonating the four explosive bolts that connect the two stages and firing other charges that propel guillotines to sever the wire bundles and plumbing that pass between them. At T-5 seconds, the DSKY flashes Verb 99, requesting that the crew authorize the ascent engine's ignition. As the Commander arms the ascent engine, the LMP presses PRO on the keyboard to confirm the ignition request. The normal Verb display returns, and the AGC sends the ignition command to the ascent engine when the countdown reaches zero.

Launch from the lunar surface
Ascent in the LM is often compared to being on a fast-moving elevator, and is a surprisingly comfortable experience. At ignition, the ascent stage rises at about 10 ft/sec^2 (3 m/sec^2), creating an accelerating force equal to about one-third of Earth's gravity; only twice what the astronauts were experiencing standing on the surface. Acceleration increases gradually until engine cut-off, when it will have built to about two-thirds that of normal Earth gravity. After about five seconds, and now rising at 40 ft/sec (12 m/sec), the ascent stage pitches 54° face down to build up horizontal velocity as it climbs into orbit. Such an abrupt maneuver may be a bit disconcerting to those used to seeing launches on Earth, where vehicles rise vertically and pitch over only gradually. Atmospheric drag creates such a performance penalty that launch profiles limit large pitchovers until the vehicle has risen above a large part of the sensible atmosphere. Ascending from the airless Moon allows a far more optimal trajectory.

During the 7½-minute trip into orbit, the crew spends much of their time monitoring the trajectory and the performance of the spacecraft systems. At the beginning of the ascent, the AGC displays Noun 94, giving the accumulated velocity, the altitude rate and computed altitude. The crew consult a cue card Velcro'd onto the instrument panel to monitor the performance of the ascent. Similar to the reference card used during the descent, the ascent cue card shows the expected velocities, rates and angles at 30-second intervals.

Because the ascent engine has no gimbal, the only means of adjusting the thrust vector is to change the attitude of the vehicle. The guidance routines continually compare the desired trajectory with the actual values, and adjust the attitude of the ascent stage by firing the RCS thrusters to keep the spacecraft on target. Weight and balance, so critical on aircraft to ensure the center of gravity is within limits, are even more important on the LM's ascent stage. The spacecraft's center of gravity must be over the thrust line of the ascent engine within a small margin, otherwise the Digital Autopilot will have difficulty in maintaining the vehicle on course. Samples and other items returned from the surface are carefully stored in the ascent stage to ensure that the vehicle's weight distribution is within limits. The actual trajectory is compared against the desired values every two seconds, and the LM's attitude is adjusted to stay on target. The corrections for the trajectory errors create a distinctive slow back-and-forth and up-and-down rocking of the vehicle as it aims for its insertion target.

With about a minute left in the burn, attention turns to Noun 85, showing the remaining velocity to be gained in all three axes. At this point, the LM is still 1,000 ft/sec (300 m/sec) short of orbital velocity, and is flying almost parallel to the surface. At 500 ft/sec to go, the LMP opens the propellant crossfeed valves to allow fuel and oxidizer from the APS tanks to flow into the RCS system. With attaining orbit practically assured, some of the propellants from the APS system can be tapped for use by the RCS jets. The propellants from the APS tanks do not replenish the RCS tanks, they flow directly to the thrusters. Nearing the end of the burn, attention turns

ASCENT

TFI	θ	OHW (0° YAW)	VI	H DOT	H	SBD
0:00			15.1	0	0	120/-38
0:10			60.0	54.0	300	
0:30	308	39	170.0	93.0	1900	
1:00	305	38	440.0	127.0	5200	148/14
1:30	302	35	730.0	153.0	9400	
2:00	299	33	1040.0	173.0	14300	152/19
2:30	295	31	1370.0	185.0	19700	
3:00	292	29	1710.0	192.0	25400	156/24
3:30	288	27	2080.0	192.0	31100	
4:00	285	24	2470.0	185.0	36800	161/29
4:30	281	22	2890.0	173.0	42200	
5:00	277	19	3320.0	156.0	47100	168/35
5:30	273	16	3780.0	134.0	51500	
6:00	269	13	4270.0	107.0	55100	176/41
6:30	264	10	4790.0	77.0	57900	
7:00	259	7	5340.0	45.0	59700	186/46
7:11	257	6	5540.6	32.2	60107	188/47

Figure 138: LM ascent cue card

to monitoring the spacecraft's velocity. Despite receiving an on-time shutdown command, the LM's ascent engine does not reach zero thrust instantaneously. The resulting tail-off, plus small attitude and guidance errors, invariably leaves the vehicle with small velocity errors. These are apparent from the Noun 85 display. Immediately after APS shutdown, the Commander uses the TTHC to eliminate the velocity errors in all three axes using the Noun 85 display as a reference. Further demonstrating the precision of the AGC and the ascent engine, the final residuals are small: usually a fraction a foot per second!

Post-insertion CSI preparations, liftoff + 8 minutes
Once the LM is in the desired orbit, several events must occur in quick succession. The guidance platform is realigned to correct for the minor misalignment introduced by the ascent burn. Once completed, the LM maneuvers to point its "face" (the + Z-axis) towards the CSM and the tracking strobe light is turned on. Using the rendezvous radar, the LM crew begins its hunt for the CSM. Throughout the rendezvous sequence, the LM maintains its face pointing towards the CSM. In this attitude, the CMP can see the LM's flashing beacon and the spacecraft structure does not obstruct the radar's view of the CSM. Program 20, Rendezvous Navigation, is started, and the computer begins to track the CSM using the radar. Range and angular position from the rendezvous radar are passed to the LM computer every minute to enable the AGC to compute the CSM's position and velocity.

Coelliptic Sequence Initiation: liftoff + 1 hour
The next maneuver in the rendezvous sequence is the Coelliptic Sequence Initiation, or CSI burn. When the LM boosted itself into lunar orbit, the insertion orbit was quite elliptical, with an apolune of approximately 45 nm (83 km) and perilune of only 9 nm (17 km). One hour after insertion, the LM has reached the high point and is positioned to raise the perilune to circularize the orbit. The CSI maneuver is an RCS burn of about 45 seconds, and its objective is to raise the perilune to 45 nm. The LM crew checks for any "out-of-plane" corrections that might be necessary for matching the orbital planes of the two vehicles. Program 32 is used to begin CSI computations.

Soon after orbital insertion, the LMP enters 410 + 10000 on the DEDA to start Coelliptic Sequence Initiation computations in the AGS. The target values for the burn, specifically the timing of the burn and its velocity changes, are derived from the CSM's state vector, which in turn was obtained by the rendezvous radar.

Although designed to compute parameters for the CSI maneuver, Program 32 also performs processing for the related Constant Delta-H maneuver. Using the simplifying assumptions that the burns will be made in-plane and parallel to the surface below, P32 computes Hohmann transfer parameters to bring the LM into an orbit 15 nm below the CSM and properly phased for the TPI maneuver. The role of P32 is only to calculate the velocity changes necessary for the burn. Although the software in the AGC can make a reasonable assumption of when the burns are to occur and the required velocity changes in all three axes, ground controllers are also able to calculate an accurate set of values. This data was radioed to the crew about 90 minutes before launch, and copied onto preprinted forms in the checklists. With three maneuvers (CSI, CDH and TPI) factored into the calculations, a large amount of data must be displayed, checked, and entered. Following the checklist, the astronauts work their way through the DSKY displays for times, velocities and other burn parameters. After starting P32, the software responds with V06N11, its estimate of the CSI ignition time, which normally occurs when the LM reaches apolune. Using the data provided by ground controllers, the crew enters an updated time for CSI ignition.

During the rendezvous process, a surprising capability of the computer comes to light. When discussing the AGC, it is common to state that the "computer is now running Program X", with the implication that the computer can run only one Major Mode at a time. Generally, this is correct, and the invocation of a new Major Mode program automatically terminates the currently running one. The exception to this situation occurs when running the Rendezvous Navigation program, P20 (or one of two related programs, P22 and P25). If another Major Mode program is running, starting P20 terminates that program and begins executing in its place. However, if P20 is active (as it is during all of the rendezvous maneuvers), it remains running in the "background" when the various targeting and maneuvering programs are started. This provides the crew with continuous range and position updates on the CSM, allowing the targeting programs to continually refine their calculations.

Ideally, the insertion burn places the LM in the same orbital plane as the CSM, eliminating any need for out-of-plane maneuvers. The small residual errors that do exist are usually too small to require a dedicated plane change maneuver, but an adjustment is usually still necessary. Much of this correction is possible during the CSI burn by adding an out-of-plane component to the maneuver. Using Verb 90, the details of this component can be viewed and modified if necessary. Program 32 concludes with the display of the velocity changes for the CSI and CDH maneuvers. P20 continues to run in the background, to refine the estimate of the CSM's position. Running P32 is an iterative process. After the computed values are presented, additional updates on the CSM's position become available from P20. By recycling the program using Verb 32, P32 performs its calculations again using the refined data, and recomputes the required velocity changes. This can be done until the computed velocity no longer changes significantly.

All the while the LM is running P32, the Command Module Pilot is running his own P32, LM Coelliptic Sequence Initiation, using the data from marks on the LM and ground updates. If the LM proves unable to perform any of its maneuvers, the CSM will assume the active role and perform the maneuvers itself. Additionally, the full suite of LM rendezvous programs are available on the CSM, not only to provide sanity checking of the LM and ground solutions, and also to provide the capability to act as a backup to the LM should its computer fail.

The CSI burn, as with all burns performed during the coelliptic rendezvous, is performed by the RCS thrusters. While the APS engine could perform this maneuver, the attitude it would require might possibly cause the rendezvous radar to break its lock with the CSM. Using only thrusters for the burn also ensures tighter control over the velocity changes, and hence a more accurate maneuver. Program 41, RCS Thrusting, is started and uses data calculated in P32 as the source of its targeting data. Although the CSI maneuver may have velocity components in all three axes, no attitude maneuvers are performed, because the LM has the ability to translate along all three axes when using the RCS system.

Program 41 requires keying into the DSKY remarkably few items of data, in large part because the burn information was calculated in P32. The first exchange is a request by the computer to automatically maneuver to the desired attitude for the burn. Using the RCS system there is little need to maneuver the LM into another attitude, but the flashing V50N18 serves as a reminder that the Guidance Control should be set to PGNS. Setting the Mode Control switch to Attitude Hold further ensures that the burn will use the current attitude. The velocities to be gained in the X-, Y- and Z-axes are obtained from the P32 calculations and displayed in V16N85. The crew has the option of accepting or modifying these values with new data. After this last piece of information, P41 becomes remarkably quiet during its execution. No display counts down the time before ignition; the only indication that the event is to happen is that with 35 seconds remaining, the DSKY blanks to indicate the Average G routine has started. Five seconds later, the display returns to its static display of velocity to be gained. Unlike burns with the larger engines (such as the ascent or descent engine) Verb 99 does not flash to request the crew's approval to fire the thrusters. At the time of ignition, the thrusters fire automatically. The crew watches the DSKY and its Noun 85 display, observing the change in velocity for the three axes. Ideally, the three registers display zeros when the burn completes, indicating a perfect maneuver. If there are any residual velocity components to be gained, the Commander uses the translational hand controller to correct the error, which is typically only a few tenths of a foot/sec (a few cm/sec).

Now that the LM has performed the CSI burn, the state vector kept in the CSM's AGC is no longer valid. For the CSM to continue tracking the LM, the CMP runs Program 76, LM Target Delta V, using the burn parameters as reported by the LM crew. From this velocity data and time of ignition, the CSM's computer is able to update its estimate for the LM's state vector. It can now extrapolate the LM's future position with sufficient accuracy to drive the sextant to where the LM should appear in its field of view, while continuing to keep the transponder pointed towards the LM's rendezvous radar antenna.

Optional plane change: liftoff +1 hour, 30 minutes
Immediately after the CSI burn, program P20 is started again to track the CSM and collect new rendezvous solutions. Twice, several radar marks are taken, solutions computed, recorded and plotted on specialized charts. All the while, the CMP continues his sextant sightings and VHF ranging. Assuming that the insertion onto lunar orbit occurred according to plan, the two spacecraft should both be orbiting in approximately the same orbital plane after the CSI maneuver. A small amount of error can be eliminated during the CSI and later burns, but a separate plane change maneuver becomes necessary if the out-of-plane component is beyond a predefined threshold. Soon after the CSI maneuver, ground controllers and the crews decide whether the out-of-plane component is sufficiently large to justify a special burn. Nearing the time for the plane-change burn, the LM emerges from the Moon's

shadow and approaches the point where the two orbital planes intersect. Now in sunlight, the tracking beacon is turned off, as its light is washed out against the lunar surface below. As a plane change is not normally performed,[10] a dedicated program is not provided in the LM's AGC. The crew obtains the necessary data from ground controllers, and inputs the values into Program 41, which uses the RCS thrusters to bring the LM's orbit into the CSM's plane.

Constant Delta Height maneuver: liftoff + 2 hours

The Constant Delta-H, or CDH, burn must place the LM in an orbit that maintains a constant 15 nm (24 km) difference in altitude below that of the CSM.[11] The maneuver is burdened with a somewhat inappropriate name. Ideally, the LM should be in a perfectly circular 45 nm orbit after the CSI maneuver, 15 miles beneath the CSM's perfectly circular 60-nm orbit, with the phasing such that the Terminal Phase Initiation maneuver will occur exactly on time. In such a case, there is no need for a CDH burn. In reality, the orbits of the CSM and LM are not perfect, and it is necessary to adjust the phasing between the two vehicles for the all-important TPI burn. As a result, the CDH burn also serves as a phasing maneuver that adjusts the LM's orbit so that TPI occurs at its optimal time. If it appears that the TPI burn is going to be early, the CDH maneuver will place the LM is a slightly higher, slower orbit to delay the point when TPI is to occur. In the same manner, compensating for a late TPI is through using the CDH to lower the LM's orbit to enable it to arrive at the designated point sooner.

In addition to the rendezvous solutions made by both the CSM and the LM, the RTCC has also been computing its own solution based on MSFN tracking. All three solutions are evaluated to determine the best values for the CDH burn. Generally, the measure of the "best" value takes into account both the accuracy of the data source and also the consistency of the solutions. Ideally, each of the several sets of marks taken on the other vehicle should arrive at similar solutions for the upcoming burn. A set of solutions for the upcoming maneuver will be discarded if they do not settle into a narrow range of values, or disagree significantly with the solutions obtained from other sources. When Program 33 is started, the exchange of data between the crew and the computer is similar to that used in the CSI program, P32. All the while that P33 is running, P20 continues to track the CSM and updates its tracking. If the solution computed onboard does not agree with the other solutions available to the LM crew (ground computations and onboard charts), then entering Verb 32 will cause P33 to return to the beginning of the program and recalculate using the most current data on the CSM's orbit. Over time, the change in velocity will converge to a small range of values as additional tracking measurements are

---

[10] In fact, none of the Apollo missions required the LM to perform a plane-change burn.
[11] Under abort rendezvous scenarios, the LM is often in an elliptical "bailout" orbit in order to reestablish the proper phasing relationship between itself and the CSM. The CSI and CDH maneuvers are used together to circularize the LM orbit at 45 nm to set up for a standardized terminal phase.

made. As with the CSI maneuver, the CDH burn is performed using P41, RCS Thrusting. Normally, the maneuver lasts only a few seconds and produces a velocity change of only a few feet per second.

An entry of 410+20000 on the DEDA starts CDH computations on the AGS. The AGS continues to receive updates from the radar and computes its own solution of the CSM's orbit from the marks taken.

Terminal Phase Initiation: liftoff +2 hours, 40 minutes
Tracking of the CSM using P20 continues, and Program 34, Terminal Phase Initiation, is started on both the LM and CSM to compute the parameters needed for the maneuver. On the AGS, 410+30000 is entered into the DEDA to begin computations for the TPI solution. The timing of the burn and its velocity changes are carefully planned to ensure that a standard approach is performed. Developed from the experience from Gemini, this maneuver starts with the CSM about 26° above the LM's local horizontal plane, and the LM will travel about 130° around the Moon before it draws alongside the CSM. The geometry of this approach allows the LM to rise to the CSM's orbit from below and behind, only to slip slightly ahead of the CSM when their altitudes match.

Terminal Phase Initiation, as its name implies, is the last and arguably the most precise of the rendezvous maneuvers. While appearing somewhat inefficient, the maneuver is designed to tolerate small errors in velocity and in the estimates for the spacecraft's position. Small errors that remain after the TPI maneuver are eliminated during two planned midcourse corrections. If allowed to continue, the nominal TPI trajectory places the LM ahead and slightly above the CSM, and the excess velocity that takes the LM above the CSM is canceled during the final braking phase. About 2 hours and 40 minutes after insertion, the TPI burn is performed. The LM's RCS thrusters are used to increase the spacecraft's velocity by 20 to 40 ft/sec (6 to 12 m/sec). Timing is critical, as an early or delayed ignition results in errors that must be compensated by larger midcourse corrections.

After the TPI burn, attention in the LM turns to fine tuning the trajectory through two midcourse corrections. While the TPI burn may have been right on time, and the velocity change is accurate to within a few tenths of a foot per second, even small errors in position, attitude or velocity can be sufficient to cause the LM to miss by a large fraction of a mile. As the LM arcs towards the CSM, the smaller distances and slower relative speeds allow highly accurate measurements of range and velocity. Program P20 continues, and new radar marks on the CSM are taken.

TPM maneuvers: liftoff +2 hours, 55 minutes and 3 hours, 10 minutes
In the 45 minutes between the TPI and TPF maneuvers, opportunities for two midcourse corrections are available if needed. With Program 20 running in the background, Program 35, Terminal Phase Midcourse, calculates the midcourse velocity changes. These small but essential maneuvers require at most a few feet per second in each axis. The first midcourse correction is performed about 15 minutes after TPI, and the second one 15 minutes later. The velocity changes are specified in all three axes, with the LM maintaining a fixed attitude throughout the terminal

phase in order to maintain a solid track on the CSM. As the LM is now on its intercept course, there is no longer any need for the CMP to take marks using the sextant, and his duties now switch to preparing the CSM for docking. He moves from his position at the sextant in the lower equipment bay, and climbs into the left-hand couch. Distance, rate and angular offset information from the VHF ranging system are displayed on his DSKY. At this point, the two spacecraft are operating on their own, as the rendezvous radar, VHF ranging system and old-fashioned eyeballs are far more accurate than any data that the ground controllers could provide.

Terminal Phase Final: liftoff + 3 hours, 25 minutes
During the terminal phase, crews often comment that the LM ascent stage appears to be "riding up on rails". By design, this is not an illusion. The terminal phase is designed so that the vehicles will appear to be fixed in space relative to each other. The LM appears to be fixed against the background of stars, allowing the CSM to track it without having to continually change its attitude. Shortly after the LM's second midcourse maneuver, the CMP selects Program 79, Final Rendezvous. P79 maneuvers the CSM from the attitude that held the radar transponder pointing towards the LM, to an attitude that aligns the CSM with the LM ascent stage for docking. Once this maneuver has completed, the program provides the range, range-rate and angular difference between the CSM's X-axis and the LM. The transponder, which was necessary when the two spacecraft were hundreds of miles apart, can be switched off because the reflected signal from the CSM skin is sufficiently strong that the radar can still provide accurate data to the AGC.

As the LM approaches the CSM, its crew have several sources of information available to monitor the progress of the rendezvous. After the final midcourse burn, only 15 minutes remain in the rendezvous. Program 47, the Thrust Monitor program, is started. It will provide data on the changes in velocity during the final braking maneuver. Both the DSKY display and the tapemeters on the Commander's panel that displayed altitude and rate during the lunar landing are now used to show the distance and closing rate of the CSM. And the cross-pointer used during the landing to show lateral velocity is pressed into service to show the relative motion between the CSM and LM in the X- and Y-axes. Together, these instruments provide the crew a complete picture of the position and velocity relationship between the two spacecraft.

Terminal Phase Final is a series of small braking burns, begun about 6,000 feet (just under 2 km) from the CSM, and completed when the two vehicles are only a few hundred feet apart. Using data from the rendezvous radar, the Commander in the LM monitors the closing rates and adjusts the braking burns to perfectly match the LM's orbit with that of the CSM. A series of four burns are made, each to achieve a specific amount of velocity reduction. Program 47 allows the crew to monitor the velocity changes with V16N83. Range and range-rate to the CSM can be displayed on the DSKY using Verb 83, and the velocity monitor can be called back up by pressing the PRO key. The crew can toggle back and forth between the velocity and radar data display by keying Verb 83 and then PRO. In these last few minutes of the approach, the LM Commander manually slows the spacecraft by 30, 20, 10 and 5 feet per second at distances of 6,000, 3,000, 1,500 and 600 feet respectively, using the two forward-facing RCS thrusters. At the end of these braking maneuvers, the LM has parked itself just slightly ahead of the CSM.

When the final braking is complete, the two spacecraft are in identical orbits, and the docking can proceed at an unhurried pace. Several minutes of "station keeping", or orbital formation flying, are performed to allow the crews to photograph each other's spacecraft and verify that the docking systems are ready. For the first time during the rendezvous, the LM turns its "face" away from the CSM and pitches down to present its docking target and drogue to the CSM. The CMP, with the advantage of better visibility, takes the active role for docking. Thrusting forward at little more than one foot per second, the CSM closes the 150 feet (50 meters) distance to the LM.

A docking target is mounted on top of the LM. It is a "T" shaped cross centered 15 inches (38 cm), above an 18 inch (46 cm) diameter disk. Sighting the docking target through the COAS, the CMP can perceive any misalignment during his approach to the LM. Immediately before contact, the CSM halts its movement towards the LM, and both vehicles set their Digital Autopilots to Free mode. This inhibits the autopilots from trying to maintain a specific attitude. While the docking itself is performed slowly and smoothly, the vehicles tend to rotate a small amount on initial contact. If the DAP on either spacecraft were active, it would sense this motion and struggle to correct the attitude error, possibly introducing stresses beyond what the vehicles were designed for. Were the DAPs on both spacecraft to remain active, they would fight each other as each tried to assume the "correct" attitude. After the DAPs are configured for docking, the CSM moves forward, the docking probe slides into the LM's drogue, and small, fingernail-sized capture latches at the tip of probe engage the small end of the conical drogue. Damping any motion resulting from the contact is handled by a set of three shock absorbers mounted on the probe. Not unlike those installed in an automobile, the shock absorbers both dampen the motions resulting from contact and align the two vehicles. Talkback indicators on the CMP's panel indicate that the LM has been captured, and that a "soft dock" has been achieved. The probe is retracted, and when it has pulled the LM into contact with the Command Module's docking ring, the twelve docking latches securely grab the LM for a "hard docking".

MINKEY processing in the CSM
By the time Apollo 15 flew, new programming in the Command Module's computer reduced the workload of the CMP during the busier periods of the rendezvous. Known simply as the MINKEY (MINimum KEYstroke) routine, it automatically sequenced the CMP through the several rendezvous programs, routines and displays that previously required entering a large number of keystroke sequences. Crews praised the MINKEY software in its ability to reduce the workload on a very busy CMP. Although the LM crew could have certainly benefited from the savings in workload available by the MINKEY routine, the lack of memory in the LM AGC prevented its implementation.

The MINKEY controller runs in the background as Routine 07 in the CSM's computer, beginning immediately after starting one of the rendezvous programs (P31 through P36). Verb 50 displays with the Noun 25 checklist code of 00017, asking the CMP whether he wants to use the MINKEY sequencer. Entering PRO starts the sequencer, while keying ENTER tells the software that all the tasks associated with the rendezvous programs are to be performed manually. Under MINKEY control, much of the interaction with the computer is reduced to verifying the most important data and accepting the previously computed data by keying PRO in response to a display. During a standard rendezvous, the ordering of programs run by the CMP follows a fixed sequence of P32 (CSI Targeting), P33 (CDH Targeting) and P34 (TPI Targeting). MINKEY automatically runs each program in sequence, and reinitializes the W-matrix at the end of each program. Depending on the amount of velocity change calculated in the rendezvous program, MINKEY calls the SPS or RCS thrusting program in case the CSM is required to assume the active role in the rendezvous. Pressing ENTER during any of the automatic routines cancels the MINKEY sequence and returns the AGC to fully manual operation.

Direct rendezvous

After the experience of several successful missions, the maturity of the rendezvous procedures and systems increased to where the direct rendezvous method was practical. Direct rendezvous, it can be said, is identical to the coelliptic method with all the unnecessary parts taken out! Beginning with Apollo 14 and continuing for the three remaining missions, the direct method reduced the elapsed rendezvous time to under two hours and required less than one orbit to perform. Remembering that the conservatively designed coelliptic technique assumed that several opportunities for orbital corrections would be needed, it was designed to include extensive position and range updates to ensure accurate navigation and a methodical set of orbital maneuvers. The direct rendezvous method eliminates both the Coelliptic Sequence Initiation and Constant Delta Height maneuvers, and makes the TPI burn shortly after orbit insertion. The sequence of events is:

0:00:	Launch from the lunar surface and insertion into a 9 × 45 nm (17 × 83 km) orbit.
0:10:	Tweak burn, if required to refine the orbit
0:40:	Terminal Phase Initiation
0:55, 1:10:	Terminal Phase Midcourse maneuvers
1:25:	Terminal Phase Final, manual braking and docking.

To assure that a rendezvous is practicable on the first orbit, three important capabilities must be demonstrated. First and most importantly, control of the engine and guidance system must be sufficiently accurate to perform the insertion and TPI burns with precision. Next, the ability for both the CSM and LM tracking systems (the rendezvous radar on the LM, and the sextant plus VHF ranging on the CSM) must be adequate for the task. Finally, streamlined crew procedures to perform all the tasks within the greatly truncated time available is essential. The first two

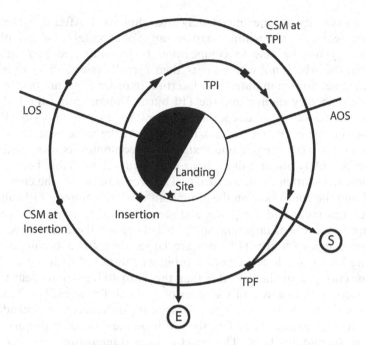

Figure 139: Direct rendezvous maneuvers

capabilities were demonstrated by Apollo 9 through 12, and the procedures were refined in ground simulators. In all cases, the systems performed as expected or better, and their capabilities were proven in situations that were not entirely nominal. For example, Apollo 11 had difficulty in establishing an accurate fix on its landing site, and Apollo 12's ascent engine burned the spacecraft into a slightly higher orbit than planned. Later, the CSM's orbit on Apollo 15 was much more elliptical than planned. All these variants had been anticipated, and the systems were designed to compensate for a certain level of error.

Sequence of events in direct ascent rendezvous

Direct rendezvous begins in the same manner as the coelliptic method, with pre-ascent updates sent to the LM by the ground, an alignment of the guidance platform, and the optional tracking of the CSM during one of its passes over the landing site. As crews stayed longer on the surface, the Moon's own rotation moved the landing site away from the CSM's orbital plane, and beyond the capabilities of the LM to execute a plane change. The task of changing planes was assigned to the CSM and usually performed on the day before the scheduled ascent from the Moon. The timing and parameters of this plane change were such that the CSM and LM would be essentially in the same plane at the nominal liftoff time, thereby reducing the propellant demands and the workload on an already packed schedule.

Liftoff and orbital insertion are the same as the coelliptic method, and the LM is placed in a 9 × 45 nm (17 × 83 km) orbit. A quick evaluation of the orbit using the accumulated velocity is done immediately after shutdown. After insertion, the LM's orbital parameters must be within a narrow range of acceptable values, otherwise the TPI burn may not be able to compensate. If ground-based and onboard data indicate that the orbit is not within limits, then a small "tweak" burn to fine tune the orbit is scheduled for ten minutes after insertion in order to ensure that the LM is on an accurate trajectory leading into the TPI burn. Assuming that the LM is in good health and the trajectory is acceptable, the crew begins program P20, Rendezvous Navigation, on the computer to track the CSM. At the same time, the CSM tracks the LM using its VHF ranging and sextant, and computes its own solutions to the TPI maneuver. Only about half an hour of tracking is available between insertion and the time when attention must turn to TPI preparations, so the crews in the LM and CSM and the controllers on the ground all work to compare TPI solutions. The LM rendezvous radar and computer, and ground-based targeting, are generally the most accurate sources of targeting, so the bias is to use these solutions.

Velocity changes for the TPI burn are larger than for a coelliptic rendezvous because the LM is in a lower, elliptical orbit, as contrasted with the 45-nm circular orbit of the coelliptic method. Rather than the 20 to 40 feet/sec maneuvers using the RCS thrusters, a TPI burn is of the order of 70 to 80 feet/sec (21 to 24 meters/sec), necessitating the use of the APS engine. Despite the differences, the intended effect of the maneuver is the same. After TPI, the LM is on the familiar trajectory to meet the CSM some 40 minutes later. The benefits of a standardized terminal phase are clearly apparent, as the two midcourse corrections, final braking, and docking are the same as discussed in the coelliptic technique.

Rendezvous after an abort

On a nominal mission, the rendezvous process begins with an on-time ascent from the surface, and maneuvers extensively planned and precisely executed. But what happens if there is a problem that is serious enough to require abandoning a landing attempt? Depending on the timing of the abort, the LM can be left in a wide variety of possible orbits, each of which might have its own rendezvous solution. Scenarios ranging from a systems failure in a critical component to simply running out of fuel without an acceptable place to land, are hardly the far-fetched dreams of a malevolent simulation supervisor; nor are landing aborts an exercise that is practiced and shelved with little expectation that they will be needed. Problems experienced by Apollo 11 and 14 involved issues that would reasonably require an abort. During the prelanding checks on Apollo 14, an errant solder ball in the Abort pushbutton was intermittently triggering the abort discrete in the computer. If the solder ball were to set the abort indicator shortly after the descent began, then the AGC would automatically abort using the descent stage. Problems late in the descent, such as Apollo 11's program alarms, or when Apollo 14's landing radar initially refused to lock on, would have forced jettisoning the descent stage and using the ascent stage engine to return to orbit.

Optimally, the crew would wish to abort with the ascent and descent stages still attached. The fuel, oxygen, power and cooling water available in the descent stage greatly extends the lifetime of the spacecraft, and provides time and maneuvering capability for a greater range of options. An abort option with an unstaged LM is available for only the first four or five minutes of the powered descent. After that point, there is insufficient fuel in the descent stage to bring the entire LM back into the proper orbit, requiring that the crew shed its dead weight and rely on the ascent stage. Normally, aborts in the LM are initiated under the control of the AGC by pressing either the Abort or Abort Stage buttons on the Commander's main instrument panel. Pressing either button terminates the currently running program, and starts one of two abort programs. Pressing the Abort button early in the descent causes the AGC to switch to Program 70, pitching the spacecraft forward and throttling the descent engine up to full thrust to guide the LM to a safe orbit. If a situation demands an abort late in the descent, with fuel, velocity and altitude low, then pressing the Abort Stage button will automatically start Program 71, which shuts down the descent engine, fires the pyrotechnic devices to separate the stages, and ignites the ascent engine to achieve orbit. Aborts can also be initiated through directly entering the abort program number through the DSKY using Verb 37. However, given the likely time pressure, simply hitting the appropriate abort button is generally the easiest means of recovering from a bad situation.

Despite its name, control does not automatically pass to the Abort Guidance System during an abort. If the AGC is still operating properly (that is, the abort was required because of other LM system problems), this will handle all of the chores necessary for orbit insertion. In the case where the computer, platform or other components of the PGNS are suspect, the Commander can flip the Guidance Control switch to AGS, which then assumes guidance and navigation duties.

The need for a standardized terminal phase is especially important in rendezvous after an abort, and depends on the CSM being "ahead" of the LM, forcing it to "catch up" from behind. In the minutes before powered descent, the LM is slightly ahead of the CSM, making a standard coelliptic rendezvous impossible. As the LM slows during its descent to the surface, its decreasing velocity allows the CSM to pull ahead in its orbit, and in the event of an immediate abort from the surface, the CSM is too far ahead for a standard coelliptic procedure. Restoring this all-important phase relationship and returning to the standard rendezvous requires that the LM adjust its orbit or make an additional circuit of the Moon. The rendezvous technique used depends directly on how far into powered descent the LM is when it declares the abort. An early abort, with the CSM insufficiently ahead of the LM for an optimal rendezvous, created the need for the LM to "fall back" to its optimal position for TPI. Manipulating the orbital period is the means to adjust this phase relationship, and it is performed by the active vehicle. Increasing the LM's orbital period by increasing its apolune causes the LM to increase the distance between itself and the CSM. Likewise, decreasing the orbital period by lowering the LM's orbital altitude allows it to catch up with the CSM.

Limitations in the AGC prevent onboard real-time calculation of the various abort options, so data for many of the abort burns is determined in the RTCC and relayed to the LM crew. They then use one of the "External Delta-V" programs to set up the maneuver and manage the burn. Most of the other programs used in the LM (such as descent, ascent and rendezvous) have the guidance and navigation solutions calculated internally; that is, using only the resources available on the spacecraft. The External Delta-V series of programs use data relayed from the ground. This includes the velocity to be gained, the time and duration of the burn, and the attitude the spacecraft. Three programs are available for External Delta-V maneuvers, and differ primarily by the engine(s) used for the maneuver. Program 40 is for burns with the Descent Propulsion System, Program 41 uses the RCS thrusters, and P42 is for burns using the Ascent Propulsion System. Whether an abort occurs early or late in the powered descent, and whether the LM has staged or not, are the determinants for which of these programs will be used.

As an example, the LM is well into its braking maneuver, and a serious problem develops onboard. Once the decision is made to abort, the Commander presses the Abort Stage button[12] (but almost certainly not before muttering something unprintable). Immediately, the descent engine is shut down, the LM is staged, Program 71 is started on the AGC, and the ascent engine is fired. If the spacecraft is less than 25,000 feet (7.6 km) above the surface, the LM will pitch over to a vertical attitude to slow its descent; otherwise, it will pitch forward and gain horizontal velocity. If the abort occurs between 12 minutes prior to PDI and about 5 minutes into the descent, the LM has a considerable distance to "fall back" in order to reestablish the proper orbital relationship with the CSM. A higher orbit is necessary to slow the LM and increase its separation from the CSM, and an early abort requires two orbits to create a sufficient distance. Rather than entering the familiar 45 x 9 nm orbit, the LM increases its apolune to approximately 115 nm. Although the high point of the orbit is well above the lunar surface, the perilune is still only 9 nm. Once the LM reaches the high point of the orbit, the Boost maneuver is performed to raise the perilune to about 45 nm. Now that the LM is in a sufficiently high orbit, attention turns to continuing the adjustments to synchronize with the CSM. As the LM reaches its perilune, it performs a Height Adjustment Maneuver (HAM). Like the Boost maneuver, the HAM burn parameters are computed on the ground and are verbally communicated to the crew. Performed at the first perilune after the abort, the HAM is a small maneuver designed to set the stage for the next phase of the rendezvous.

As the LM reaches its second apolune, the elegance of the rendezvous architecture again becomes apparent. After the HAM, the LM crew begins to track the CSM using Program 20, Rendezvous Navigation. This data is used to calculate the CSI maneuver, which is now used to begin the two-step process of lowering the LM into the standard 45 nm circular orbit. Despite the fact that the LM performs the CSI at a much higher altitude during an abort than after a normal ascent from the surface, the objective remains the same. Using the combination of CSI and CDH burns, the LM transitions from an elliptical orbit to a circular orbit that is properly phased for the TPI maneuver.

---

[12] If this example were for an abort much earlier in the descent, the Commander would select the Abort button rather than Abort Stage, and the LM would retain its descent stage.

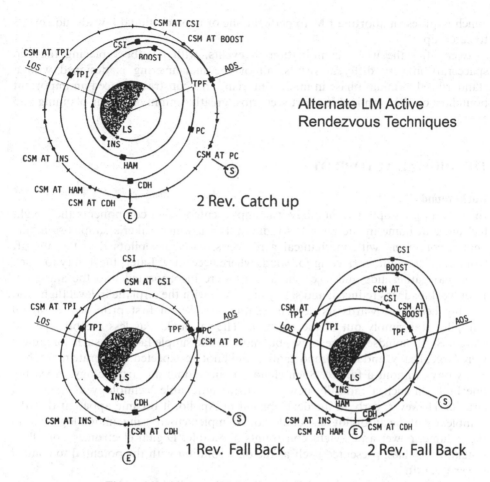

Figure 140: Alternate LM active rendezvous techniques

An abort late in the descent, such as after the LM pitchover maneuver, finds the CSM pulling ahead of the LM, which is a more desirable position with respect to phasing. Unlike an abort early in the descent, an abort at this point always requires jettisoning the descent stage. Only one fall-back orbit is necessary to produce the desired separation between the LM and CSM. As with aborts early in the descent, a late abort brings the LM into an orbit much higher than the CSM in order to create a longer orbital period. As the LM approaches the apolune of the first orbit after insertion, the CSI burn adjusts the perilune to 45 nm. As the LM nears its new perilune, the CDH burn lowers the apolune to place the LM into a circular orbit with the proper phasing for the terminal phase.

If the LM aborts very late in the descent, the CSM is very well placed for a standard coelliptic rendezvous. The CSM is sufficiently ahead of the LM so that a fall-back maneuver is no longer necessary; rather, the LM now must "catch up" with its target. As time passes after landing, the CSM moves further and further ahead, which requires an aborting LM to perform one or two additional low altitude orbits to catch up.

Even after the worst combination of events, with an abort placing the two spacecraft in very different orbits, all of the maneuvering is designed with a standardized terminal phase in mind. Enforcing this constraint placed an important boundary on the permissible abort scenarios, greatly simplifying their planning and execution.

THE DIGITAL AUTOPILOT

Background
In the 1960s, autopilot technology was implemented using components that might feel more at home in the aircraft's radio and in a watchmaker's shop. Essentially analog computers with mechanical gyroscopes, early autopilots flew the aircraft from point to point, correcting for winds where necessary. Later, the ability to track radio navigation aids and steer the aircraft were incorporated into the autopilot. Pilot workload was reduced, and safety improved with the airplane in level flight and reliably heading towards its next navigation fix. When first proposed, a digital autopilot was simply out of the question. The rationale was less the oft-repeated "Right Stuff" mindset of the crews, but more of the simple fact that such computers were large, heavy, and most importantly, could not be expected to operate more than a few days without a failure. Such characteristics doom any technology, especially one built on such a radically new and (then) unreliable technology as integrated circuits. However, early in the development of Apollo, it became apparent that the complexity of the autopilot task could not be implemented using analog logic within the volume, power and weight constraints allowable. Digital electronics, for all its potential troubles, presented itself as the only solution with the potential to control the spacecraft.

For this to be successful, entirely new technologies had to be created in very short order. Apollo pioneered the concept of a "fly by wire" spacecraft, in which no mechanical connection exists between the hand controller used by the astronaut and the jets that actually maneuver the vehicle. The Digital Autopilot software in the AGC interprets the motions of the hand controller using mathematical formulas that define how the vehicle must respond to forces acting upon it. These formulas, called "control laws", use the displacement of the control stick and the rate of its motion to decide on the appropriate amount of thruster jet activity. The thrusters are small rocket engines in their own right, with rapidly acting valves allowing the engine to be started and shut down quickly and precisely. Additionally, the thrusters are capable of a wide variety of burn times. Long firings of the thrusters, often lasting several seconds, are necessary for maneuvers to separate the two spacecraft or to settle fuel in a tank. For precise control of the spacecraft, very short bursts of less than 0.02 seconds can precisely maintain the vehicle's attitude.

Control laws create a specific set of handling qualities that can be altered to create different flying characteristics. For example, a given amount movement on a control stick can be interpreted to mean a large or small deflection of a control surface. This amount of deflection, and the speed it moves, determines the controllability of an aircraft. Flying qualities, or the "feel" of an airplane, are often described as harmonious, sensitive, heavy, or any number of adjectives. Pilot expectations also help define how an aircraft handles. A fighter or aerobatic aircraft should be highly responsive to only a slight motion of the stick. In a commercial airliner, stability and smooth coordinated control is the goal. By adjusting the sensitivity, or "gain", of the controls, different senses of controllability are created. This does not imply a Boeing 747 can be modified to maneuver like the latest fighter jet, only that the handling qualities are not immutably fixed. A classic case is the modified Gulfstream G-II used for Space Shuttle approach and landing training. In the "normal" flying mode, the aircraft behaves like the standard business jet that it is derived from. In the shuttle simulation mode, thrust reversers deploy and the handling characteristics are altered through software to reproduce a Shuttle making a steep gliding approach to the runway.

Digital autopilot requirements

The Digital Autopilot, or DAP, is an essential software component in both the Command Module and Lunar Module computers. Unlike an autopilot in an aircraft, the DAP does not perform any navigation tasks; rather, it focuses on the maneuvering and control characteristics of the spacecraft. Specifically, the DAP is tasked with several important functions:

- The DAP must hold the spacecraft attitude to within an angle specified by the crew. This angle, called a deadband, typically has a value of between 0.3 degrees and 5 degrees. Maintaining a small deadband is not always desirable. The slightest drift from venting or other small force will cause frequent jet firings, making holding a given attitude more fuel intensive than a wider deadband.
- Virtually all of the CSM's and LM's maneuvering is done under the Digital Autopilot's control. A wide range of programs calculate the attitude changes which the DAP must perform: powered flight, inertial platform alignments, and rendezvous all require maintaining a specific attitude. While tracking landmarks on the surface, the attitude continuously changes at a constant rate. Powered flight is especially demanding on the DAP, as a large part of mission success depends on accurate maneuvering to achieve position and velocity targets.
- It is easy to design an autopilot for situations where all of the hardware is working well. However, a thruster jet may fail or the crew might need to disable specific jets. The DAP must adapt to these configuration changes in order to ensure controllability of the vehicle.
- During powered flight using the CSM's Service Propulsion System or the LM's Descent Propulsion System, thrust must be directed through the spacecraft's center of mass. A misaligned engine will cause the spacecraft to rotate, forcing the DAP to use additional fuel as it tries to correct for this

error. Jackscrews under control of the DAP move the engine on its gimbals to ensure that it fires through the center of mass.

While similar in concept, the Digital Autopilots in the two vehicles have little software in common. Spacecraft thruster configuration, fuel systems and vehicle symmetry are only a few of the major differences. Control of the CSM must be over the full spectrum of the mission, from boost through entry, including orbital, translunar and atmospheric flight. The LM's DAP has far different needs, coming alive only in lunar orbit with much of its focus on the lunar landing and ascent. Still, there is much hardware in common between the two spacecraft. The Inertial Measurement Units and hand controllers are identical, and the propulsion system gimbals are conceptually similar.

RCS jets: providing the force to maneuver

Looking closer at the different configurations of the Command Module, Service Module and Lunar Module gives a perspective on the different DAP configurations that are required. Maneuvering the CSM is performed using four sets of four-jet "quads" mounted on the Service Module. Each thruster provides 100 pounds of thrust (445 Newtons) in steady-state operation. These thruster quads are mounted at 90-degree intervals around the circumference of the Service Module, and at approximately the center of mass when the CSM is loaded with fuel. Each quad is supplied by a dedicated propellant system, which itself contains two independent sets of propellant tanks and valves. Thruster quads are offset -7.25 degrees along the X-axis, requiring the DAP to make a small rotational adjustment in its calculations between the physical location of the jets and their rotation axis. This is not a particularly difficult exercise, as the CSM is reasonably symmetrical in pitch and yaw; the roll axis is unaffected. The jets themselves direct their thrust 10 degrees away from the Service Module's skin to limit heating from plume impingement. Thrusters on the CSM can be enabled or disabled individually, allowing considerable control over the RCS configuration.

The Command Module thruster arrangement is completely different from the Service Module, and the thrusters themselves are of a different design. The Command Module is conical, with the exterior covered in an ablative heat shield. Twelve thrusters, each rated at 93 pounds of thrust (414 Newtons) are imbedded within the heat shield, which protects them from the heat of entry. For redundancy, two thrusters are available in each axis for rotations in the positive direction, and two for negative rotations. The Command Module's thrusters perform only rotational motions, as there is no need for translations during the short time between jettisoning the Service Module and entry. Each set of jets is supplied by an independent propellant system, and either system can adequately control the CM. Individual jets cannot be enabled or disabled directly. The only level of control is to enable or disable a propellant system that affects all six jets that it supplies. A loss of one system will reduce control authority, but the CM is still maneuverable.

The LM is a far more complex case. Here, the thruster quads are rotated 45 degrees along the X-axis, with the jets aligned along the translational axes. With the thrusters in the "corners" of the ascent stage, the thruster arrangement still provides for full translational and rotational control. Two independent fuel systems supply the RCS, each of which connects to two thrusters in each quad. The thrusters and the fuel systems that supply them are arranged such that a failure in one fuel system will still allow full control of the vehicle. As with the CSM, failed or disabled jets are compensated by the DAP, using a table of thruster combinations to ensure controllability. Control of the LM is not significantly different in the ascent or descent configuration. The DAP also includes a capability in which the LM can maneuver the docked CSM/LM. While inefficient and not especially precise, this feature allows maneuvering the entire stack in the event of the Command Module AGC being unable to exercise control.

Hand controllers

Manual rotations of the spacecraft are commanded through the Attitude Controller Assembly (ACA), a pistol-grip controller capable of rotation in three axes. Only a small motion, about 1.5 degrees from the neutral, or detent, position, is needed to send a rotation command to the AGC. A much larger deflection of 11 degrees fires the jets directly, without computer control. Each valve (fuel and oxidizer) in an RCS jet has two solenoid coils that open and close the valve. One coil is sufficient to operate the valve and fire the engine. When thrusters are commanded by the AGC, the primary coil is used. Moving the hand controller to the direct position disables the primary coils in that axis, and energizes the secondary (direct) solenoid coil. This allows the crew to fire a thruster directly, outside of computer control in the event of the automatic systems being unusable.[13] Direct mode also creates a small problem in nomenclature. Although direct commanding of RCS jets is exclusively under the astronaut's control, the term "manual control" is never used for this mode.

Reconfiguring thrusters and the effects on the DAP

When thruster configurations in either spacecraft are changed, the DAP can compensate by modifying its jet selection rules. While it may seem unwise to disable thrusters intentionally, there are several operational requirements for this. Of course, thruster failures or leaking propellant valves certainly demand isolating a jet. More commonly, the need to prevent a thruster from firing is the concern over the effects its plume will have when it impinges on the other vehicle. Most obvious is the upward facing thrusters on the Service Module and their effect on the LM ascent stage when the two spacecraft are docked. In addition, disabling some jets and forcing others to perform the work is useful for equalizing propellant usage across the different fuel systems.

Phase planes: defining the autopilot state

Using an aircraft as a familiar example, engineers use mathematical models and wind

---

[13] In the CSM, the Stabilization and Control System (SCS) is completely independent from the AGC, and it is able to fire the thrusters. Thrusters in the LM can also be controlled by the Abort Guidance System.

tunnel testing to develop equations of how an airplane will perform in response to a pilot input. These expressions become the control laws, which take into account the aircraft's configuration and the external environment to define how to move the control surfaces. Combined, these control laws become the core of an autopilot, and changes in any of these variables can cause vastly different results. For example, moving the airplane's elevator up by 10 degrees might cause a gradual rate of climb at maximum weight during takeoff, but might pitch the aircraft at a dangerous rate when flying at cruising speed and altitude. An autopilot must constantly maintain an understanding of the variables that affect the aircraft's controllability.

The autopilot in Apollo has a somewhat easier job than that found in aircraft. The control laws in the AGC's Digital Autopilot consider the set of physical characteristics (mass and moment of inertia) of the vehicle, the position and authority of the RCS jets, and the attitude and rotational rate of the spacecraft. Interaction with the external environment can be ignored while the spacecraft is operating in a vacuum. The laws that govern how the spacecraft is controlled can be reduced to a concept called a "phase plane". A phase plane is the mathematical set of rules that define when, and in which direction, the RCS jets will fire. Several variables are involved in defining the boundaries of the phase plane, such as rotational rates, attitude error and the tolerances for maintaining an attitude. Like most mathematical entities, a phase plane is best understood by using a graph that depicts the relevant decision points and actions used by the autopilot.

One key area of the phase plane is where no thruster activity occurs. Known as the deadzone, it is the region where the spacecraft attitude is correct within the limits of deadband or is improving towards the correct attitude. While the basic idea of a DAP is very simple, the actual mechanization of a phase plane introduces a bit of complexity. Consider the example of creating an autopilot for a self-driving car. Traveling down a straight road is not particularly difficult – all that is required is to stay in the middle of the lane. The autopilot only needs to ask two key questions. First, has the car moved out of the driving lane? Next, is the car still in the lane, but drifting (essentially, turning gently) left or right, and by implication will leave the

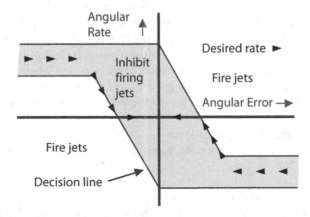

Figure 141: Generic phase plane

driving lane shortly? Decision-making now involves three variables – position, angular rate and time. Because roads have a finite width, and the car is only able to turn at a fixed rate, the position and angular rate variables need to have acceptable boundaries defined. We shall assume that the width of a highway lane is 12 feet (4 meters), and the car's maximum turning rate is no more than 1°/sec. As the car continues down the road, the tires pass over the painted stripe on the right side of the road, meaning that the car has exceeded its "deadband" (the roadway width). A flick of the steering wheel to the left brings the car back towards the center of the lane. Of course, rapidly turning back to the center of the lane is uncomfortable, and staying off the edge of the road for long is unsafe, forcing the compromise of a 1°/sec turning rate. Unfortunately, in this autopilot model, the turn does not stop once the car returns to the center of the driving lane. Eventually, the car finds itself at the left side of the lane where the correction process will repeat itself. In the end, the car will endlessly cycle back and forth between the left and right sides of the lane. This simplified description easily translates to the issues in spacecraft attitude control.

In the phase plane diagram, the horizontal axis defines the attitude error of the spacecraft, and the rotational rate error is defined on the vertical axis. The shaded area is the "deadzone" where we will not perform any thruster firings, and the boundaries of the deadzone are defined by the attitude deadband and maximum maneuvering rates. These boundaries, called "decision lines", define the transition between drifting flight and the region where thruster firings adjust the vehicle's state on the phase plane.

In Figure 142 the spacecraft state is plotted on this diagram at point A, with positive rotation and negative attitude error. In this highly simplified diagram, the DAP control logic varies its objectives as a function of attitude error. Following the path to Point A in Decision Area 1 (DA1), the spacecraft has a relatively large negative attitude error and a large positive rate. If left alone, the spacecraft will eventually reach the deadzone, but ideally the time outside the deadzone should be as brief as possible. This is the same case as a car driving outside of its lane, and very slowly returning. By increasing the rotational rate in a positive direction to fall within the center of the deadzone, the vehicle returns to the desired attitude in a timely manner. Decision Area 6 is very similar, but here the rotational rate increases in a more negative direction towards the lower deadzone. When entering the deadband region (zone 4), a perfect world would have the rates driven to zero. However, the uncertainties of vehicle mass and moment of inertia make attempting to calculate the exact amount of torque in order to simply reach the edge of the deadzone impractical. If the mass estimates are too low, the DAP will apply too much of a correction, which creates serious controllability problems. Reducing the torque necessary to zero the rates by 20 percent prevents overcontrolling. Without the rates completely nulled, the rotation of the vehicle will cause it to leave the deadzone, and jets will fire again (and once again, reduced by 20 percent) in the attempt to null the rate. After several cycles, the rates on the spacecraft slow to a barely perceptible value.

To follow this in an example, assume the spacecraft is in DA3 heading towards the state (attitude and rate) of point A, with a high positive rate and a negative attitude error above the rate limit deadband. The first order of business is to reduce the rotation to the maneuvering rate defined when the DAP was initialized. Slowing the rotation brings the state to point B, just outside the deadband but still far from the desired attitude and rate. The positive rotation continues to bring the state across the deadband to point C, which lies just inside the decision line of the phase plane in DA4, with the intention of zeroing out the rotational motion. Firing times for nulling the rotation are retarded to avoid overcontrolling, and the jet firing establishes it within the deadband with a small positive rate still remaining. The vehicle drifts towards point D, and the jets fire to push the state towards the deadband and zero the rates. A small negative rate still exists at point E, and the state slowly moves to point F, where the rate will be slowed even more by additional jet firings. The decisions and actions at points G and H are the same as points E and F, and the spacecraft cycles back and forth between the deadband limits. Eventually, the corrections are so small that the DAP reaches the minimum limit of thruster firing time of 14 ms, and the spacecraft is comfortably within its rate and attitude deadbands.

For a vehicle rotating back and forth between the deadband limits, wouldn't it make sense to stop the rotation just as it passed the zero attitude error point? Certainly it would appear to reduce fuel consumption and improve attitude accuracy. The DAP has the ability to sense the crossing of the phase-plane line into the deadband, and can project when it will reach zero attitude error. Continuing with this reasoning, a thruster could fire to hold the spacecraft exactly on attitude. This scheme would indeed work to hold the vehicle at the center of the deadband, but

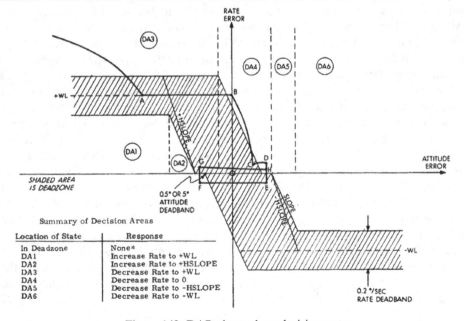

Summary of Decision Areas

Location of State	Response
In Deadzone	None*
DA1	Increase Rate to +WL
DA2	Increase Rate to +HSLOPE
DA3	Decrease Rate to +WL
DA4	Decrease Rate to 0
DA5	Decrease Rate to -HSLOPE
DA6	Decrease Rate to -WL

Figure 142: DAP phase-plane decision areas

another issue arises. This technique has effectively defined another deadband zone, one that is smaller than that originally requested by the crew. The result is that the spacecraft will be constantly correcting small attitude errors, a problem sometimes called "chattering", and ultimately use more fuel than staying within the original, wider deadband. Small disturbing forces occur continually in a spacecraft, especially with a crew aboard, and a very small deadband requires thrusters to fire frequently for precise attitude control.

In combination with the DAP settings, the attitude controller provides different capabilities as a function of how much the attitude controller is moved. Normally, the crew will select the fully automatic mode, where a small deflection in any axis will produce a "rate command" rotation that is very similar to how one would use a stick to control an aircraft. A slight deflection produces a small rotation rate, while more substantial motions of the controller generate greater rotation rates. This control is fully managed by the DAP, which enforces maximum rotation rates and ensures coordinated rotations about each axis. A second option is to fire the thrusters once each time the controller is moved, using their "minimum impulse" mode. This allows for very fine control over the vehicle's attitude, but is hardly the most efficient mode of maneuvering. A thruster firing in minimum impulse mode shuts down just as the ignition transient begins, peaking at 80 pounds of thrust and only 60 percent of the jet's efficiency. Finally, a hardover of the controller bypasses all DAP control, and directly fires the thrusters for as long as the ACA is deflected. None of the fly-by-wire intelligence of the DAP is available in direct mode, forcing the astronaut to manually compensate variations in vehicle mass and the coupling effects from the various thrusters.

Mass and moment of inertia
Mass and moment of inertia (also called the mass moment of inertia) are the two variables that directly affect a vehicle's reaction to a thruster jet firing. Mass and inertia are two concepts that are familiar to all. If a force is applied to the center of mass of a body, it will accelerate in a straight line. To understand how a body will rotate when an off-center force pushes against it, consider two flywheels of the same diameter and mass. The first is of a traditional design, where most of the mass is concentrated on the rim of the wheel. The second flywheel is the opposite, with most of the mass concentrated at the center with a lightweight rim. Pushing along the rim to rotate a traditional flywheel is more difficult than the second wheel, because the mass distribution is concentrated further away from the rotation axis. The mass moment of inertia quantifies this property, and is defined using the mass of the object and its distribution throughout the rotating body.

Newton's second law, $F = ma$, applies to a body moving in a straight line. To determine how much a rotating body will accelerate when a given force is applied, the first step is to find the center of mass, and then add up the masses in terms of distance from the center of mass. As this is effectively a calculus exercise, the mathematics are beyond the scope of the book. However, the goal is to simplify the problem to a model with a point mass located a fixed distance from the axis of

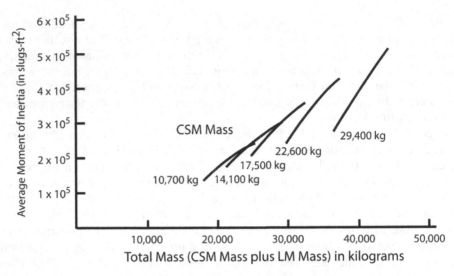

Figure 143: CSM and LM moment of inertia

rotation. When a thruster fires in a spacecraft, the force of the jet and its distance to the center of mass produces a known torque. Knowing this torque and the moment of inertia we can calculate the angular acceleration and ultimately the angular rate of the body. Recalculating the moment of inertia of a complex body like a spacecraft can take a long time. The AGC, already limited in memory and CPU power, simply does not have the resources for this task. At the same time, there is no need for such complex calculations. Because the spacecraft uses fuel and other consumables in predictable ways, the effect on the moment of inertia is also predictable. A single expression relating the vehicle's configuration, mass and the resulting moment of inertia is created as part of the engineering analysis of the spacecraft. While not rigorously accurate, this expression is more than adequate for the digital autopilot. An example is seen in Figure 143 showing the graph of mass versus moment of inertia for the CSM when docked with the Lunar Module.

Unintended consequences: thrusters and coupling
The mass distribution of the CSM or LM is not symmetrical, and the thruster jets do not fire exactly along, through or perpendicular to the center of mass. This results in an effect called coupling, where a thruster used for a rotation in one axis creates unwanted motions in another axis. This problem is most easily seen with the CSM configuration in Figure 144. Here, the center of mass of the CSM is assumed to be forward of the thruster quads and centered along the X-axis. As expected, all rotations of the vehicle are around the center of mass. The coupling problem becomes apparent when diagramming the components of the thrusters force with respect to the center of mass. Firing a single right roll thruster also causes a small amount of pitch down and left yaw. Eventually, the DAP will sense that the vehicle is outside of the pitch and yaw deadbands, and perform jet firings to correct the

Single thruster jet firing

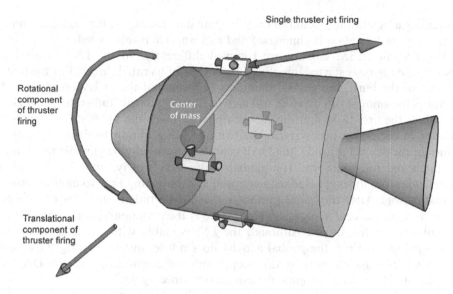

Rotational
component
of thruster
firing

Center
of mass

Translational
component of
thruster firing

Figure 144: Rotational coupling in the CSM

attitude. Firing thrusters in pairs with the jets on opposite sides of the CSM reduces (but does not eliminate) the coupling problem, as the undesirable effects tend to cancel one another out.

Coupling is not just an issue with rotation. Single-jet maneuvering is often used for fine attitude control or times where fuel conservation is especially important. In Figure 144, the thruster is not aligned with the rotation axis of the vehicle. In the resulting force diagram, a small translational component is created when the jet fires. While this component is small, the velocity changes from these translations add up, and usually result in the need to made regular corrections to the trajectory or orbit.

The DAP and the gimbal trim system

Gimbal trim for the main propulsion engine performs the same function as trimming does in an aircraft. In an aircraft, the pilot has two options for maintaining a specific attitude or compensating for an out-of-balance situation. First, the pilot can simply hold the control wheel to continuously exert a correcting force, but this becomes tiring very quickly. The other option is to manipulate small aerodynamic surfaces, called trim tabs, that force the larger control surfaces to move. This reduces the workload on the pilot. The pilot can adjust the trim to maintain the desired attitude without exerting any force on the control wheel. The purpose of trimming the SPS of the CSM or the DPS of the LM is similar, to direct the engine's thrust through the center of mass and reduce the work that the DAP must perform to maintain a fixed attitude. If the engine is not thrusting through the vehicle's center of mass, the offset force causes the vehicle to rotate. Using information from the IMU, the DAP senses this change in attitude, and fires jets to correct the error. Especially during long burns, an out-of-trim vehicle requires a substantial amount of thruster activity to maintain a fixed attitude, consuming a significant amount of fuel. By gimbaling the engine through the center of mass, unwanted rotation is eliminated and fuel wastage is minimized.

The LM and CSM engines are trimmed differently. In the LM during lunar descent, the engine is started initially at 10 percent of its rated thrust. For the first 26 seconds of the burn, the engine maintains this minimal thrust level and the DAP monitors the amount of pitch and roll activity. Nulling the attitude rates through the gimbals is the first task, and during each pass through the DAP it determines the amount of correction necessary in the two axes. In addition to firing thrusters to maintain the desired attitude, the DAP uses the descent engine gimbals to eliminate the pitch and roll rotations. The gimbals move very slowly, only 0.2 degrees per second, allowing for precise adjustments and also providing time to handle runaway gimbal motors. After the initial trimming, the LM attitude is steady with its engine firing through the center of mass. The engine is then commanded to throttle to 94 percent of its rated thrust. Although the LM is stable with the engine correctly positioned at this time, the gimbal activity does not terminate. During the descent, the center of mass changes as the propellants are consumed, and the DAP will compensate by sending the gimbals into motion once again.

During the preparations for firing the CSM's SPS engine, the crew refers to a checklist entry or an update from controllers on the ground for the best estimate for the initial gimbal setting. When the gimbal motors are powered up and the DAP initialized, the engine is manually positioned to its trimmed position. After a short test of the gimbal motors by the computer, the engine returns to its trimmed position, where it remains until the engine is ignited. As with its Lunar Module counterpart, the DAP in the Command Module trims the SPS engine to ensure it is thrusting through the center of mass.

For the SPS and DPS engines, gimbal use is normally associated with eliminating rotational forces while the engine is burning. Yet, during the lunar descent, the spacecraft's pitch continuously changes throughout the landing profile. Just as the engine gimbals are designed to arrest any unwanted rotation, they can also be used for the opposite task: to aid in maneuvering the vehicle. The DAP uses both the engine gimbals and the RCS jets to share control during the descent. Every other DAP processing cycle, the gimbal trimming code is executed to allow the DAP to use both the thrusters and the gimbals to alter the vehicle's attitude. Each time a new attitude is calculated by the descent targeting program, the DAP fires the thrusters in one 0.1-second timing cycle, and adjusts the gimbal position in the next 0.1-second cycle. Precious maneuvering fuel is saved by using the gimbals to assist in pitching the LM over during the descent.

A similar system is used in the CSM. The SPS is rated at 21,900 pounds of thrust (97,500 Newtons) and is capable of gimbaling up to 6 degrees in pitch and yaw. Without centering the thrust along the center of mass, the resulting rotation would have the potential to overstress the vehicle's structure. Based on the mass and configuration of the CSM or CSM/LM stack, the gimbal angles are estimated and manually set using controls on the Command Module's main display console as part of the preparation to fire the SPS. During long burns, the DAP will adjust the gimbals to ensure that the thrust remains directed through the center of mass.

Autopilot control modes

Three different modes for the CSM and LM DAP are available. If Auto is selected, this provides fully automatic maneuvering, allowing attitude-hold functions or vehicle maneuvering under the direction of another program in the AGC. Attitude Hold maintains the vehicle's current attitude, unless a crewmember uses the hand controller to initiate a maneuver. If Free is selected, this places the DAP into an idling mode, inhibiting any thruster activity and allowing the spacecraft to drift. This is used in both spacecraft during docking, immediately after the capture latches engage for a "soft dock". Small misalignments when the two vehicles make contact result in rotational motions that are damped out by the docking probe's shock absorbers. If the DAP were to attempt to correct for these motions, the force from the thruster firings could damage the docking mechanism.

When the PGNS Mode Control is in Attitude Hold, the DAP responds to a hand controller as a request to override an existing maneuver or to change the attitude being held. This mode has an unusual side effect at lunar touchdown. At an altitude of about 500 feet (160 meters), the Commander has selected a landing site and wishes to steer the LM towards it. Switching to Attitude Hold from Auto allows the LM to be flown "manually", using the DAP to maintain the last commanded attitude. While the Attitude Controller Assembly (ACA) is not being moved (i.e. it remains in the "detent" position) the DAP holds the LM at the last commanded attitude. At contact, the LM settles onto the surface, but will almost certainly not come to rest in the same attitude it maintained during descent. Now on the surface, the LM is stuck on the ground with no way to rotate, yet the DAP is focused on returning the vehicle back to its descent attitude, with the thrusters constantly firing in a futile effort to move the vehicle. Although the Lunar Contact light is illuminated for the crew, the AGC is not aware that the LM has landed and the crew must inform the DAP that the current attitude is acceptable. When the Commander nudges the hand controller out of its detent position, the DAP interprets the action as a request for a new attitude to maintain. With the LM solidly on the surface, the DAP sees the new attitude is holding and no longer needs to fire the thrusters.[14] The ACA does not have to move out of detent in all three axes to set the new attitude. The "out of detent" discrete is applicable for motion in any one axis, and is sufficient to force the DAP to hold the current attitude in all three axes.

DAP processing overview

In both the CSM and the LM, the DAP is the single largest program in the computer, with about 10 percent of the code or about 4,000 storage words required to hold the software. Despite its size and complexity, implementing the DAP in the AGC is a much easier and far lighter solution than a comparable analog system. The

---

[14] This situation will not last forever if the DAP remains enabled long after landing. As the Moon rotates, the "attitude" of the LM relative to the inertial frame of reference also changes, and the DAP will once again try to bring the vehicle back to the "correct" attitude.

high level processing flow of the DAP is similar in the LM and CSM, but many implementation details are considerably different, especially concerning jet selection. Processing occurs in three phases. First, every 0.1 seconds the T5RUPT interrupt occurs and the DAP reads the CDU counters to determine the spacecraft's attitude. Comparing the current attitude with the previous value, the rotational rate is determined, and the rate estimation filters are updated. Translations from the IMU axes to body axes are performed, and the FDAI error needles are updated. In plotting the attitude and rate on the phase plane, important questions must be answered. Is the spacecraft's attitude within the deadband and the rotational rate within the limits specified by the crew? Is that rotational rate bringing the spacecraft towards or away from the desired attitude?

The second phase of DAP processing is to determine whether the vehicle state is outside the phase plane deadzone, and if so, the necessary torques are calculated to move the state back towards the deadzone. Using basic variables of mass, moment of inertia, jet thrust and moment arm length, a thruster firing duration is calculated. A check is made of the DAP control mode, which influences thruster firing schedules. For example, fully automatic control by the DAP provides variable-length thruster firing times, while minimum impulse (as its name implies) generates short fixed-length firings. Allowable rotation rates are specified in the DAP parameters and impose a limit on the firing times of the jets. In the CSM, firing times are limited to between 14 and 100 milliseconds – with 14 ms being the minimum impulse for the thruster. Thruster firings cannot exceed 100 milliseconds as they would otherwise overlap the next DAP cycle. This limitation is different in the LM DAP, which extends the limit to 150 ms. To avoid the problem with overlapping the subsequent DAP cycle, the LM DAP will skip the next cycle when performing a long thruster firing.

Now that any necessary rotational movement has been calculated, the DAP moves onto the third phase, which is to handle translational requests. Translation is never commanded automatically by the DAP; the only interface is the astronauts' translational hand controller. In both the CSM and LM, translation is performed using either two or four jets, depending on the translation axis. In most cases, the translational thrusters will not be firing through the center of mass, inducing an unwanted rotational motion. If calculations determine that the undesired rotation produced during translation is more than the rotational jets can counteract, the translational request will be scaled back. As it is possible for a translation request to overwhelm a pre-existing rotation command, the DAP honors rotational requirements, which have priority over translation. As an example, the docked CSM/LM combination has a center of mass considerably forward of the Service Module thrusters. If an astronaut in the CSM uses the translational controller to translate the entire stack upwards along the -Z axis, the location of the Service Module thrusters also creates a pitch down rotation. Here, the rotation induced by the translational thrusters could be more than the pitch thrusters can compensate for, so the translation request is scaled back or ignored. Assuming that combining the rotational and translational maneuvers is possible, the required firing times are combined when the thruster firings are scheduled. Note there is no indication in the computer or other displays that the translation has been limited.

Failed thrusters and jet selection

Before the final selection of the jets to be fired is made, any thrusters that have failed or are disabled by the crew must be identified. Jet failures can have a serious impact on controllability, and so there are several mechanisms in place to handle problems. Most problems are sensed by instrumentation hardware outside of the AGC, known as the Caution and Warning Electronics Assembly (CWEA). A failure detected by the CWEA illuminates a failure light and sets a red talkback flag on the vehicle's console. Thrusters can be rendered unavailable for a number of reasons: failed on, failed off, failed opposing, or disabled by the crew. In the LM, the CWEA monitors the jets for failure conditions. Sensing a rise in chamber pressure without seeing a command from the AGC is a "failed on" situation. It automatically fires jets opposing the failed thruster to prevent the spacecraft from rotating out of control. The crew, seeing the red talkback, must promptly disable the affected thrusters manually. A "failed off" thruster is detected in a similar way, by comparing the AGC commanded signal with the thruster chamber pressure. Minimum impulse thruster firings, being of such a short duration, can fail to register a chamber pressure high enough to be considered "on". This situation is infrequent, but if the instrumentation detects a lack of chamber pressure several times in minimum impulse mode, or once in a longer firing, the jet is considered to have failed off. Two jets firing in opposite directions also generates a warning indication that requires the crew to disable thruster quads.

Failed jets sensed by the CWEA are only one way to lose thrusters. Varying levels of manual control of thrusters are also available, as each of the three spacecraft (CM, SM and LM) has its thrusters and fuel supplies arranged differently. Service Module thrusters can be controlled individually, but jets on the Command Module cannot. Only by closing the propellant valves for an entire set of six jets can a problem thruster in the CM be shut down. Two independent fuel supplies are used in the Lunar Module, and each jet is connected to one fuel system only. Isolating jets by fuel system or by pairs within a thruster quad is possible, but enabling or disabling individual thrusters is not. The goal is to isolate the failure to the smallest number of jets, preserve as much control as possible and simplify the work the DAP must undertake to determine alternative thruster configurations. Figure 145 shows that the assignment of jets, fuel systems and shutoff valves provides a significant amount of redundancy.

When confronted by failed jets, the DAP is not able to produce balanced rotations or translations. In the case of a forward translation in the LM, two aft-facing thrusters are necessary. If one of the two jets were to fail, firing the remaining jet would cause the LM to yaw around the X-axis. Translation is still possible, but requires an additional thruster firing in another axis to counteract the adverse yaw. Although the translation is inefficient, it is still possible to perform the maneuver. Adjusting the jet selection logic for a loss of a rotation thruster is similar. The DAP is provided with a table of "alternative" jets, and it recalculates the maneuver using these substitutes. If ever the LM DAP finds itself with no jets available for a rotational or translational maneuver, it will issue a program alarm with a code in the range 2001 through 2004.

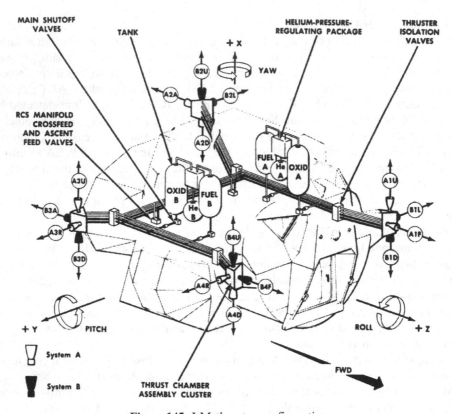

Figure 145: LM thruster configuration

After all the requests, limitations and alternative jets (if necessary) are identified, the specific thrusters required for the maneuver are selected. Only at this point is it possible to calculate the thruster firing times for each axis. Since all combinations of rotation and translation are expressible in terms of firing jets in each rotational axis, the task of timing jet firings is also simplified. To track the timing for each set of thrusters, a three-entry queue is maintained for the TIME6 interrupt, or T6RUPT. Entries in the list are sorted by the duration that the thrusters in each axis are scheduled to fire. Like the waitlist, times in T6RUPT queue are not the absolute times of the thruster firings but the intervals between the thruster shutoff times. For example, assume the X-axis thrusters will burn for 90 ms, the Y-axis for 65 ms and the Z-axis for 20 ms. The T6RUPT queue contains the values of 20 (the duration of the Z-axis impulse), 45 (the difference between the Z-axis and Y-axis), and 25 (the difference between the X-axis, and the last interval). At the beginning, the T6RUPT is enabled and set to 20 ms, and the required X-, Y- and Z-axis jets are ignited. After 20 milliseconds the T6RUPT timer expires and the Z-axis jets are turned off, leaving the Y- and X-axis thrusters burning. T6RUPT is now restarted with a wait time of 45 ms. When T6RUPT expires this second time, the Y-axis jets are shut down, leaving the X-axis jets to complete their work. Once again, the

T6RUPT counter is set, this time for 25 ms, and when it counts down to zero the final jets are turned off.

Autopilot axes in the Lunar Module
A top view of the LM shows the expected Y-axis and Z-axis, which intersect at the presumed center of mass. However, the thruster quads do not lie along either of these axes; rather, they are at a 45 degree angles to them. Obviously, this complicates a DAP implementation. To simplify the control law problem, the DAP breaks the problem down into two independent sets of RCS control systems. The P system is composed of eight jets and thrusts along the X-axis. The U and V system has four jets apiece, and thrusts in the Y-Z plane such that U and V are offset by 45 degrees from Y and Z and are aligned with the thruster quads (Figure 146a). This is not a particularly complex change, and the transformation between the traditional X-Y-Z axes to the P-U-V axes is not difficult to calculate.

A larger problem is that the LM's mass distribution is not symmetrical,[15] and that approximately 70 percent of the LM is fuel and other consumables. If the propellant mass were distributed evenly, the DAP's task would not be difficult. Mass distribution during powered flight changes significantly as over 18,000 pounds of propellant (over half the weight of the LM) is burned during the 12 minutes of lunar descent. Complicating the matter is that the majority of the propellant in the descent stage is below the center of gravity. As the fluid is settled to the bottom of the tanks, a significant change to the moment of inertia occurs, especially in pitch and roll.

In the ascent stage, the size and location of the propellant tanks give rise to a particularly difficult situation. As the ascent stage lifts off from the lunar surface, much of its mass is distributed along the Y-axis (Figure 146a). At about 5,200 pounds (2,360 kg), the weight of the fuel and oxidizer is much larger than the basic ascent stage structure, itself which is about 3,000 pounds (1,360 kg). Further, the mixture ratio of the propellants, 1.6:1 (oxidizer to fuel) means that the oxidizer tank is significantly heavier than the fuel tank. Displacing the lighter fuel tank outboard in order to counterbalance the heavy oxidizer tank has created the familiar asymmetrical "chubby cheek" face of the LM. Burning all of this propellant during ascent creates an even more complex problem in managing the moment of inertia. The mass of the propellants and the majority of the ascent stage mass is aligned along the Y-axis (Figure 146b). With the RCS thrusters at a 45-degree angle from the Y-Z plane, a firing of a jet along the V-axis causes a coupled rotation with the U-axis rather than in line with the V-axis (Figure 146c).

A solution might be to move the RCS jets forward and backward like the wings on a jet fighter, so that rotations are aligned to the U- or V-axis. Given the effort to minimize LM weight during its development, and the complexity of such a movable system, this method could never be considered. However, it is possible to "move" the rotational axis in the calculations for the thruster firings, creating the same effect as

---

[15] The symmetry of the LM is something that only another LM could love.

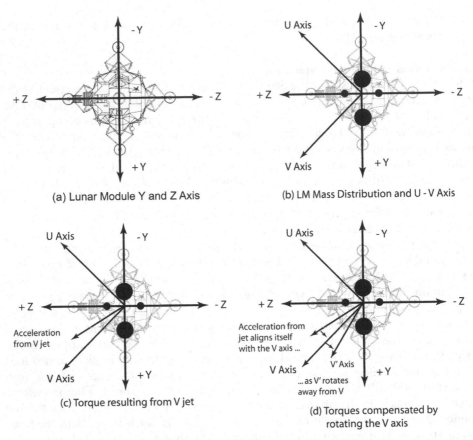

(a) Lunar Module Y and Z Axis   (b) LM Mass Distribution and U - V Axis

(c) Torque resulting from V jet

(d) Torques compensated by
rotating the V axis

Figure 146: Non-orthogonal axes for U, V in the Lunar Module

having the RCS jets move physically (Figure 146d). We have seen that transforming the Y-Z axis to the U-V axis system can simplify the DAP logic. Immediately after liftoff from the lunar surface, the stage is full of propellant. Due to the lack of LM symmetry, a jet firing along the U- or V-axis does not produce a torque only through that axis. Consider the motions along the U-axis when firing one of the upward or downward thrusters on the forward U quadrant.[16] The ascent fuel tank is a large mass, and is approximately behind the thruster quad. The resulting torque has a component of forward rotation, rather than purely along the U-axis. This asymmetry causes an undesirable torque around the V-axis. Without a means to physically eliminate this effect, the task is to compensate for it in software. By mathematically "rotating" the U-axis aft to define the U'-axis, the calculations for jet firings are made with this rotated, nonorthogonal axis. The DAP's calculations

---

[16] A similar process is used for the V-axis, but the angles for U' and V' are not related to each other and are calculated independently.

see the thrusters firing in the U'-axis and the effects of its torque in other axes, prompting the need for additional thrusters to fire to compensate. This rotation of the control axis is not static. As fuel is burned and the ascent stage lightens, the torque vector rotates backward towards the U-axis. As such, the U'-axis needs to move forward in order to maintain consistent control. The 1/ACCS routine calculates the amount of fuel burned, and subtracts that amount from the mass of the LM. Now, with the new value for mass, a new moment of inertia must be calculated for all three axes. This series of computations produces new values for the angles of U' and V'.

Routine 60/62: automatic maneuvers
Space flight is rarely a static environment, and maneuvers to new attitudes occur frequently. While the crew can maneuver the spacecraft manually using the hand controllers, the DAP can perform these maneuvers more accurately and use far less fuel in the process. Many of the Major Mode programs require a specific attitude to accomplish their task, and incorporate an automatic maneuver request in the dialog with the crew. In these programs, the crewmember is presented with a Verb 50 Noun 18 display, asking if the crew wishes to perform an automatic maneuver to the new attitude. If authorized, the program calls the routines to place the vehicle in a new attitude, and then continues after the new orientation has been achieved. Ad hoc maneuvers requested by the crew are also possible, using Verb 49 to begin the dialog for the maneuver. Both the automatic and crew initiated maneuvers operate identically in the AGC.

Conserving fuel is always important, making efficient maneuvering a priority with the Digital Autopilot. One technique for maneuvering to a new attitude is possible by rotating the vehicle in each of the three axes, one after another. While this is the simplest way of altering the orientation of the vehicle, it is also inefficient in terms of time and fuel. A far more optimal way is to rotate directly in all three axes to the new attitude. Routine 60 maneuvers the spacecraft using input from the crew or another program. Verb 49 calls Routine 62, which asks the crew for final attitude gimbal angles. Once the angles are entered, Routine 62 calls Routine 60, which in turn calls on two major subroutines to perform the maneuver, VECPOINT and KALC-MANU.

On entry to Routine 60, VECPOINT takes the requested attitude and computes the desired gimbal angles for the maneuver. After VECPOINT returns to Routine 60, the desired gimbal angles are passed to the KALCMANU maneuvering routine, whose first task is to calculate the most efficient maneuver, namely a rotation around a single vector. Defining this vector begins with the spacecraft's initial and final attitude. Both attitudes are represented as vectors, and when joined together they form a geometric plane. The third vector, used for the axis of rotation, is calculated as the vector cross-product of the first two vectors, and is therefore perpendicular to the plane. Figure 147 shows how this process works. Vectors A and B define the current and desired attitude, respectively. The vector normal to the plane defined by the cross product of A and B is the rotation vector R. The maneuver is now defined by rotating the vehicle along the R-axis.

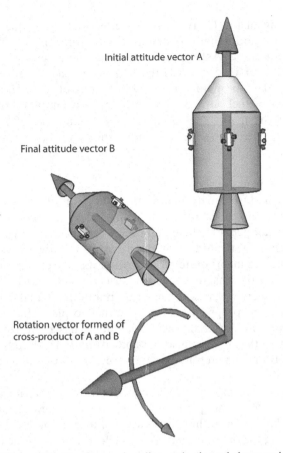

Initial attitude vector A

Final attitude vector B

Rotation vector formed of
cross-product of A and B

Figure 147: Starting and ending attitude and the rotation axis

It is natural to think that the DAP's attitude-hold logic is just that – ensuring that the vehicle does not move within its frame of reference. However, what happens if the frame of reference itself is rotating? It was earlier presumed that if the state of the vehicle is along the phase plane's rate error line (the vertical axis), then this implies that the rotational rate is zero. The DAP, seeing no rotation and presumably within the attitude deadband, assumes that no corrections are necessary. Now, imagine the rate error is biased clockwise at 1°/sec by shifting the phase plane right by one degree. Just as in the attitude hold case, the DAP checks the rotation rate to see if we are within the deadzone. Starting with a motionless vehicle, the DAP sees the state as being outside the phase plane and "rotating" in a negative direction. Thrusters are fired to bring the spacecraft's state to the middle of the deadband and null the rotation. In doing so, a positive rotation is set up. We have created a constant-rate rotation autopilot from the attitude-hold DAP.

After KALCMANU calculates the axis and angles of the maneuver, its second task is to take advantage of the constant-rate rotation technique to turn the vehicle to its new attitude. The angles and rates for the first second of the maneuver are calculated and are passed to the DAP to begin the maneuver. Once the maneuver is underway, KALCMANU sets a timer for 1 second in the future. At the end of this interval, KALCMANU recalculates the rotation axis and the amount of rotation using the latest gimbal angles, and creates new values for the next second of rotation, which are passed to the DAP. This process repeats until the spacecraft is close to the new attitude and the rotation must be halted. The bias previously applied to the rate error in the phase plane is removed and the final attitude angles are sent to the autopilot, returning to the familiar attitude-hold phase plane.

But what happens if the rotation along the new axis travels through the gimbal lock region? VECPOINT is responsible for avoiding gimbal lock, but it only checks the final angles to see if they fall within the gimbal lock zone, not the intermediate angles along the maneuvering path. If the final attitude is within the gimbal lock area, the maneuver is modified to avoid this region. Nor does KALCMANU perform any gimbal lock monitoring or active avoidance. In cases where gimbal lock cannot be avoided, manual procedures, including realigning the platform to another REFSMMAT, are necessary. During the maneuver, the crew observes its progress through the error needles on the FDAI. These needles reflect the vehicle's attitude error with respect to its phase plane, and should not deflect more than the maneuver deadband. If a significant error develops, or is progressing towards gimbal lock, the astronauts can stop the spacecraft by nudging the ACA out of its detent position. Sensing the ACA motion, the DAP terminates the automatic maneuver, stops the rotation in all axes, and holds the spacecraft in the current attitude.

Constant maneuvers: PTC, landmark tracking and orbital rate mode
In contrast to the times when the spacecraft must maintain a fixed attitude, several tasks require the spacecraft to rotate at a constant rate. Certainly the most common is the Passive Thermal Control, or PTC, maneuver, also known as the "Barbeque Mode". The Command Module heat shield is sensitive to extended cold soaking during the voyage to and from the Moon. PTC is designed to slowly (about 1°/min) rotate the entire vehicle. This rotation exposes the ablator to heating during the one half hour in sunlight and cooling for the remaining half hour, to prevent cracks forming in this material. Although originally conceived for thermal control of the heat shield, PTC has the added benefit of maintaining the spacecraft temperature within a comfortably narrow range.

Passive Thermal Control begins with a new REFSMMAT loaded into the CSM's computer. This is defined with the X- and Y-axes in the plane of the ecliptic and the Z-axis perpendicular, facing towards the southern celestial pole. The Y-axis parallels the Earth-Moon line at the time of Translunar Injection, facing towards the Earth. The X-axis completes the right-handed system. The spacecraft is maneuvered until its X-axis is oriented parallel to the Z-axis of the REFSMMAT, and is held in that attitude for 15 to 20 minutes for the attitude to stabilize. For much of the Apollo program, PTC was started manually. This created a marginally acceptable situation, as without active control the spacecraft gradually diverged from the rotation axis. Crews frequently had to be awakened to reestablish the proper attitude and restart

PTC, a task that could easily take another 30 minutes. Later versions of the CSM software introduced Option 5 to Program 20, the Universal Tracking program, to provide the ability to perform fixed-rate rotations. No longer having to manually establish PTC, the crew workload reduced to selecting P20 and then entering the starting attitude and the rotation rate through Nouns 78 and 79. Using the same DAP logic as the automatic maneuvering, a constant-rate rotation began. With the DAP now controlling all three axes, the spacecraft maintained PTC attitude quite accurately. In even later versions of the Command Module software, Verb 79 provided automatic Passive Thermal Control and Orbital Rate maneuvers. Now integrated into the DAP, Verb 79 eliminated the requirement to start Program 20. This improvement reduced the task of establishing PTC to entering Verb 79 and providing the rotation rate and the deadband in Noun 79.

Two additional tasks use the automatic maneuvering capability of the DAP in the Command Module. Several times during the mission, the crew will track a particular landmark on the lunar surface using the optics. The most familiar application of landmark tracking is sighting the Lunar Module on the surface, so that its location can be precisely determined. Tracking other landmarks, such as a prominent crater or mountain, provides important spacecraft navigation information, as well as defining the locations of sites that are not charted with the highest degree of accuracy. Option 2 of Program 20 starts a fixed-rate rotation to keep the sextant pointing at the target. Program 24, Rate Aided Optics Tracking, starts next. In P24, the computer drives the optics to keep them pointing towards the selected landmark. If the spacecraft were not rotating as part of this tracking exercise, mechanical limitations of the sextant would prevent it from tracking the landmark for the entire time that the vehicle is above the horizon. While the fixed rotation rate keeps the sextant pointing in the general direction of the landmark, it is not sufficient for tracking. Program 24 drives the optics with a high degree of accuracy, and keeps the sextant centered on the target.

When a spacecraft is in a fixed attitude in orbit, it will maintain that attitude with respect to the stars. This may not be completely desirable. Pointing down directly at the surface at one point will lead to pointing directly away from the surface half an orbit later. If the goal is photography, it is much more useful to maintain a fixed attitude with respect to the horizon, in a reference system called Orbital Rate. The vehicle is set constantly rotating with respect to the stars, such that the rotation rate keeps pace with the orbital motion. Option 2 of Program 20 is also used to establish Orbital Rate rotation. Like landmark tracking, the crew specifies a rotation rate, and the DAP pitches the vehicles at a rate that matches the orbital period. Note that the DAP is simply rotating the spacecraft blindly, without any further references to the orbital state. If the orbit is very elliptical, or the orbital period is not well synchronized with the rotation rate, the spacecraft will not accurately track the horizon.

CM entry in the DAP
The final task of the DAP is maneuvering the Command Module during entry. Although the CM does not have wings, its shape and offset center of gravity give it a

modest amount of aerodynamic lift. This stability, plus the small amount of lift, allows the vehicle to be "flown" through the atmosphere up- or down-range and to either side of the ground track. Rolling the lift vector upwards stretches out the entry path and reduces the g-loads on the crew, and directing the lift downwards digs the spacecraft deeper into the atmosphere, shortening its trajectory and increasing the deceleration. In the same manner, pointing the lift vector right or left provides crossrange maneuvering.

Command Module pitch thrusters are roughly aligned with the center of mass, and do not produce a significant amount of unwanted torques in other axes. Yaw control is somewhat more complicated. Yaw thrusters are offset from the center of mass and create a roll coupling in the process, requiring roll thruster firings to offset the torque. The pitch and yaw attitudes are maintained to within ± 2 degrees. Because the CM is remarkably stable until about Mach 2, the aerodynamic forces hold the spacecraft in its proper attitude. As a result, the only control logic for pitch and roll are rate dampers. A wide deadband during entry is reasonable because of the buffeting the spacecraft experiences, and in any case precisely maintaining an attitude is unnecessary.

Saturn takeover function

Primary guidance and control of the Saturn V is not performed by the AGC, but uses an independent inertial platform, computer and related support electronics in the Instrument Unit (IU) located atop the S-IVB stage. This computer, known as the Launch Vehicle Digital Computer (LVDC), is responsible for performing the calculations and steering commands necessary to meet the guidance targets. Reliability is assured through triple-redundant electronics, and voting logic to detect and circumvent hardware and software errors. In the Command Module, the only indication of IU failure is a single indicator light, which illuminates if there is a failure in the LVDC computer or its platform. The IU provides no other information to the crew about the ascent or the condition of its systems.

AGC Program 11, Earth Orbit Insertion Monitor, independently monitors the ascent, and indirectly the IU's performance, by performing the same guidance calculations as the LVDC. During ascent, the DSKY displays the vehicle's velocity, altitude and altitude rate, and the error needles on the FDAI indicate any deviation from the AGC's computed trajectory. The Commander's attention is never far from two status lights that reflect the condition of the launch vehicles guidance: LV RATE and LV GUID. The LV RATE light illuminates when the vehicle exceeds predetermined pitch or yaw rates ($\pm 4°$/sec) or roll ($\pm 20°$/sec). A failure in the IU's inertial platform or the LVDC will illuminate the LV GUID light. In either case, switching control of the Saturn to the AGC from the failed IU is accomplished by setting the LV GUIDANCE switch to CMC from IU. At that point, the AGC assumes control of the launch vehicle, and sends steering commands to guide the stack into orbit. Like the LVDC, the AGC will perform open-loop polynomial guidance for the first few minutes, and then switch to closed-loop iterative guidance. If the AGC isn't controlling properly, or if an engine or gimbal failure creates a control problem that the AGC cannot handle, the Commander can assume manual control of the booster. Entering Verb 48, DAP Initialization, and changing the first octal digit in the configuration code to a '3' in Register 1, enables him to literally "fly" the booster using his attitude controller. Polynomial guidance is terminated, and the pilot relies on an attitude versus time "cheat sheet" Velcro'd to the panel for steering information. In this mode, the rates are limited to 0.5°/sec. While this sounds like a rather sluggish rate, it is born of a necessity. All launch vehicles are flexible, given their lightweight structure and lack of internal bracing. Bending of the Saturn V vehicle is problematic, especially during the first few minutes when the first stage is thrusting. Maneuvering at any faster rate can set up dangerous bending oscillations that are not easily damped and can potentially lead to the breakup of the vehicle. While the image of an astronaut steering 3,300 tons of rocket by hand may seem incredible, simulations demonstrated that the crews could guide the stack into orbit with reasonable accuracy.

ERASABLE MEMORY PROGRAMS

A feature that the AGC shares with virtually all other computer systems is that critical programming is protected in read-only memory. Whether using the core rope of the AGC, or the BIOS of a PC, the integrity of the system depends on assuring that the software is not changeable. This has important implications during error recovery. In the event the system experiences a fatal error, the only option might be reinitializing the system to an known, idle state. A fresh copy of software is loaded into memory and started, banishing any trace of the errors that caused the original problem. In a PC, this software originally starts in the BIOS, quickly turning to the hard drive as the source of software to load. Without secondary storage like disk or tape, the core rope in the AGC assumes the dual role of the software repository as well as memory for the processor. However, unlike the flash BIOS found in PCs, the programming in core rope is permanent, preventing even emergency changes to the software.

Unsurprisingly, the AGC software was designed to exacting mission requirements. Given the perennial shortage of memory, every piece of programming code that was not essential for mission success was discarded. But what can be done about situations which require programming that is not included in the flight version of the software? The AGC is noteworthy for its lack of contingency programs, yet workarounds to some potentially serious problems are only possible through software. The solution requires that the crew or ground controllers have the ability to load new software by manually entering the binary codes directly into memory.

While almost unheard of in the 21st century,[17] entering binary code directly into

---

[17] Well into the 1980s, small programming changes were often made by entering binary instructions byte by byte, bypassing the need to recompile large amounts of software. This technique continues in the 21st century with a far more sinister motive. Many computer viruses load binary codes as they exploit security flaws in a system.

memory was a normal practice for the first decades of computing. Rows of lights and switches provided computer operators the means to load instructions and verify the contents of memory. The most familiar example was initializing a computer for its day's work, where a small "bootstrap" program was entered using the switches, and then a button was pressed to execute it. This bootstrap program loaded additional programs, which in turn initialized the operating system. Though tedious and famously error-prone, this manual process was an integral part of early computing. From this experience, it was only natural that urgent updates to the AGC would occur in the same way. Loading erasable storage with temporary programs is a reasonable solution, but an apparently insurmountable problem arises: virtually every word in erasable storage is used, and in many cases by multiple programs. With no free memory available, where will ad hoc software be loaded? On closer inspection, there are several areas available for holding such data: the Vector Accumulator (VAC) areas, the Core Sets and unused temporary variable areas.

VACs are commonly pressed into service for these temporary programs, formally known as Erasable Memory Programs, or EMPs. While normally used to hold data such as an interpretive program's stack, there is nothing to prevent AGC instructions from being stored and executed within a VAC. With 43 words available in each VAC, small but complex programs are possible. VAC areas are ideal locations for erasable memory programs because their use does not interfere with the normal operation of the Executive. Once a VAC is marked as "in use" by an EMP, no other job will be able to allocate or delete it. However, an unintended consequence of using VACs for EMPs does exist. In heavy computing situations where large numbers of interpretive programs are running, there is an increased likelihood of experiencing a 1201 Executive Overflow program alarm indicating that there are no more VAC areas available.

The Core Set area is also available for storing an EMP. Unlike the Vector Accumulator areas and their 43 words of storage apiece, the 12 words of a Core Set cannot hold anything but the most trivial of programs. Making full use of the Core Set area requires abandoning the concept of the Core Set table as a collection of 12 word entries, and looking afresh at the entire table as one contiguous area. With almost twice the storage of a single VAC area, the Core Set table area now makes it possible to load and execute large EMPs. Contrasted with VAC areas, which are effectively ignored by the Executive and so do not substantially affect the overall operation of the AGC, the use of the Core Set table for EMPs places a serious restriction on the operation of the AGC. During normal operation, Core Sets move about in storage as they are copied in and out of Core Set 0. Unless actively running, the location of a program's Core Set is unpredictable, and can be in any entry of the table. This behavior prevents jobs from running concurrently with the EMP which has overlaid much of the Core Set table. A movement of Core Sets would quickly corrupt both the EMP and the job's Core Set, and this limitation imposes the requirement that the AGC be set to idle in Program 00 when running an EMP from the Core Set area.

EMPs can also exist in a large area that is reserved for temporary variables. Data in this area is located in the upper range of erasable storage, and is assigned to specific mission programs. By carefully selecting the variables overwritten by the EMP, it is possible to continue to run many of the AGC programs at the same time.

EMPs can execute as any type of schedulable work in the AGC, either as a job or a waitlist task. When running as a job, the program can run under the control of the Interpreter or as basic AGC instructions. However, while an EMP can obtain a VAC area for its processing, the Interpreter cannot process instructions in erasable storage. A transfer of control to fixed storage is the only option for an EMP to execute interpretive programming. Given an EMP's temporary nature, there is no attempt to protect its code against program aborts or restarts. An incorrectly keyed word of data will likely result in a program alarm in another part of the system, and the recovery activity might clear the memory the EMP resides in. At this point, the crew must not only address whatever issues result from the restart, they are also faced with the task of keying the EMP back in from scratch.

Because EMPs are nonstandard, defining the parameters and loading the program must be performed carefully. The location for the EMP must be consistent with other AGC requirements. Will using a VAC area to store the EMP be sufficient? If more storage is required to hold the program, should the Core Set area be used? Using a VAC area does not prevent other programs from executing, and is often the only option if the program cannot be run "standalone" with the AGC idling in Program 00. The other option, using the Core Set area, precludes any other jobs from running. The specific location then drives which banking registers are used. Ideally, to minimize the program's complexity, there should be no bank switching.

For the program to be entered into storage and executed, parameters for Noun 26 must be drawn up, specifying the characteristics that are necessary to schedule the EMP. The three parameters entered with Noun 26 contain the same information that is passed to FINDVAC/NOVAC and WAITLIST routines. The format of Noun 26 differs when starting a job or waitlist task. Waitlist tasks require that the first word contains the delay time in centiseconds before the task is to run. Scheduling a job requires a much more complex set of information, with two fields defining the characteristics of the job. Bits 14 through 10 contain the job's priority, and bit 1 indicates whether the EMP is an interpretive program or not. The second and third words of Noun 26 specify addressing information and are the same for both types of EMP. The second word of Noun 26 is the 12-bit entry address of the EMP program. As with all 12-bit addresses, this value is insufficient without banking information to completely define a location in storage. The third word contains this banking register data, with the FBANK in bits 15 through 11, the Superbank data in bits 7 through 5, and the EBANK defined in bits 3 through 1. With the three words of Noun 26 prepared, and the software converted to binary, the EMP is ready for loading and execution. Given the possibility of a data entry error using the DSKY, manually keying the data is practical only when entering small programs or at times when communication with Earth is not possible.

When data is loaded through an uplink, Verb 71, Block Update, is started and up to 20 words can be entered directly into memory in a single update. If necessary, multiple updates may be used to complete the EMP load. The Noun 26 data must be entered prior to starting the EMP. Two different methods are available. For ground-based updates, an additional Verb 71 is added to the uplink sequence, updating the Noun data directly. If this data is not uplinked, it must be entered manually using Verb 25. Starting the EMP is done by entering Verb 30 when the program runs as a job, otherwise Verb 31 schedules the program as a waitlist task. After this point, the program runs like all other jobs or tasks in the AGC. Once the EMP has completed, the crew can clear out the memory manually, or force a restart of the AGC and clean up the memory as part of this processing.

AGC DATA UPLINK AND DOWNLINK

From the very beginning of software development, programmers have always needed to observe how their programs are operating, to ensure that they are running efficiently and correctly. Simply reviewing the output of a failed program after it completes is hardly a useful method for identifying the root cause of a problem. At that point, evidence of the error is usually obscure, leaving few clues to test the forensic skills of the programmer. Many errors are detectable only when they are in the act of causing their mayhem, requiring the programmer to continually display critical areas of memory, hoping to be able to identify the moment of failure. Elaborate debugging environments, universally used in modern software development, were decades away in the early 1960s when the AGC software was written. Still, a wide range of tools were available for debugging software in the AGC and other computers of that era. On a dedicated computer, starting, halting a program, or even executing one machine instruction at a time was a completely reasonable way to search for errors. Complementing the printed contents of memory on paper, programmers displayed data on a row of lights, each light corresponding to one bit of storage or a register. Monitoring software execution often extended to tracing individual signals in the computer circuitry, where specialized hardware analyzed the types and frequencies of the instructions in real-time. Despite being slow and tedious, these methodologies worked extremely well and gave developers essential views into the dynamics of their software. The lack of elaborate tools makes diagnosing issues in a piece of software such as a mission program very difficult. This task is even more complex to perform when trying to troubleshoot an operating system such as the Executive in the AGC. Unlike a program that has a well-defined sequence of events, the flow of the Executive is driven by interrupts, timers and the unpredictable nature of the mission programs.

Given the high-stakes nature of spaceflight and the critical role that the AGC plays in its success, it is reasonable to insist on constant monitoring of all vital parameters in the system. This may appear to be a bit of overkill, as the software may perform error-free throughout the mission. Still, a critical error may occur without warning, and the data that indicates the computer's state to that point is invaluable. Engineers have long been accustomed to a steady, real-time flow of data from the spacecraft. Much of the work of the ground controllers and their support staffs involves monitoring temperatures, pressures and voltages in order to assess the health of their systems. To complement this insight on the mechanical and electrical systems, a comparable level of detail is necessary for the computer. Using a level of detail that is not seen in ground-based computing, critical memory areas are dumped to the ground on a regular basis, to provide real-time debugging data for the engineers. As with much of the telemetry sent to Earth, the quantity of the data and its level of detail is far beyond the needs of the crew onboard the spacecraft. However, instrumenting a computer is a very different task to instrumenting an environmental system, whose hardware is unchanging. Different software is running in the computer at any given time, and the data downlinked should reflect only the active programs. Continually dumping all of erasable memory is unnecessary, as it requires a considerable amount of time to accomplish and only a small set of data must be continually monitored. Dumping only the relevant parts of memory is not only a more efficient use of the available telemetry channel, it allows that data to be transmitted more frequently.

Data downlinks
How does one monitor the health and internal execution of a computer from 240,000 miles away? Clearly, monitoring the AGC requires extensive and invasive tools to capture every relevant word of data and send it to the ground for monitoring and analysis. Although the amount of data transmitted to Earth is limited, 200 words or about 10 percent of all of erasable storage is transmitted. Much of the data reflects the overall status of the computer and contains the data words such as state vectors, flagwords and DSKY displays. About 100 words are tailored to the Major Mode that is currently running. During rendezvous in the CSM, for example, the downlink contains different data than during launch. During the startup of each Major Mode, and, in the case of the LM, also Routine 47, AGS Initialization, an identification code is set to establish which downlist is used. There are six downlists used in LM, and five in CM. Downlists are grouped into several related categories, each of which transmits the same data (Appendix Q contains a breakdown of the downlists for the Major Modes).

The 200 AGC words in the downlist are transmitted to the Earth as 100 "downlink words", each of which is 40 bits long. The first downlink word contains two codes: the ID code of specific downlist being sent, and a synchronization bit string whose value is always 77340_8. Each downlink word starts with a "Word Order Code", set to "0" except for the first and 51st word in which it is set to '1', indicating the beginning and middle of the list. Next in the string of the downlink word is the contents of Channel 34 and 35, each of which is 15 bits of data and 1 bit of parity. Finally, the 40-bit string is concluded with a repeat of the first seven bits of Channel 34. This is more for the fact that the downlink system operates internally with 8 bit segments with the seven bits acting as little more than "filler". Unlike the uplink word which repeats bits strings for redundancy, these duplicated bits are not used for error detection.

The downlink process begins with the hardware generating an interrupt every 20 ms, with the DOWNRUPT interrupt entry point at 04040_8. Processing in the interrupt determines which of the 200 words must be downlinked next, and the next

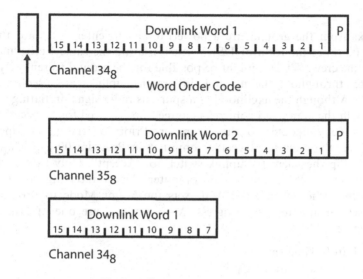

Figure 148: Downlist word format

two words are stored in channels 34 and 35. Writing to these registers actually begins the process of sending the data to the downlink hardware, which then formats the data and arranges for it to be radioed to the ground. Downlinking data is an ongoing process, and is not subject to a particular schedule. Formatting and transmitting all 200 words takes about 2 seconds, after which the cycle starts again with another set of 200 words. Over the time it takes to transmit one downlist, it is unlikely that all 200 words will remain unchanged. Consistency of the data is important for accurately interpreting the state of the AGC, so a "snapshot" of particularly important variables is necessary. Each time the downlink list begins, data that might be particularly volatile is saved to ensure it reflects the state at the beginning of the downlist.

A special case is found when dumping all of erasable storage, which is performed by Verb 74. Complete dumps of memory are rarely required except before major mission events, such as before launch or lunar landing. In these cases, sampling only the important words in the AGC is not sufficient to verify that the software is ready for that event. A minor problem, such as a variable that is not quite within expected tolerances might have a serious impact later in the mission. When Verb 74 is entered, 2,048 words of data for all eight erasable banks are transmitted to the ground, not once, but twice. With such a large amount of raw data, a set of identifiers are necessary to organize the data. Unlike the downlists, which are organized into sets of 200 words, a storage dump is composed of eight 130-word sequences, the first two words of which are header information, beginning with an identification word containing 01776_8. The second word contains the erasable bank number and a bit indicating whether this is the first or second copy transmitted to the ground. The remaining 128 words are the contents of the erasable storage bank. Transmitting eight banks takes 20.4 seconds to complete, and the second copy of storage is transmitted immediately after the first copy completes. During the storage dump, the standard downlist is not transmitted.

Data uplinks

Data uplinks from the ground are a powerful way to enter data into the AGC. Uplinked data takes the form of DSKY entries, using the same Verbs and Nouns available to the crew. While it might be possible for the ground to remotely operate the computer throughout the mission, in practice uplinks are used for specific updates only. Although the likelihood of a spurious radio signal initiating an update was small, both the crews and controllers agreed on the need for a special interlock to permit or deny updates to the computer. Prior to sending an update, the controllers asks the crew to place the computer in Program 00 (no Major Mode running) and flip the telemetry uplink switch to "Accept". Only then is data sent from the ground able to reach the computer. Upon receiving the first word transmitted from the ground, the AGC sets the Major Mode to Program 27 to establish that the update is in progress. Most updates use one of four different extended Verbs:

Verb 70: Liftoff Time Increment
Verb 71: Contiguous Block Update
Verb 72: Scatter Update
Verb 73: Octal Clock Increment

As uplinks to the AGC take the same form as manual DSKY inputs, it was reasonable to combine these two operations into a single set of software routines. As such, uplink data is almost indistinguishable from manual input, and the details of its processing are found in the DSKY section. Although key codes are not monitored directly, the downlist software is still running and contains the contents of the uplinked buffers, enabling the uplink to be verified.

With the data received by the AGC in the INLINK register, all that is visible are the three 5-bit key codes. Seven other bits are part of the update word, for a total of 22 bits per key uplinked. Three bits specify the subsystem to receive the data, and another three bits identify the spacecraft for which the update is intended.[18] The remaining 16 bits contain the three copies of the key code, plus a leading '1' to force an overflow to trigger the uplink interrupt. Data is not transmitted blindly by the ground, because there is an actual dialog between the computer and the ground controllers. After receiving each uplink word, the hardware sends an 8-bit acknowledgment that the data has been successfully received and decoded. When the AGC recognizes that all the data has been entered, the update program flashes V21N02, which is visible on the ground via the downlist. If everything is acceptable, the controllers will transmit Verb 33, Enter, to proceed with the update without any more DSKY inputs.

[18]   One argument says that this format was developed with the expectation that systems other than the AGC would be updated by ground controllers, and that conceivably more than one pair of Apollo spacecraft might be aloft at any time.

Error detection

Although the reliability of the data link between the AGC and ground controllers was excellent, any of a number of problems could prevent a successful exchange of information. Static on the radio channel or a momentary delay in processing a bit string could result in lost or corrupted data. Any situation where the data might be in error is serious, and if it occurs, the entire data stream becomes suspect. Digital error correction codes were not used in Apollo, leaving the only alternative to mark the transmission as failed and start over. The hardware receiving uplink from the ground or downlink from the AGC was sensitive to the data rate. Large memory buffers were not a feature in these electronics, so all the components had to operate in a carefully synchronized manner. Two program alarms were reserved for data transfers that fell out-of-sync. An 1105 alarm occurred when the data being sent from the AGC was faster than the downlink hardware could accept, and an 1106 alarm was raised when the ground was transmitting faster than the spacecraft could accept.

While errors in the downlist might be troublesome, it was rarely a critical issue as another set of data would be transmitted shortly. An error in data uplinked to the AGC was a far more serious matter, as an invalid update could disrupt program operations or cause serious errors. While error correcting codes were not used, the uplink key codes were designed to reliably detect errors. By sending three copies of the 5-bit key code, with one copy sent with its bits inverted, any data corruption was usually apparent. If the validation routine in the uplink program detected a problem, it would illuminate the OPR ERR light on the DSKY. Observing this light, both the crew and ground controllers would cancel the current update, and have the data retransmitted.

COMMAND MODULE ENTRY

Entry back into the atmosphere is perhaps the most dynamic series of events that occur during a spaceflight mission.[19] Unlike the series of thrusts that combine to launch the spacecraft from Earth, or even the dynamic maneuvers necessary for landing on the Moon, entry requires transitions from vacuum to atmospheric flight (and in some cases, not once but twice) in a vehicle whose flying characteristics change substantially over just a few minutes. With the mission nearing the end, the crew begins the process of preparing the spacecraft for its return through the Earth's atmosphere, ending with a splashdown in the ocean. During the ascent from Earth, the Saturn rocket converted the chemical energy of millions of pounds of volatile

---

[19] In most descriptions of a spacecraft returning to Earth, the terms "entry" and "re-entry" appear interchangeably with little harm done. In the case of returning from the Moon, a nominal trajectory involves the spacecraft making a "double dip" into the atmosphere, rendering the use of "entry" and "re-entry" confusing. Here, the term entry is used exclusively, with each penetration of the atmosphere being carefully noted.

chemicals into heat, to send the spacecraft to the Moon. During entry into the atmosphere, this process reverses, and the tremendous velocity accumulated on the journey home is converted back into heat. In the same manner as an air compressor becomes hot as it pumps air into a tire, the spacecraft heats the air ahead of it through compression. With the CM moving at hypersonic velocities, the air ahead of it cannot get out of the way fast enough, and the molecules are squeezed together with incredible energy. As the kinetic energy is converted into heat, the temperatures reach thousands of degrees.

Several options exist for designing an entry profile, each of which is dependent on the velocity and angle of the CM when it hits the atmosphere. Returning from a lunar mission creates the most dynamic entry profile and a complex sequence of events, but also serves to demonstrate the capabilities of the AGC and spacecraft. Much of the data in the following example comes from Apollo 11, but is typical of the lunar missions.

The most important goal of the entry software is to ensure that the spacecraft will be decelerated to less than 25,500 ft/sec (7,770 m/sec), because this will ensure it cannot remain in an orbit around the Earth. Known internally to the AGC as VCIRC, this velocity is an important boundary between the different phases of entry. Slowing below this velocity will guarantee capture by the Earth's gravity. The second goal is to make sure the spacecraft is guided accurately to the splashdown zone. Balancing the need to reduce velocity is the need to limit the amount of deceleration to about 8 g, the highest humans can withstand for a short time while maintaining consciousness. Upon hitting the upper fringes of the atmosphere, the vehicle's velocity of 36,000 ft/sec (10,970 m/sec) is very near the velocity required to leave the Earth and journey to back the Moon. If not slowed, the CM will whip around the Earth, and likely enter an orbit around the Sun. Most of the velocity accumulates during the final hours of "falling" to the Earth. Only 8 hours prior to entry, the spacecraft is traveling at a relatively leisurely pace of 6,500 ft/sec (1,980 m/sec). The formal start of entry is at the Entry Interface, a boundary defined as 400,000 feet (122 km) above the Earth's surface. Once the entry begins, the spacecraft and crew are entirely on their own. The high temperatures create a sheath of ionized gas around the spacecraft that blocks both incoming and outgoing radio signals. Not only is voice communication impossible, but ground controllers have no way to monitor telemetry to assess the health of the crew and their spacecraft. This blackout begins at the point when the crew begins to feel their first g-forces. The duration of this radio blackout can vary considerably depending on the entry profile used, and may range from less than three minutes to almost a minute longer.

Deceleration begins quickly as the spacecraft slams into the upper fringes of the atmosphere, rising from a fraction of 1 g to almost 7 g's in less than 30 seconds. Two minutes after Entry Interface, the harsh deceleration has paid off, and the spacecraft is now traveling at less than orbital velocity. While there is no longer any chance that it will return to space, much more energy needs to be dissipated before reaching the denser layers of the lower atmosphere. From this point on, the AGC is tasked with steering the spacecraft towards the targeted splashdown point while it performs the tasks necessary to slow the vehicle. If necessary, the AGC will reverse the downward trajectory and fly back toward space. This maneuver is required if the AGC needs to "stretch" the entry by adjusting the flight path to direct itself towards the recovery zone. This maneuver has the added benefit of reducing the heat and g-loads on the spacecraft and crew. Any such relief is short lived, however, because the entry concludes with another, longer spike of nearly 6 g's designed to bleed off much of the remaining velocity. During this last phase, the computer is rolling the spacecraft, adjusting its lift vector to steer as close as possible to its targeted landing point. Almost home, the last critical event of the mission occurs. A tightly choreographed sequence deploys a series of parachutes to lower the spacecraft into the ocean. Apollo has arrived back on Earth.

The entry corridor
The entry corridor is a path that describes the optimal trajectory back into the Earth's atmosphere. The corridor is narrowly defined, using both the point where the spacecraft reaches the Entry Interface and the angle above the horizon that the spacecraft penetrates the atmosphere. This level of precision is necessary, for it is carefully chosen to ensure that the spacecraft enters the atmosphere and slows enough to be captured by the Earth's gravity, yet ensures that the deceleration forces are tolerable for the crew. Entering the atmosphere head-on is certainly a fatal proposition, as to attempt to punch through such rapidly thickening air would produce tremendous heating and deceleration forces. An entry angle too shallow, or originating too high above the Earth's surface, will not decelerate the vehicle sufficiently to ensure that it will return home.[20]

The comfort of the crew is not the only consideration. Engineering limits for the thermal protection provided by the ablative heat shield are also important. In a steep entry, not only are the deceleration forces high, the temperature of the plasma surrounding the vehicle may be beyond what the ablator can survive. A long slow entry limits heat shield temperatures and g-forces on the crew, but also increases the total heat load on the spacecraft. While the vehicle will not burn up during such a shallow entry, the crew runs the risk of being baked. A more controlled way to pass through the atmosphere is a gentle, nearly grazing path, where the tradeoff of heat and velocity are far more manageable. Using an entry angle of 6.48 degrees permits a deceleration rate that produces tolerable g-forces and a manageable temperature and heat profile. This entry angle is measured at 400,000 ft (122 km), and if the Earth were an airless body, this trajectory would pass through a perigee at an altitude of 20 nm (38 km) and then climb back to a high apogee. However, the sensible atmosphere exists far above that point, so that much of the deceleration from lunar velocity occurs before reaching this altitude.

A final midcourse correction maneuver is performed if needed in the final hours before entry. It is the last opportunity to tweak the flight path before the crew turns

---

[20] The overused term, "skipping off of the atmosphere" is misleading, as the spacecraft does not skip like a stone across a smooth lake. Rather, it continues to travel through the atmosphere without slowing enough to be captured by the Earth's gravity.

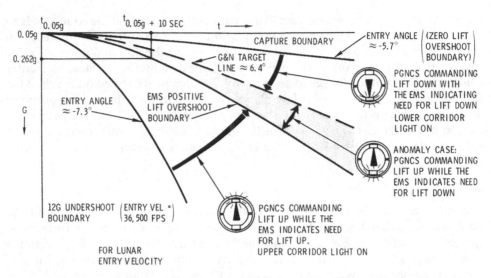

Figure 149: Command Module entry corridor

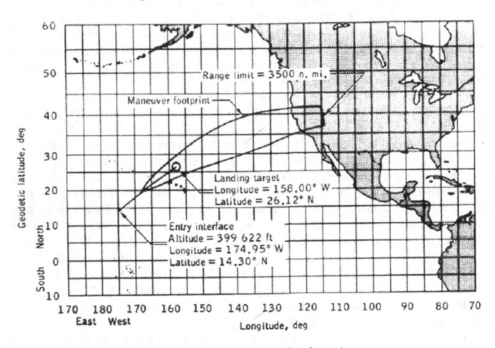

Figure 150: Entry maneuvering footprint

its attention to preparing the spacecraft for entry. While the ideal entry corridor is about 6.48°, both the spacecraft and the software can safely accommodate a wider range of values. The spacecraft can return at steeper angles of up to 7.3°, but at the cost of a higher g-load and greater peak temperatures on the heat shield. Any angle steeper than this creates deceleration forces that could seriously injure the crew or damage the spacecraft. At the opposite end, the entry angle can be as shallow as 5.7°. Such entries are longer as the spacecraft decelerates less over time, and yield a higher total heating load. Entry at less than 5.7° will not decelerate the vehicle sufficiently for the Earth's gravity to capture it, and will result in the spacecraft slipping through the upper atmosphere into an elongated orbit. When entry does finally occur, the angle will be substantially steeper and the deceleration will likely destroy the spacecraft.[21]

In steep entries, peak heating is the primary issue from the tremendous amount of kinetic energy that must be dissipated quickly. To ensure crew survival, this must be done without exceeding 12 g's. A shallow entry produces lower temperatures but the heating lasts for a longer period, increasing the total heat load. In this case, more of the accumulated heat from the entry flows through the ablative material and heats the stainless steel structure, which is limited to a maximum temperature of 600° F (365° C).

Entry Monitor System description

Monitoring the progress of the spacecraft's entry is a task involving a complex set of variables, many which must be maintained within a narrow range of values. The AGC, whose only interface with the crew is the DSKY with its numerical registers, is woefully inadequate to present such a composite view of the data to the crew. With velocities, altitudes and g-forces changing so rapidly, a human cannot read, interpret and act upon numerical data quickly enough. This task is difficult enough when training in a simulator, but during an actual entry the buildup of g-forces makes this task even more challenging.

During the first two minutes, the most important task is to slow the spacecraft sufficiently that it is assured of returning to Earth. A little more than two minutes past Entry Interface, attention turns to balancing the deceleration profile with accurate targeting to the splashdown zone. Unlike the initial deceleration to suborbital speed, which was performed with the roll attitude fixed, in this phase the CM must be actively steered to its target. Designed to provide a graphical picture of the spacecraft's velocity versus acceleration, the Entry Monitor System, or EMS, gets the data from its internal accelerometers and the Stabilization and Control System. Being independent of the AGC, it provides an essential cross-check to the data the DSKY is providing, and if a problem occurs with the AGC or the IMU, the EMS is capable of guiding the spacecraft through the atmosphere. Since the Command Module is very stable in pitch and yaw during entry, the only axis that needs to be managed is roll. The roll angle, which also directs the lift vector, is displayed on the EMS and frees the astronauts from having to scan and interpret the FDAI for attitude information. A scrolling progress strip graphically shows the progress of the spacecraft as it passes through the entry corridor.

---

[21] However, the demise of the spacecraft in this case is of little concern – the oxygen in the spacecraft will be long gone by the time the Command Module enters the atmosphere again.

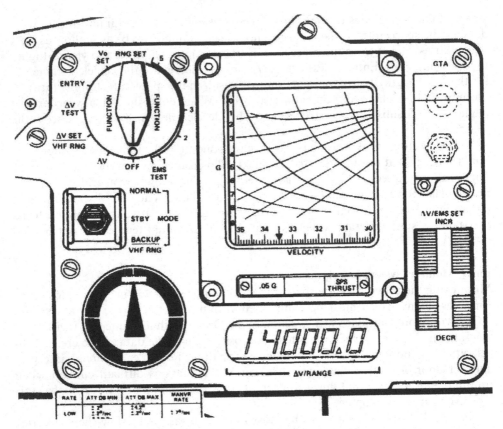

Figure 151: Entry Monitor System panel

EMS components: corridor lights, monitor scroll, roll indicator and velocity
Three instruments dominate the EMS panel, as well as three lights. At the bottom is the velocity meter, which is set by the crew either from information read up from ground controllers or from the Noun 63 display in Program 61 of the AGC. Once the EMS senses 0.05 g, the velocity indicator displays the velocity of the spacecraft, counting back from its initially set value. It provides an important cross-check to the AGC calculated values.

Without graphical displays of the sort taken for granted in 21st century computers, the Monitor Scroll is able to provide a real-time display of the entry profile of the spacecraft. To depict how the entry is progressing, the Monitor uses a long strip of Mylar preprinted with entry control parameters defined in terms of both acceleration and velocity. The backside of the strip is coated with a soft emulsion, and a scribe presses against its surface. The scribe is driven vertically from the EMS accelerometer, moving downward as the g-forces increase. As the spacecraft decelerates, the scroll moves to the left, with the current velocity aligned with the scribe. The effect is similar to drawing on an Etch-a-Sketch toy, with its stylus on the

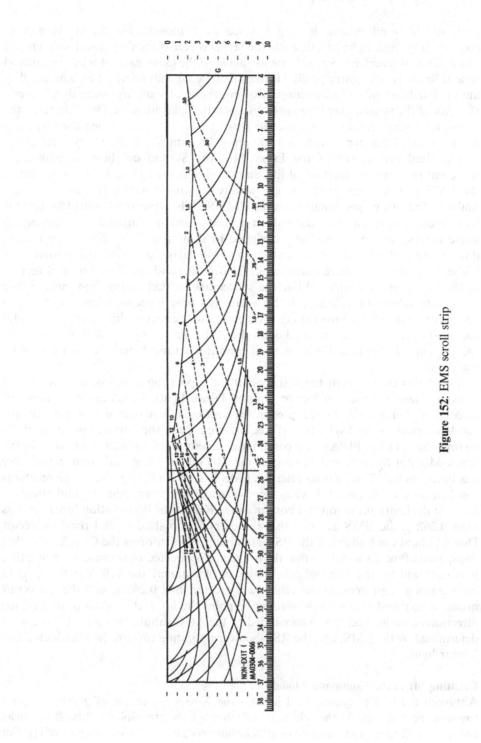

Figure 152: EMS scroll strip

back of the window and leaving a trace as it moves. The Mylar is moving continuously from right to left, while the stylus is constrained to travel only up and down. Two different EMS scrolls are available to the crew, each of which is tailored to a different type of entry profile. Rather than being characterized by whether they are for Earth orbital or lunar return missions, the scrolls are organized by whether a skip out of the atmosphere is required to reach the splashdown target. Normally the non-exiting scroll is used, but a second pattern is available for when the trajectory has to be extended to as much as 3,500 nm (6,480 km) from the Entry Interface.

The final instrument of the Entry Monitor System combines a pointer to represent the current direction of lift, and two indicator lights. As the only part of the EMS that requires hardware not wholly contained within the unit, the Roll Stability Indicator uses signals from the CM's Stabilization and Control System for its reference. With an oversized indicator needle, and uncluttered by intermediate angle marks, the RSI is perfectly designed to show the vital information needed during the earliest part of the entry. In these early moments, the g-forces are highest (to the point where vision might be affected) and decisions on the direction of the lift are most critical. While the spacecraft is decelerating from near-escape velocity to suborbital velocity, the lift vector is a binary decision: up or down. The simplicity of the RSI allows the crew to determine the current lift vector even under the most difficult conditions. Although the RSI is driven by the SCS, it will reflect the current roll attitude. Under normal conditions, this attitude is commanded by the AGC.

Two lights complement the RSI, one at the top of the indicator and one at the bottom. These are the Lift Vector Up and Down lights, which are illuminated to direct the Commander to the proper lift vector. When 0.05 g is sensed by the accelerometers in the IMU, this will be detected when the Servicer performs its 2-second check of the PIPAs. A second, independent accelerometer is installed in the Entry Monitor System, and illuminates the 0.05 g light when that point in the entry has been reached. Ten seconds after the EMS has sensed 0.05 g, the accelerometer is checked for the amount of deceleration the spacecraft is experiencing, and attention turns to the lights at the top and bottom of the RSI. If the deceleration forces are less than 0.262 g, the EMS assumes that the entry is too shallow and the Lift Vector Down light at the bottom of the RSI illuminates. This informs the Commander that the spacecraft needs to roll so that the lift vector is directed downward to ensure that it is captured by the atmosphere. In the same manner, the Lift Vector Up light illuminates if the forces on the vehicle are greater than 0.262 g, and the spacecraft needs to be rolled heads down in order to point the lift vector upward. Ideally, the direction of the lift vector as commanded by the AGC should match the lift vector as determined by the EMS, and the RSI should be pointing toward the illuminated Lift Vector light.

Creating lift in the Command Module
Although the CM is symmetrical around the X-axis, its center of gravity is offset towards the +Z axis. As the vehicle travels through the atmosphere, this offset causes the spacecraft to trim at an angle of attack that produces a modest amount of lift. For

a given velocity, the attitude the spacecraft trims itself in and amount of lift that results is fixed. As the direction of the lift vector cannot change, the only means to control the trajectory of the spacecraft is by rolling the CM and thus the direction of the lift vector. With the vehicle in a heads-down attitude, the lift vector is directed upward. In a heads-up attitude, the lift vector is down. Manipulating the lift vector enables the CM to be steered in much the same manner as rolling an aircraft alters the lift vector causing a turn left or right. Crossrange maneuvering is possible by rolling a little to either side of full lift up or full lift down. A significant amount of crossrange control is possible by rolling only 15 degrees to either side of full lift up or full lift down, yet the impact on the vertical component of entry is small. The CM is very stable at hypersonic speeds, and very little RCS activity is needed in the pitch and yaw axes. Roll jets perform the majority of thruster activity, being used to alter the lift vector. Once the CM gets below Mach 2 the stability decreases, and RCS activity increases significantly and continues until drogue chute deployment.

The digital autopilot during entry

Control of the Command Module during entry is a complex task, given the changes to spacecraft configuration and maneuvering in and out of the atmosphere. By far, the most important axis for the DAP to manage is rolling on the X-axis. Maintaining the Command Module in the center of the entry corridor becomes an exercise in managing the lift vector. Starting at the Entry Interface at 400,000 feet (122 km) the lift vector is pointed toward the Earth in order to "dig down" into the atmosphere and prevent any tendency to skip back out into space before slowing to suborbital velocity. G-forces build as the spacecraft rapidly decelerates. The lift vector is rotated up when the calculated trajectory will bring the spacecraft within 25 nautical miles of its target. Deceleration is maintained at between 3 to 4 g, continuing to reduce the velocity. If necessary, a controlled skip-out increases the altitude, back into thinner air, possibly even right out of the atmosphere. By this point, the vehicle is below orbital velocity, and a more conventional entry profile is begun. Through these phases, the DAP is maintaining attitude as commanded by the entry targeting routines, rotating the spacecraft to orient the lift vector. If a skip-out is required, the aerodynamic forces may no longer be sufficient to stabilize the vehicle, in which case the DAP will assume active control of all three axes. Aerodynamic flight resumes when the CM digs into the atmosphere for the second and final time, and the DAP maintains attitude through the pitch and yaw dampers. Finally, as the vehicle slows to less than 1,000 ft/sec (300 m/sec, or just under Mach 1) at an altitude of about 65,000 feet, there is no longer sufficient aerodynamic force to maneuver, and active steering is terminated.

Command Module entry sequence of events

Entry interface -4 hours

Despite the technical capabilities in the spacecraft, few tasks are more important than recording the PAD updates relayed from the ground. All critical entry parameters, such as times, velocities and attitudes are read to a crewman, who verbally confirms their contents and records the data on a preprinted form. Not all the data is destined for entry in the computer. Data such as the time of the horizon check or the selection of the entry profile in the Entry Monitor System are outside of the computer's scope, but are just as necessary as the entry attitude. Much of the PAD data is used to update or confirm the data presented in Program 61, the Entry Initialization program.

Ground controllers provide an update of the state vector and entry REFSMMAT in preparation for a platform alignment. An opportunity for one last midcourse correction exists at three hours before Entry Interface, and if required, will fine-tune the entry angle. With two hours remaining, the pace inside the spacecraft quickens with a final P52 platform alignment. Logic circuits, critically important to manage the firing of the pyrotechnics responsible for the parachutes, are powered up and verified by ground controllers. The Entry Monitor System is powered up and a number of tests are conducted to ensure that its circuits and the scroll assembly are working correctly, then the scroll for the type of entry is positioned in the window.

Powering up all of the systems in anticipation consumes a large amount of power, generating lots of heat in the process. Both the primary and secondary cooling systems, located in the Service Module, are started to handle this thermal load. In fact, there is more cooling than strictly necessary at this time, in order to "cold-soak" the electronics and structure as much as practicable without endangering the equipment. This additional cooling is necessary because once the Service Module is jettisoned, there is no active cooling available in the Command Module. While the electronics are cooling, heaters are turned on to maintain the CM thrusters at an operating temperature of at least 28° F (-2° C). At temperatures below this value, it is possible that the nitrogen tetroxide oxidizer will freeze on contact with the injector valves, preventing the thruster from firing.

With all the systems activated, the crew completes their final stowage tasks and strap themselves into their couches. Batteries for the pyrotechnics are connected to the sequential events circuits and are tested, and the Command Module RCS system is activated. With less than 45 minutes before reaching the atmosphere, the EMS is initialized by selecting the appropriate scroll to use, entering the entry velocity, and aligning the RSI to the Stabilization and Control System. Now sufficiently warmed, the Command Module thrusters are tested to ensure that they respond to hand controller inputs. With its mission completed, many of the systems in the Service Module are shut down, isolating the Command Module in preparation for their separation. Power is still available in the Service Module, to run the control sequencers that will separate the two spacecraft and fire thrusters to pull the Service Module away. With the hardware configured for entry, the first of the entry programs is started in the AGC.

Program 61 Entry Preparation: entry interface -25 minutes

The AGC does not calculate most of the entry parameters used. This data is calculated in the RTCC in Houston, and read up to the crew several hours prior to entry. About 25 minutes before Entry Interface, the crew starts Program 61, Entry Preparation. This performs the preliminary entry navigation computations and checks the IMU to ensure it is up and aligned. A final integration of the state vector is performed, and Average G, the Servicer routine used to update the state vector during accelerated flight is also started. Much of the dialog between the AGC and the crew consists of presenting data calculated by the AGC for the crew to confirm or update. In addition to confirming basic entry parameters, the crew uses much of this data for initializing and operating the Entry Monitor System. Three sets of Nouns are displayed. Nouns 61, 60 and 63 are used to provide a wide range of entry-related data. Beginning with Noun 61, the predicted latitude and longitude of the splashdown point are entered, plus the entry roll attitude. The next two Nouns supply the velocities, angles, g-loadings and ranges relating to the entry. Pressing PRO to accept the data entered in Noun 63 terminates P61 and automatically calls up Program 62.

Program 62 CM/SM Separation and Preentry Maneuver: entry interface -15 minutes
With the mundane but essential task of entering entry data complete, the AGC starts Program 62, CM/SM Separation and Preentry Maneuvering. P62 performs yet another check of the platform to ensure that it is operating and properly aligned, and generates a program alarm if the IMU is not ready. At this late stage, it is unlikely there will be time for troubleshooting or realignment of the IMU, and a manual entry becomes necessary. Assuming all is well with the platform, P62 will display V50N25, Perform Checklist Item, with Register 1 containing 00041_8 to request that the crew begin the CM/SM separation process. The CSM first maneuvers to an attitude which is perpendicular to the flight path to ensure that the two vehicles will not recontact after separation. Separating the Command Module from the Service Module and its lifegiving supplies of power, oxygen and propulsion is such a critical maneuver that it is not trusted to the computer. In addition to the steps necessary to arm the pyrotechnics that will cut the two modules apart, the switches that fire those charges are protected by special covers to prevent their inadvertent operation. Springs or other mechanical devices are not sufficient to push the two modules safely apart, so the Service Module Jettison Controller on the Service Module commands the thrusters to fire and push the now derelict hardware away.

At this point, the Command Module is completely on its own. After separation, the entry Digital Autopilot adopts a heads-down attitude to enable the crew to monitor the horizon as a final attitude check. More than an exercise to verify the correct entry angle, the automatic maneuver to the horizon check attitude verifies that the AGC and IMU are able to correctly command the CM to a given attitude. A reference line on the Commander's rendezvous window indicates a 31.7° angle between the horizon and the spacecraft X-axis. If the horizon does not align itself within ±5° of this mark on the window, either the AGC or the IMU is assumed to have failed, and the crew will employ the EMS and manual procedures for entry guidance.

P62 once again displays V06N61 to confirm the splashdown point and the lift orientation, allowing a last minute change if necessary. Pressing PRO accepts the data and advances the display to V06N22 with the gimbal angles for entry. The CM maneuvers to the entry attitude with the lift vector pointing down to assure capture in the atmosphere. When the CM is within 45 degrees of the intended attitude, the AGC automatically advances to P63.

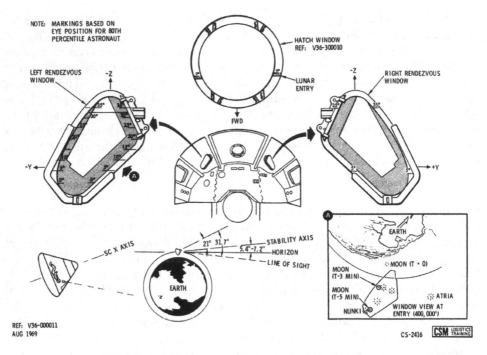

Figure 153: View from CM of Earth's horizon prior to entry

P63 Entry – Initialization: entry interface -2 minutes
Now configured and committed to entry, Program 63 initializes the entry guidance equations used to control the spacecraft as it decelerates as well as steering it to the recovery zone. The entry DAP holds the entry attitude steady, as the first wisps of the atmosphere will be encountered shortly. Displaying Noun 64, the AGC presents the current acceleration, velocity, and range to go to the splashdown point. Little more needs to be done by the computer and the crew, and both wait for the first onset of gravity, marked by the 0.05-g light. Meanwhile, the Earth pulls the CM closer and faster, and the effects of the rather shallow entry angle become apparent, bringing the spacecraft around the Earth nearly to the side opposite the Moon. In the minutes before reaching the atmosphere, the crew observes the Moon setting behind the Earth, finally disappearing at two to three minutes before the spacecraft reaches the top of the atmosphere.

Although defined as the boundary between Earth and space, the 400,000 foot (65 nm, 122 km) Entry Interface altitude is somewhat arbitrary. Few air molecules exist at this altitude and the spacecraft continues to accelerate for the next 30 seconds. As a practical matter, the entry point is when spacecraft reaches deep enough into the atmosphere to experience a deceleration force of 0.05 g. The AGC expects this to occur at 300,000 feet (91 km), but it can vary by thousands of feet. Apollo 11, for example, experienced 0.05 g at 310,000 feet. Regardless of the altitude at which it occurs, when 0.05 g is sensed the AGC switches to Program 64 and the real work of entry begins.

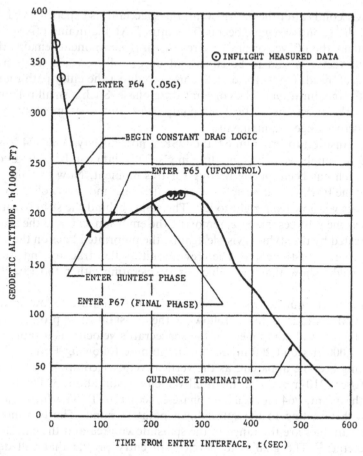

Figure 154: Entry program number and altitude

P64 Entry – post 0.05g

Now traveling through the atmosphere, the three-axis attitude-hold mode of the DAP ends. Assuming that the spacecraft is near the center of the entry corridor, the CM rolls to a heads down, lift vector up attitude to limit the deceleration loads on the crew. The atmospheric DAP begins with a wide pitch and yaw deadband, and the hypersonic stability of the CM requires only pitch and yaw dampers for attitude control. Roll, and the resulting lift vector, continue to be under the active control of the guidance routines. The view out the windows is often described as being inside a neon light, with flows of brightly lit plasmas coming off of the spacecraft. These plasmas are the result of the heating as the spacecraft compresses the air in front of it and are composed of ionized gases. Carrying an electrical charge, the plasma is also responsible for the radio blackout that occurs for approximately the first 3½ minutes of entry.

Early in the entry, the overriding goal is to slow the spacecraft to a value below orbital velocity, with steering towards the splashdown area a secondary concern.

Within a few seconds after the 0.05-g event, the spacecraft is experiencing 1.4 g's and the constant-drag control regime begins. The entry DAP maintains a fixed roll angle, which maintains the lift vector and ensures the drag component remains the same. While the lift vector is said to be either full up or full down, a small crossrange adjustment is allowed. Up to 15° of roll is used to direct the entry path right or left, and is held for the duration of the constant-drag phase. Such a small roll angle does not significantly affect the vertical lift component, but does measurably help in meeting any crossrange requirements.

With the transition to Program 64, the DSKY now displays V06N74 to show the commanded roll angle, velocity, and drag in g's. This data can be compared against the EMS, which has been quietly waiting for this moment. Now that the vehicle is decelerating, the EMS velocity display starts to decrease and the scroll strip begins to move across its window from right to left. The stylus behind the strip begins moving downwards as the g-forces increase, removing the emulsion as it and the scroll move. The trace created by the stylus is visible against the preprinted lines on the scroll, and shows the progress of the entry. The crew monitors this tracing, and watches for trends that might show that the vehicle is decelerating either too rapidly or too slowly.

Deceleration very quickly rises to over 6 g's in the minute following the 0.05-g event. The CM attitude is fixed, following the constant-drag profile, and P64 is monitoring the vertical component of the spacecraft's velocity as it rapidly declines from about 4,000 ft/sec (1,200 m/sec). In the minute following Entry Interface the effect of lifting become apparent, as the vertical velocity component has slowed to about 700 ft/sec (210 m/sec), yet the total velocity is still about 32,000 ft/sec (9,800 m/sec). At this point, P64 enters a new phase called HUNTEST which evaluates the trajectory with targeting to the splashdown point in mind. The two most critical HUNTEST variables are the range to the splashdown area and the current velocity of the spacecraft.[22] The goal is to develop an entry profile that will deliver the spacecraft to within 25 nm (46 km) of the target. Several solutions are evaluated, each of which might require a different set of entry programs in the AGC. Most importantly, HUNTEST must consider if the entry solution requires Program 65, Entry – UPCONTROL, and if so, whether P65 should be followed by Program 66, Entry – Ballistic (Kepler). Regardless of the selection of either or both of these programs, the last program to manage entry is always Program 67, Entry – Final Phase.

Rather than streaking through the atmosphere in a constant downward motion, entries from the Moon usually require a phase where the downward rate is paused or, more likely, reversed. Such an "UPCONTROL" phase by P65 serves several purposes. Reducing the drag on the spacecraft also means that its velocity is no longer decreasing, and it will travel further downrange. By increasing the vehicle's

---

[22] Note that the altitude of the spacecraft is not considered in any of the entry programs. Altitude becomes relevant only when all guidance is terminated and the parachutes are ready to be deployed.

altitude, the drag and heating are reduced, both of which serve to reduce the stress on the human crew and the heat shield. With the primary task of guaranteeing the spacecraft's capture in the Earth's atmosphere now past, HUNTEST concentrates on meeting the range requirements of entry using UPCONTROL to control the distance traveled. This is done now because as the amount of energy left in the spacecraft decreases, its ability to shorten or extend its range also diminishes, and by P67 may no longer be adequate. By quickly entering UPCONTROL and lofting the trajectory soon after HUNTEST begins, the range can be extended considerably. If the range to the splashdown target is short, no UPCONTROL solution will exist and the AGC will enter P67 directly. HUNTEST evaluates the solutions every 2 seconds, shortly after the Servicer reads from the accelerometers and updates the vehicle's position and velocity. Constant drag is maintained until HUNTEST's calculations show that an UPCONTROL solution will bring the spacecraft's estimated range to within 25 nm of the target. The velocity calculated by HUNTEST for when UPCONTROL is entered determines the range that can be achieved by the maneuver, and also whether the spacecraft will exit the atmosphere. If a skip-out, or ballistic phase is necessary, HUNTEST continues the constant-drag phase until the skip-out velocity is safely below VCIRC, the minimum orbital velocity around the Earth. If the skip-out velocity is greater than VCIRC, then the spacecraft will fly off in a highly elliptical orbit that will doom the crew.

P65 Entry – Upcontrol

P64 transfers control to P65 when it has arrived at an UPCONTROL solution, and is automatically selected when the HUNTEST range estimate is within 25 nm of the targeted splashdown area and velocity is less than VCIRC. P65 is necessary when the spacecraft needs to dissipate excess energy and extend the trajectory to meet range requirements. As its name implies, UPCONTROL rolls the spacecraft to a heads-down attitude, which directs the lift vector upward. Although the Command Module does not produce a significant amount lift, it is still sufficient to take the vehicle back out of the atmosphere. However, entering P65 does not automatically imply that the spacecraft will leave the atmosphere. The duration of this phase may be brief, keeping the entire trajectory within the atmosphere. As an example, UPCONTROL was entered on Apollo 11 at about 182,000 feet (55 km), and brought the spacecraft up only to about 220,000 feet (67 km) before descending again. During the highest part of this phase, the spacecraft still experienced 0.6 g's, and there was enough air to continue slowing the vehicle.

P65 initially displays V16N69 for the crew, with bank angle, drag in g's at the point of skip-out, and the skip-out velocity. This is not the only display that is available, as several other Nouns are able to provide velocity, range and altitude rate. There are a number of ways to exit from P65, depending on the amount of UPCONTROL needed for the entry. If a skip-out is required to extend the range of the entry, P65 will wait until it senses an exit from the atmosphere. As the spacecraft climbs and the density of the air falls, the drag it experiences also decreases. Once the deceleration reaches 0.2 g's, the spacecraft is considered no longer to be in the atmosphere and P65 hands over to P66 for the ballistic phase of the entry.

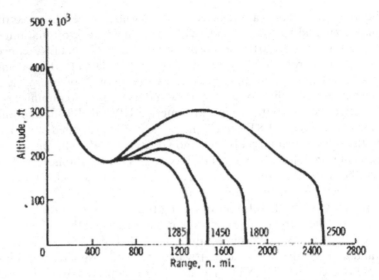

Figure 155: Entry altitude and range profile

HUNTEST's decision to require a skip-out trajectory and the transition to P66 depends on the distance between the splashdown point and the point of the Entry Interface. Unless the splashdown target lies near the limit of the spacecraft's downrange capabilities, a ballistic phase is not required. While the UPCONTROL phase still exists, the spacecraft remains in the atmosphere where drag continues to slow the vehicle. With the vehicle still experiencing drag, there is no need to transition to the program for purely ballistic flight. UPCONTROL ends when the overall velocity has reached a sufficiently low value and the spacecraft has resumed decending towards the Earth.

All entries from lunar trajectories will execute P65, but not all will direct the trajectory upward or skip out of the atmosphere. If the range to the target area is short, HUNTEST will extend the constant-drag phase to slow the vehicle and rule out an overshoot. When HUNTEST arrives at a targeting solution, P64 terminates and P65 is automatically started. P65 may find that the additional slowing of the spacecraft has brought its velocity to under 18,000 ft/sec (5,500 m/sec) where any significant upward deflection of the entry path is not possible. Recognizing this, P65 will immediately terminate and start P67 to manage the final phase of the entry.

P66 Entry – Ballistic

Program 66 is entered only when P65 had detected that deceleration has dropped to under 0.2 g's, and the spacecraft has left most of the sensible atmosphere. Up until this point in the entry, atmospheric forces held the Command Module in a stable attitude. Now effectively back in the vacuum of space, the Digital Autopilot can no longer rely on pitch and yaw dampers to maintain the desired attitude. As the atmosphere continues to thin, the DAP returns to its three-axis attitude-hold mode when the deceleration force reaches 0.05 g. V06N22 comes up on the DSKY when

P66 begins, and displays the entry attitude gimbal angles. The crew monitors their attitude indicators and the error needles to ensure that the spacecraft is in the proper attitude when it enters the atmosphere for the second time.

In P66, the spacecraft is on a purely ballistic path, with no means to alter its trajectory and little deceleration force to slow its progress. There is little for the crew to do but wait until gravity pulls the spacecraft back into the atmosphere. If the drag has dropped to below 0.05 g the DAP will have switched to three-axis control. Once this condition is sensed again, the pitch and yaw dampers are reengaged. P66 terminates completely when the atmospheric drag increases to 0.2 g's, switching to Program 67.

Program 67 Final Phase

True to its name, Program 67 controls the final phase of the spacecraft's entry through the atmosphere. Although the initial entry is highly dynamic with the choice of specific trajectory based on a large number of variables, the final phase is fairly standardized. Each of the three predecessor programs has the option of exiting to P67. P64 will transfer to P67 directly if the entry is from Earth orbit, or if the range to the splashdown target is short and the velocity is sufficiently low. Whether a skip-out follows the UPCONTROL phase also dictates when P67 will be entered. If a skip-out does not occur after UPCONTROL, P65 will exit to P67 as soon as the velocity is sufficiently low and the vehicle begins descending again. If a skip-out is needed, P67 is entered when the deceleration force reaches 0.2 g's. In all cases, the spacecraft is still high in the atmosphere and is traveling at a large fraction of orbital velocity.

When P67 is entered, most of the maneuvering to the splashdown area has already been accomplished. P67 manages the final steering to the target by rolling the lift vector to eliminate the remaining range error. Upon entry, P67 presents the crew with V16N66, showing the roll angle and the downrange and crossrange errors. As the spacecraft returns to the thicker air below 200,000 feet, g-forces quickly build a second time, peaking at 6 g's before falling back. The velocity has reduced sufficiently for the ionization around the spacecraft to dissipate and end the communications blackout. While not a complete blackout, communications during the second heating phase is still "ratty", often with heavy static and frequent dropouts. Around 65,000 feet (19.8 km), the velocity has fallen to 1,000 ft/sec (305 m/sec) and active steering is terminated because the parachute deployment sequence will be occurring shortly. At this point, control of the spacecraft goes to the SCS, and the role of the PGNS has ended for the mission

Parachute deployment sequence

Once the Command Module falls below 65,000 feet (19,800 meters), there is no active steering towards the targeted splashdown point. The spacecraft continues to slow as it enters thicker air. At this point, the AGC is no longer involved with flying the spacecraft, but continues to report the current latitude and longitude and the range to the targeted splashdown point. Altitude is not displayed on the DSKY, nor is there a radar onboard to measure altitude above the surface. Rather, the CM has a barometric altimeter which displays the approximate altitude. From this point onwards, the Earth Landing Sequence Controller manages the various elements of the Earth Landing System. Using a collection of timers and barometric switches, the ELSC is responsible for jettisoning the forward heat shield and deploying the parachutes. Most of these events are initiated through firing pyrotechnic charges. Unlike the vacuum of space, where springs are generally sufficient to push two pieces of hardware apart, the aerodynamic forces around the vehicle are complex and exert significant pressures. Mortars, similar to their combat namesakes, actively propel parachutes upwards and outward from the descending spacecraft.

At 24,000 feet (7,300 meters), barometric switches close and mortar charges fire to separate the forward heat shield, or apex cover, from the rest of the spacecraft. A small parachute built into the apex cover assists in separating the two and ensures that they will not recontact. Now exposed to the air, two drogue parachutes are fired from their mortar canisters 1.6 seconds later. These are only 16.5 feet (5 meters) in diameter, and they stabilize the CM as it descends. As the spacecraft descends through 11,000 feet (3,350 meters), the drogues are cut loose and three pilot parachutes are fired from their mortars. On opening, each pulls an 83.5 foot (25 meter) main parachute from its container. If allowed to completely open immediately, such large parachutes would likely suffer damage or injure the crew inside the spacecraft. To prevent such a shock, reefing lines prevent the parachutes from opening all at once. Over the next 10 seconds, the reefing lines are cut in two stages to allow the canopies to inflate completely. By 9,000 feet (2,750 meters), the parachutes are fully opened, slowing the CM's rate of descent to about 25 ft/sec (7 m/sec). Less than two minutes later, it splashes down in the ocean and the crew is finally home.

COMPUTER PROBLEMS DURING APOLLO 11 AND APOLLO 14

Apollo 11 executive overflows

If the drama of the first lunar landing was not enough, five program alarms occurred during the last seven minutes of the descent. All the available Core Sets and VAC areas, the necessary requisite for starting work in the AGC, were being used and new work could not start in the computer. A restart cleared the error, and the landing continued, but the problem surfaced again four more times.

Only once had an Executive Overflow alarm appeared in simulation, and an abort was called as a result. A review of the simulation revealed that the controllers were too quick to call an abort, and a successful landing was likely if the simulation had been allowed to progress. From this experience, the support staff reviewed all the possible program alarms and documented the appropriate responses.

The cause of the alarms began innocently enough. Shortly after the Descent Orbit Insertion burn that placed the LM in a 60 nm by 50,000 ft orbit (110 km by 15 km), the rendezvous radar was turned on to verify its ability to track the CSM, as it would be required to do during the rendezvous after leaving the lunar surface. However, an abort during the descent would also require the radar to search for the

CSM. The radar operated in three modes: SLEW, which allowed the crew to manually position the radar; AUTO TRACK, which used the radar's own electronics to track the CSM without control from the AGC; and LGC.[23] Only in the LGC mode was the computer actively positioning the radar and accepting its range and range-rate data.

To lighten the workload slightly in the event of the abort, the rendezvous radar was left powered up. (If the radar had been turned off, and an abort was called for, the crew would have to perform the power-up and orientation of the radar in addition to their other post-abort tasks.) During the descent, the relative positions of the two vehicles and the changes in the LM's attitude would make it likely that the rendezvous radar would lose lock. Early in the powered descent, the crew switched the radar mode to SLEW, a reasonable choice since it would prevent the radar from moving on its own. Although the radar was powered up, it was not providing data to the AGC. Unknown to the crew and controllers, the radar Coupling Data Units were flooding the AGC with counter interrupts. When in AUTO TRACK or SLEW mode, two 800Hz power supplies provided the voltage to the radar CDUs. These sources were to be synchronized in phase, using a signal generated by the AGC. An oversight in the power supply design specifications allowed the frequencies of the two 800Hz supplies to be locked (which was desired), but not in phase. The CDU, expecting the signals to be in phase, was unable to cope with the incorrectly phased power, and began to send counter interrupts to the AGC. Counter interrupts are notable in that they use only a small amount of computing resources, stealing only 11.7 microseconds of time from the AGC. In this case, each of the CDUs was sending 6,400 counter interrupts per second. With two CDUs affected, 12,800 interrupts per second were consuming approximately 15 percent of the available processing time of the AGC.[24] During the lunar landing, by far the busiest part of the mission for the AGC, it was actively performing work for only approximately 85 percent of the time.

Jobs in the AGC rarely run for long periods of time, and usually start, perform their work, and terminate with the expectation that they will be scheduled to run again if necessary. When jobs terminate, the Executive clears the Core Sets and VAC areas, which are now available for reuse by the next job. Each time a periodic job starts, one of its first tasks is to schedule the next iteration of itself. Perhaps the best example is the Servicer, which regularly runs every two seconds and is responsible for integrating velocity data into the state vector.

With the counter interrupts stealing the little cushion of available processing time during the descent, jobs were not able to finish executing. The Executive, lacking the sophistication to detect jobs running longer than anticipated, faithfully tried to run the next job at its scheduled time. Without previous jobs completing in a timely manner, the programs continue to hold their Core Sets and VAC areas. As the next

---

[23] Lunar Module Guidance Computer, the name given to the LM's AGC.

[24] This stream of counter interrupts can be compared to a "hot I/O", where a hardware failure in a device creates a number of unexpected I/O interrupts.

iteration of a program started, the Executive looked for an available Core Set (or VAC area) and eventually found them all occupied. Unable to continue, the Executive responded with a 1201 (Executive Overflow – No Core Sets) or 1202 (Executive Overflow – no VAC areas) program alarm and began the process of a software restart. All running work was canceled without regard to its priority, and programs were restarted according to their restart table entries. In just a few seconds, the system was able to resume processing without losing guidance and navigation data.

This restart capability was the key to continuing the mission despite the steady stream of unwanted interrupts. If the overflow occurs only occasionally, as was the case in Apollo 11, the situation is tolerable because a restart can clear the error and the computer can resume its control of the descent. Continuous restarts are another matter. A software or hardware error that prevents the computer from performing basic guidance, navigation or control operations is a serious issue, and an abort is the reasonable choice. The fact that the Executive Overflows are software restarts is also fortunate. Software restarts are more benign than a P00DOO abort or fresh start, both of which would have left the computer in an idle state, and would have required an immediate abort using the Abort Guidance System for control.

Most of the program alarms were concentrated the middle of the lunar descent. Why were they not more evenly spaced, or why did they not occur in the final parts of the descent? The first alarm, a 1202, occurred about five minutes into the descent while the AGC was running Program 63, Braking Phase. The spacecraft was is in a face-up attitude, and the engine was firing directly against the direction of travel to slow the spacecraft as much as possible. At this point, Buzz Aldrin entered a V16N68 to monitor the range to the landing site, the time remaining in P63 and the total velocity of the LM. The DSKY routine, Pinball, scheduled a job to monitor this data. Pinball is not an interpretive program, and hence does not use a VAC area, but the additional processing prevented other interpretive jobs from completing. Although V16N68 does not place a particularly heavy load on the computer, when added to the existing load, it was sufficient to generate the program alarm. At this point, all the VAC areas were used and the next job to start failed. The fact that the first Executive Overflow resulted from a lack of VAC areas was a chance occurrence, as both Core Sets and VAC areas were full. Aldrin correctly noted that entering V16N68 was generating the program alarms, and stopped keying the request into the computer.

After Program 64, Visibility Phase, began and the LM pitched over to give the crew their first view of the landing site, the amount of processing in the AGC reached it peak. Counter interrupts were still flowing into the AGC, whose processing time competed with guidance targeting, attitude control and radar updates. Overflows continued each time a job attempted to start, whether the result of a DSKY entry or the timing of when the jobs were attempting to start. Twice while in P64, Armstrong switched the attitude control of the LM to Attitude Hold from AUTO, which had the effect of reducing the amount of processing required in the computer. The first occasion was to get a feel for the handling of a much lighter spacecraft and to verify that the LM was responding to commands. The second time was to fly over "Little

West" crater, which was where the LM was headed. When over a safe landing site, Armstrong entered Program 66, Final Phase, where the processing requirements were considerably reduced. By this point, the impact of the counter interrupts was not sufficient to keep the jobs from completing.

The extent of the phase error was random, and was simply the result of the moment when the AGC sent out its first synchronization signal. Thus, the problem in the CDU was not the result of an incorrect switch setting or procedural error. It could have occurred in any prior mission. More importantly, it could have occurred at any time the rendezvous radar mode switch was set to AUTO TRACK or SLEW, regardless of whether the radar was actually powered up. Remarkably, it was possible for the problem to go undetected if processing in the AGC was able to complete. A permanent solution was that the CDUs would be zeroed any time the mode switch was in AUTO TRACK or SLEW, preventing any counter interrupts from being generated. Crew procedures were updated to advise that if an Executive Overflow occurred as a result of keying in a command, it was probably a temporary situation and the command should be retried.

Apollo 14 and the abort discrete bit
Computer issues were not restricted to processing errors or program alarms. In the case of Apollo 14, a potentially mission-ending computer error was due to hardware external to the computer. Less than four hours before the scheduled landing of the Lunar Module, Antares, ground controllers noticed from the telemetry downlink that the abort bit, Channel 30_8 bit 1, was set. This bit was connected directly to the crew's abort button, and was used to command the AGC to enter Program 70, Descent Propulsion System Abort. At this point, there was little risk of this bit causing any trouble. Another bit, LETABORT (bit 9 of flagword 9) effectively locked out the routines in charge of terminating the current program and switching to one of the two abort programs. When LETABORT was set to 1, it allowed the abort programs to run normally when they received either of the two abort discretes in channel 30_8. When the LETABORT bit is not set, abort requests are not processed.

The problem was most likely a floating solder ball making contact inside of the abort switch. Floating around in weightlessness, it could lodge itself in the switch contacts at random times. If the abort discrete became set in the AGC input channel, the result would be disastrous when the descent began. Part of the LM Servicer operation is to call Routine 11, the Abort Discrete Monitor, every 0.25 seconds. R11's logic flow was to test if LETABORT was set, and if it was, it would then see if the Major Mode was already set to one of the abort programs, P70 or P71. If the Major Mode already reflected an abort in progress, R11 would exit, assuming that if an abort was already underway there would be no need to start it again. Simply to inhibit the execution of R11, even if it were possible, would have effects that would be felt in other parts of the AGC. After R11 completes, it calls Routine 10, the LM Analog Display Routine. These displays include the cross-pointers and the altitude and altitude-rate tapemeters, which are essential to the crew during the landing.

Figure 156: Abort and Abort Stage pushbuttons in the LM

The normal sequence for powered descent was that Program 63, Braking Phase, would be started about 15 minutes before the time of ignition, allowing it to calculate the targeting values it would use. As a standard part of all Major Mode processing, the program number (in this case, 63) is loaded into the MODREG erasable storage location. One of the early steps in master ignition routine BURNBABY,[25] is to set the LETABORT bit in case an abort is necessary shortly after ignition. Once this flagword bit has been set, BURNBABY will schedule Routine 11 to perform the actual testing of the bits. For 26 seconds after ignition, the descent engine gimbals trim the engine so that it directs its thrust through the LM's center of mass. Only if Program 63 is in MODREG will BURNBABY automatically throttle up the engine to full thrust, where it will nominally remain for about 6 minutes and 40 seconds.

---

[25] Only in the 1960s could an ignition routine be named, "Burn, Baby, Burn", or more formally, BURNBABY.

Also during P63, the LM's landing radar data requires a different scaling factor than in the other programs used during landing. The Servicer tests MODREG for a value of 63, and if this is the case, then sets the appropriate radar scaling value.

At any time after ignition, the software is ready to sense whether either of the abort buttons have been pushed, and if so will return the LM to a safe orbit ready to rendezvous with the CSM. This created the dilemma. At any time after the master ignition routine began executing, the solder ball could make contact in the abort switch, irrevocably causing a P70 or P71 abort. Manually entering the commands to reset the LETABORT flag immediately after descent engine ignition would take several seconds, an unacceptable exposure to risk when the success of the mission was at stake. As all programming in the AGC resides in fixed storage, there was never an option to modify the software. A solution had to be found that involved using only data in erasable storage.

MODEREG is only a housekeeping value in the AGC's erasable storage, and is used for displaying the program number on the DSKY. Many programs reference MODREG, as their processing might be dependent on a particular Major Mode. While MODREG nominally held the value of the Major Mode, there was no reason why it could not hold another program number. This small fact became the basis of the kludge for the Apollo 14 abort discrete. Once Program 63 starts, Pinball sets MODREG to reflect the fact that program number 63 is the current Major Mode. Engineers at MIT quickly devised a elegant fix, which was to set MODREG to 71, to imply that the Program 71 abort program was running, not P63. At ignition, the LETABORT bit would be set and Routine 11 would check for the existence of an abort discrete in the input channel. If the solder ball had migrated back into the switch and set the abort discrete, R11 would then check MODREG to see if either of the two abort programs was running. With MODREG set to indicate this, R11 would assume that an abort was already in progress and would perform no further action.

However, MODREG could not be left indicating that P71 was in progress, as the throttle up and landing radar routines depended on Program 63 running. The final solution consisted of several steps:

- Program 63 was started normally, and allowed to perform it calculations.
- Before ignition, MODREG would be set to 71 in order to "spoof" R11 into thinking that an abort was underway.
- BURNBABY would perform a normal ignition and engine gimbal trim. LETABORT would be set automatically, and would be ignored by R11 when it saw P71 in MODREG.
- Throttle up would be performed manually using the Commander's Thrust/ Translational Controller Assembly.
- The LETABORT flagword bit would be reset to 0, inhibiting aborts through the Abort or Abort Stage pushbuttons.
- MODREG would be set back to 63, allowing the radar scaling routine to execute properly. R11 would not notice this, because, having found LETABORT not set, it would have exited without performing any other tests.

- Now with the correct Major Mode in MODREG, the Commander would return his TTHC to its minimum position, thereby allowing the throttle control routine to manage engine thrust.

If an abort had been necessary, the crew could have initiated the appropriate abort program manually using Verb 37, or used the longer procedure of manually setting the LETABORT bit back to 1 and then pressing either Abort or Abort Stage.

After Apollo 14, a software solution to the problem of an invalid abort discrete was devised. Crews were provided with a means to lock out the two abort programs, for use in the event that either of the abort discretes became inadvertently set.

Epilogue

Our first journey to the Moon took place at a unique junction of political and technological history. The stories of Cold War politics, a youthful President and the need to assert national superiority have been told many times. What made this journey truly audacious was that it occurred at precisely the time when the scientific and engineering capabilities to attempt it were only just beginning to emerge.

The environment in which the race for the Moon took place was not dissimilar to that of the Wright brothers, as they attempted to build their first flying machine. Creating such a machine only a decade earlier would have been extraordinarily difficult because the immaturity of internal combustion engines and the lack of aerodynamic knowledge presented formidable barriers to success. By the beginning of the twentieth century, however, the science and engineering came together. Scientists were discovering the physical laws that describe the forces on a wing as it travels through air. Gasoline engines were constantly improving. Tools and techniques for building strong, lightweight structures superseded the previous impossibly heavy designs. However, the greatest success of the Wright brothers was to combine these fledgling capabilities into a machine that, whilst not exactly elegant, demonstrated the feasibility of powered, controlled flight. Nowadays, the technology used by Orville and Wilbur is primitive, but it was a significant engineering feat for its time.

The same can be said of the Apollo Guidance Computer. The knowledge and skills required to design and build the gigantic flying machines of the space age converged during the 1960s. Although much of the technology did not exist at the start of that era, engineers nevertheless recognized the hurdles they faced. Constantly pushing the state of the art to build structures and systems never before attempted, they systematically overcame the obstacles. The development of the AGC also came at a unique point in computer history. Early computers relied on a bewildering array of technologies to store and process data, all of which were bulky, slow, and consumed large amounts of power. By the time development of the AGC began, transistorized circuits and core memory were reliable technologies. Mounting several transistors and other components onto a single chip was the only practical way to satisfy the AGC's space, weight and power requirements. Contrary to urban myth, NASA did not actually invent the microchip, but the contribution that both NASA and the Department of Defense made to the microelectronics industry may well have been even more important. Chips ordered for military and spaceflight applications demanded reliability that was then tremendously difficult to obtain. At that time, placing three transistors on a chip was an exciting achievement, but failure rates were maddeningly high. Semiconductor manufacturing was still an art, and manufacturing processes that could repeatedly manufacture highly reliable devices simply did not exist. As a result, the Apollo and military programs worked closely with the manufacturers. Companies invented new processes to improve reliability, investing heavily in clean rooms and automated assembly techniques. Component testing had to be rigorous. In fact, for chips destined for the AGC, a failure in one unit resulted in the entire manufacturing batch being rejected. An investment of this magnitude was made possible only by government orders for hundreds of thousands of chips. Gradually, yields improved. As manufacturing methods improved, doubling the number of components on the original three-transistor, single NOR-gate chip, made the final version of the AGC possible. In the end, the need to create chips of the qualities required for spaceflight gave manufacturers the skills to start the computer revolution.

Over the operational lifespan of the AGC, commercial computers evolved at a breathtaking pace, becoming ever faster and able to hold ever greater amounts of data. By the time of the final Apollo mission in 1975, processors routinely operated at millions of instructions per second, multi-megabyte memories were commonplace, and hundreds of sophisticated, high-speed tape and disk storage systems moved data in and out of the processor. In comparison to the extremely constrained internal architecture of the AGC, the new computers now allowed a rich variety of features, and considerable scope for growth. Operating systems grew in complexity, allowing sophisticated memory and workload management. Formal design methodologies were introduced to develop application programs, and high-level languages were replacing machine-level programming. Even the venerable punched card was heading for extinction, as online systems provided access via remote terminals. Ultimately, Apollo was a program with a finite end. With its primary mission accomplished and its mission architecture stretched to the limit, it was impossible to justify the time and expense for further upgrades. It is ironic that the AGC, which pioneered so many technologies, did not itself benefit from the advances that its development prompted.

By the final Apollo flight the AGC was hopelessly obsolete. Nevertheless, it would have one last mission to perform. Precisely flying an Apollo spacecraft was difficult because of the variety of vehicle configurations, masses and control options. A traditional autopilot design based on analog components would have been unacceptably heavy, complex and of limited capability. Using a computer was a daring challenge. A software-based digital autopilot was a risk, yet the AGC would have to operate flawlessly for the entire duration of the flight. As Apollo began its missions to the Moon, NASA researchers at the Dryden Flight Research Center were studying computer-control of aircraft. The AGC was an ideal basis for the project. Already paired with an inertial measurement unit, the hardware and software were reliable and the development team at MIT was still in place. A Navy

F-8 Crusader fighter was extensively modified to conduct the tests. Stripped of all manual control cables and interfaces, an AGC, IMU and a DSKY were installed in the aircraft. When the F-8 Digital Fly-by-Wire aircraft flew, the pilot was completely reliant on the AGC because there was no manual backup. Over a period of several years, this program explored a wide variety of flight regimes. With the flight software permanently in the core rope, changes to the control laws were entered into erasable memory. With the DSKY mounted in an avionics bay, the only controls available to the pilot were on a panel in the cockpit. This highly successful program paved the way for the next phase in the development of digital flight control systems.

Few periods in history have experienced a growth in technology as rapid as during the 1960s, and the unique combination of events that made it possible are unlikely to recur. Over the period of just a few short years, the combined talents of a nation created the machines for flying to another world and returning safely to the Earth. When you visit the National Air and Space Museum, pause at the AGC display and reflect on the key contribution that it made in enabling Apollo to reach the Moon.

Appendixes

APPENDIX A: AGC INSTRUCTION SET

Mnemonic	Description	Extracode?
Sequence Changing		
TC	Transfer Control	No
TCF	Transfer Control to Fixed	No
CCS	Count Compare and Skip	No
BZF	Branch Zero to Fixed	Yes
BZMF	Branch Zero or Minus to Fixed	Yes
Reading and Writing		
CA	Clear and Add	No
CS	Clear and Subtract	No
DCA	Double Clear and Add	Yes
DCS	Double Clear and Subtract	Yes
TS	Transfer to Storage	No
XCH	Exchange A and K	No
LXCH	Exchange L and K	No
QXCH	Exchange Q and K	Yes
DXCH	Double Exchange	No
Instruction Modification		
INDEX	Index Next Instruction	No
INDEXE	Index Next Instruction Extended	Yes
Arithmetic and Logic		
AD	Add	No
SU	Subtract	Yes
ADS	Add to Storage	No
MSU	Modular Subtract	Yes
INCR	Increment	No
AUG	Augment	Yes

Mnemonic	Description	Extracode?
DIM	Diminish	Yes
DAS	Double Add to Storage	No
MASK	Mask A by K	No
MP	Multiply	Yes
DV	Divide	Yes
I/O Channel		
READ	Read KC	Yes
WRITE	Write Channel KC	Yes
RAND	Read and Mask	Yes
WAND	Write and Mask	Yes
ROR	Read and Superimpose	Yes
WOR	Write and Superimpose	Yes
RXOR	Read and Invert	Yes
Miscellaneous		
XXALQ	Execute Extracode Using A, L and Q	No
XLQ	Execute Using L and Q	No
RETURN	Return From Subroutine	No
RELINT	Enable Interrupts	No
INHINT	Inhibit Interrupts	No
EXTEND	Set Extracode Flag	No
EDRUPT	Ed Smally's Interrupt	Yes
RESUME	Resume Interrupted Program	No
Involuntary		
RUPT	Interrupt	
PINC	Counter Increment	
MINC	Counter Decrement	
DINC	Diminish Absolute Value by 1	
PCDU	CDU Increment	
MCDU	CDU Decrement	
SHINC	Counter Shift	
SHANC	Counter Shift and Add 1	

Implied Address Instructions

Assembled Instruction	Special code	Description
COM	CS A	Complement value of accumulator
DCOM	DCS A	Complement double precision value in accumulator and L register
DDOUBL	DAS A	Double the double precision value in the accumulator and L register
DOUBLE	AD A	Double value in accumulator
DTCB	DXCH Z and BBANK	Double Transfer Control Switching Both Banks

Assembled Instruction	Special code	Description
DTCF	DXCH FB and Z	Double Transfer Control Switching the F Bank
NOOP	TCF I + 1	No operation by branching to the next instruction in fixed storage
NOOP	CA A	No operation by clearing and reloading the accumulator
OVSK	TS A	Overflow Skip
RETURN	TC Q	Return to Calling Subroutine
SQUARE	MP A	Square the value in accumulator
TCAA	TS Z	Transfer Control to the Address in A
XLQ	TC L	Execute using L and Q registers
XXALQ	TC A	Execute Extracode using A, L and Q registers
ZL	LXCH	Zero the L Register
ZQ	QXCH	Zero the Q Register

APPENDIX B: AGC INTERRUPT VECTORS

Vector Address (octal)	Interrupt Name	Trigger Condition	Description
4000	Startup	AGC power on	Starting address after AGC power up
4004	T6RUPT	TIME6 decremented to 0	Timer for RCS jets, used by the Digital Autopilot
4010	T5RUPT	TIME5 timer overflow	Digital Autopilot timer
4014	T3RUPT	TIME3 timer overflow	WAITLIST task scheduler
4020	T4RUPT	TIME4 timer overflow	DSKY monitoring and updating
4024	KEYRUPT1	Keystroke received from DSKY	Key code from main DSKY is available in channel 15
4030	KEYRUPT2	Keystroke received from secondary DSKY	Key code from Navigation DSKY is available in channel 16 (CSM only)
4034	UPRUPT	Data ready in INLINK register	Used for DSKY uplinks
4040	DOWNRUPT	Downlink registers ready for more data	AGC downlink telemetry
4044	RADARUPT	Data in RNRAD register is ready	Data from rendezvous radar
4050	RUPT10		LM P64 redesignations

APPENDIX C: LAYOUT OF THE SPECIAL AND CONTROL REGISTERS

Name	Address	Function
A	00000_8	Accumulator
L	00001_8	Low order product of the accumulator
Q	00002_8	TC instruction return address
EBANK	00003_8	Erasable Memory Bank
FBANK	00004_8	Fixed Memory Bank
Z	00005_8	12-Bit Program Counter
BBANK	00006_8	Both Bank Register
ZERO	00007_8	Hardwired to the value zero
ARUPT	00010_8	Interrupted contents of A
LRUPT	00011_8	Interrupted contents of L
QRUPT	00012_8	Interrupted contents of Q
SAMPTIME	00013_8	SAMPLED TIME 1
SAMPTIME	00014_8	SAMPLED TIME 2
ZRUPT	00015_8	Interrupted contents of Z (hardware)
BANKRUPT	00016_8	Interrupted contents of BB
BRUPT	00017_8	Interrupted contents of B Internal Register
CYR	00020_8	Cycle Right
SR	00021_8	Shift Right
CYL	00022_8	Cycle Left
EDOP	00023_8	Edits Interpretive Operation Code Pairs
TIME2	00024_8	T2: Elapsed time
TIME1	00025_8	T1: Elapsed time
TIME3	00026_8	T3RUPT: Wait list
TIME4	00027_8	T4RUPT
TIME5	00030_8	T5RUPT: Digital Autopilot
TIME6	00031_8	T6RUPT: Fine scale clocking
CDUX	00032_8	Inner IMU Gimbal
CDUY	00033_8	Middle IMU Gimbal
CDUZ	00034_8	Outer IMU Gimbal
CDUT, OPTY	00035_8	Rendezvous Radar Trunnion CDU, or Optics Y axis
CDUS, OPTX	00036_8	Rendezvous Radar Shaft CDU, or Optics X axis
PIPAX	00037_8	Velocity measurement – X axis
PIPAY	00040_8	Velocity measurement – Y axis
PIPAZ	00041_8	Velocity measurement – X axis
BMAGX	00042_8	RHC input – Pitch
BMAGY	00043_8	RHC input – Yaw
BMAGZ	00044_8	RHC input – Roll
INLINK	00045_8	Telemetry Uplink
RNRAD	00046_8	Rendezvous and Landing Radar Data
GYROCTR	00047_8	Outcounter for gyro

Name	Address	Function
CDUXCMD	00050_8	Outcounters for X CDU
CDUYCMD	00051_8	Outcounters for Y CDU
CDUZCMD	00052_8	Outcounters for Z CDU
CDUTCMD,		
TVCYAW,		
OPTYCMD	00053_8	Outcounter for optics (Y) or radar, SPS Yaw command in TVC Mode
CDUSCMD,		
TVCPITCH,		
OPTXCMD	00054_8	Outcounter for optics (X) or radar, SPS Pitch Command in TVC Mode
THRUST	00055_8	LM DPS Thrust Command
LEMONM	00056_8	Unused
OUTLINK	00057_8	Unused
ALTM	00060_8	Altitude Meter

APPENDIX D: COMMAND MODULE I/O CHANNELS

CM Input and Output Channels

INPUT CHANNEL 3
Bit 1-14 Contains the HIGH ORDER SCALER: 23.3 hours = maximum capacity in increments of 5.12 seconds.

INPUT CHANNEL 4
Bits 1-14 Contains the LOW ORDER SCALER: 5.12 seconds = maximum capacity in increments of 1/3,200 second.

OUTPUT CHANNEL 5

Bit	Jet Designation SM Quad No./CM Ring No.	TRANS/ROT	SM Command CM Command
1	C-3/1-3	+X/+P	+P
2	C-4/2-4	-X/-P	-P
3	A-3/2-3	-X/+P	+P
4	A-4/1-4	+X/-P	-P
5	D-3/2-5	+X/+YW	+YW
6	D-4/1-6	-X/-YW	-YW
7	B-3/1-5	-X/+YW	+YW
8	B-4/2-6	+X/-YW	-YW

OUTPUT CHANNEL 6

Bit	Jet Designation SM Quad No./CM Ring No.	TRANS/ROT	SM Command CM Command
1	B-1/1-1	+Z/+R	+R
2	B-2/1-2	-Z/-R	-R
3	D-1/2-1	-Z/+R	+R
4	D-2/2-2	+Z/-R	-R
5	A-1	+Y/+R	
6	A-2	-Y/-R	
7	C-1	-Y/+R	
8	C-2	+Y/-R	

OUTPUT CHANNEL 11

Bit	
1	ISS Warning
2	Light Computer Activity Lamp
3	Light Uplink Activity Lamp
4	Light Temp Caution Lamp
5	Light Keyboard Release Lamp
6	Flash Verb arid Noun Lamps
7	Light Operator Error Lamp
8	Spare
9	Test Connector Outbit
10	Caution Reset
11	Spare
12	Spare
13	Engine On/Off (1-On,0-Off}
14	Spare
15	Spare

OUTPUT CHANNEL 12

Bit	
1	Zero-Optics CDU's
2	Enable Optics COU Error Counters
3	Not Used
4	Coarse Align Enable
5	Zero IMU CDU's
6	Enable IMU CDU Error Counters
7	Spare
8	TVC Enable
9	Enable S-IVB Takeover
10	Zero Optics

11	Disengage Optics DAC
12	Spare
13	S-IVB Injection Sequence Start
14	S-IVB Cutoff
15	ISS Turn-on Delay Complete

OUTPUT CHANNEL 13

Bit	
1	Range Unit Select c
2	Range Unit Select b
3	Range Unit Select a
4	Range Unit Activity
5	Not Used
6	Block Inlink
7	Downlink Word Order Code Bit
8	Not Used
9	Spare
10	Test Alarms
11	Enable Standby
12	Reset Trap 31-A
13	Reset Trap 31-B
14	Reset Trap 32
15	Enable T6RUPT

OUTPUT CHANNEL 14

Bit	
1	Not Used
2	Spare
3	Spare
4	Spare
5	Not Used
6	Gyro Enable
7	Gyro Select b
8	Gyro Select a
9	Gyro Sign (1 = minus)
10	Gyro Activity
11	Drive CDU S
12	Drive CDU T
13	Drive CDU Z
14	Drive CDU Y
15	Drive CDU X

Channel 14: Gyro Selection

a	b	Gyro
0	0	–
0	1	X
1	0	Y
1	1	Z

INPUT CHANNEL 15

Bit	
1	Key codes from Main DSKY
2	Key codes from Main DSKY
3	Key codes from Main DSKY
4	Key codes from Main DSKY
5	Key codes from Main DSKY
6	Spare
7	Spare
8	Spare
9	Spare
10	Spare
11	Spare
12	Spare
13	Spare
14	Spare
15	Spare

INPUT CHANNEL 16

Bit	
1	Key codes from Navigation DSKY
2	Key codes from Navigation DSKY
3	Key codes from Navigation DSKY
4	Key codes from Navigation DSKY
5	Key codes from Navigation DSKY
6	Mark button
7	Mark Reject button
8	Spare
9	Spare
10	Spare
11	Spare
12	Spare
13	Spare
14	Spare
15	Spare

INPUT CHANNEL 30

Bit	
1	Ullage Thrust Present
2	CM/SM Separate
3	SPS Ready
4	S-IVB Separate, Abort
5	Liftoff
6	Guidance Reference Release
7	Optics CDU Fail
8	Spare
9	IMU Operate
10	S/C Control of Saturn
11	IMU Cage
12	IMU CDU Fail
13	IMU Fail
14	ISS Turn-On Request
15	Temperature in Limits

INPUT CHANNEL 31

Bit	
1	+ Pitch Manual Rotation
2	− Pitch Manual Rotation
3	+ Yaw Manual Rotation
4	− Yaw Manual Rotation
5	+ Roll Manual Rotation
6	− Roll Manual Rotation
7	+ X Translation
8	− X Translation
9	+ Y Translation
10	− Y Translation
11	+ Z Translation.
12	− Z Translation
13	Hold Function
14	Free Function
15	G&N Autopilot Control

INPUT CHANNEL 32

Bit	
1	+ Pitch Minimum Impulse
2	− Pitch Minimum Impulse
3	+ Yaw Minimum Impulse
4	− Yaw Minimum Impulse
5	+ Roll Minimum Impulse
6	− Roll Minimum Impulse
7	Spare
8	Spare
9	Spare
10	Spare
11	LM Attached
12	Spare
13	Spare
14	Proceed (Standby button)
15	Spare

INPUT CHANNEL 33

Bit	
1	Spare
2	Range Unit Data Good
3	Spare
4	Zero Optics
5	CMC Control
6	Not Used
7	Not Used
8	Spare
9	Spare
10	Block Uplink Input
11	Uplink Too Fast
12	Downlink Too Fast
13	PIPA Fail
14	AGC Warning
15	AGC Oscillator Alarm

OUTPUT CHANNEL 77 – (RESTART MONITOR)

Bit	
1	Parity Fail (E or F memory)
2	Parity Fail (E memory)
3	TC Trap
4	RUPT Lock
5	Night Watchman
6	Voltage Fail
7	Counter Fail
8	Scalar Fail
9	Scalar Double Frequency Alarm
10	Spare
11	Spare
12	Spare
13	Spare
14	Spare
15	Spare

APPENDIX E: LUNAR MODULE I/O CHANNELS

LM Input and Output Channels

INPUT CHANNEL 3
The high order bits of the LGC scaler. The least significant bit is worth 5.12 seconds. The maximum value is equivalent to approximately 29.5 days.

INPUT CHANNEL 4
The lower order bits of the LGC scaler. The least significant bit is worth 1 centisecond. The maximum value is equivalent to approximately 5.12 seconds.

OUTPUT CHANNEL 5

Bit	Jet Designation	Jet No.	Rotation Effect	(RCS Jet Firings) Translation Effect
1	B4U	1	+V	−X
2	A4D	2	−V	+X
3	A3U	5	+U	−X
4	B3D	6	−U	+X
5	B2U	9	−V	−X
6	A2D	10	+V	+X
7	A1U	13	−U	−X
8	B1D	14	+U	+X

Bits 9–15 not used

OUTPUT CHANNEL 6

Bit	Jet Designation	Jet No.	Rotation Effect	(RCS Jet Firings) Translation Effect
1	B3A	7	+P	+Z
2	B4F	3	−P	−Z
3	A1F	15	+P	−Z
4	A2A	1	−P	+Z
5	B2L	12	+P	+Y
6	A3R	8	−P	−Y
7	A4R	4	+P	−Y
8	B1L	16	−P	+Y

Bits 9–15 not used

OUTPUT CHANNEL 11

Bit	
1	ISS Warning
2	Light Computer Activity Lamp
3	Light Uplink Activity Lamp
4	Light Temperature Caution Lamp
5	Light Keyboard Release Lamp
6	Flash Verb and Noun Lamps
7	Light Operator Error Lamp
8	Spare
9	Test Connector Outbit
10	Caution Reset
11	Spare
12	Spare
13	Engine On
14	Engine Off
15	Spare

OUTPUT CHANNEL 12

Bit	
1	Zero Rendezvous Radar CDU's
2	Enable Rendezvous Radar Error Counters
3	Not Used
4	Coarse Align Enable
5	Zero IMU CDU's
6	Enable IMU Error Counters
7	Spare
8	Display Inertial Data
9	+Pitch Vehicle Motion (Pitch Gimbal Trim, Bell motion)
10	−Pitch Vehicle Motion (Pitch Gimbal Trim, Bell motion)
11	+Roll Vehicle Motion (Roll Gimbal Trim, Bell motion)
12	−Roll Vehicle Motion (Roll Gimbal Trim, Bell motion)
14	Rendezvous Radar Enable Auto Track
15	ISS Turn-on Delay Complete

OUTPUT CHANNEL 13

Bit	
1	Radar Select c
2	Radar Select b
3	Radar Select a
4	Radar Activity
5	Inhibit Uplink, Enable Crosslink
6	Block Inlink
7	Downlink Word Order
8	RHC Counter Enable
9	Start RHC Read
10	Test Alarms
11	Enable Standby
12	Reset Trap 31-A
13	Reset Trap 31-B
14	Reset Trap 32
15	Enable T6RUPT

Channel 13: Radar Selections

a	b	c	Function
0	0	0	–
0	0	1	Rendezvous Radar range
0	1	0	Rendezvous Radar range rate
0	1	1	–
1	0	0	Landing Radar X velocity
1	0	1	Landing Radar Y velocity
1	1	0	Landing Radar Z velocity
1	1	1	Landing Radar range

OUTPUT CHANNEL 14

Bit	
1	Outlink Activity
2	Altitude Rate Select
3	Altitude Meter Activity
4	Thrust Drive
5	Spare
6	Gyro Enable
7	Gyro Select b
8	Gyro Select a
9	Gyro Sign Minus
10	Gyro Activity
11	Drive CDU S
12	Drive CDU T
13	Drive CDU Z
14	Drive CDU Y
15	Drive CDU X

Channel 14: Gyro Selection

a	b	Gyro
0	0	–
0	1	X
1	0	Y
1	1	Z

INPUT CHANNEL 15

Bit	
1-5	Key codes from Main DSKY
6-15	Spare

INPUT CHANNEL 16

Bit	
1-2	Spare
3	Mark X
4	Mark Y
5	Mark Reject
6	Descent +
7	Descent -
8-15	Spare

INPUT CHANNEL 30

Bit	
1	Abort with Descent Stage
2	Descent Stage Attached
3	Engine Armed
4	Abort with Ascent Stage
5	Auto Throttle
6	Display Inertial Data
7	Rendezvous Radar CDU Fail
8	Spare
9	IMU Operate
10	G/N Control of S/C
11	IMU Cage
12	IMU CDU Fail
13	IMU Fail
14	ISS Turn-On Request
15	Temp in Limits

INPUT CHANNEL 31

Bit	
1	+EL (LPD), + PMI
2	−EL (LPD), − PMI
3	+ YMI
4	− YMI
5	+AZ (LPD), +RMI
6	−AZ (LPD), −RMI
7	+X Translation
8	−X Translation
9	+Y Translation
10	−Y Translation
11	+Z Translation
12	−Z Translation
13	Attitude Hold
14	Auto Stabilization
15	Attitude Control Out of Detent

INPUT CHANNEL 32

Bit	
1	Thruster 2 – 4 Disabled by Crew
2	Thruster 5 – 8 Disabled by Crew
3	Thruster 1 – 3 Disabled by Crew
4	Thruster 6 – 7 Disabled by Crew
5	Thruster 14 – 16 Disabled by Crew
6	Thruster 13 – 15 Disabled by Crew
7	Thruster 9 – 12 Disabled by Crew
8	Thruster 10 – 11 Disabled by Crew
9	Descent Engine Gimbals Disabled by Crew
10	Apparent Descent Engine Gimbal Fail
11–13	Spare
14	Proceed
15	Spare

INPUT CHANNEL 33

Bit	
1	Spare
2	Rendezvous Radar Auto-Power On
3	Rendezvous Radar Range Low Scale
4	Rendezvous Radar Data Good
5	Landing Radar Range Data Good
6	Landing Radar Position 1
7	Landing Radar Position 2
8	Landing Radar Velocity Data Good
9	Landing Radar Range Low Scale
10	Block Uplink Input
11	Uplink Too Fast
12	Downlink Too Fast
13	PIPA Fail
14	LGC Warning (repeated hardware or software alarms)
15	Oscillator Alarm

INPUT CHANNEL 77 (RESTART MONITOR)

Bit	
1	E Memory or F Memory Parity Fail
2	E Memory Parity Fail
3	TC Trap
4	RUPT Lock
5	Night Watchman
6	Voltage Fail
7	Counter Fail
8	Scaler Fail
9	Scaler Double Frequency
10	Spare
11	Spare
12	Spare
13	Spare
14	Spare
15	Spare

APPENDIX F: INTERPRETER INSTRUCTION SET

Interpreter Instruction Set

Store, Load, and Push-Down Instructions

Mnemonic		Description
STORE	X	Store MPAC
STODL	X	
	Y	Store MPAC, and Reload in DP
STOVL	X	
	Y	Store MPAC, and Reload a Vector
STCALL	X	
	Y	Store MPAC and CALL a Routine
DLOAD	X	Load MPAC in DP
TLOAD	X	Load MPAC in TP
VLOAD	X	Load MPAC with a Vector
SLOAD	X	Load MPAC in Single Precision
PDDL	X	Push MPAC onto Stack and Reload DP
PDVL	X	Push MPAC onto Stack and Reload DP
PUSH		Push MPAC onto Stack
SETPD	X	Set Push-Down Stack Pointer

Arithmetic Instructions

Mnemonic		Description
DAD	X	DP Add
DSU	X	DP Subtract
BDSU	X	DP Subtract From
DMP	X	DP Multiply
DMPR	X	DP Multiply and Round
DDV	X	DP Divide By
BDDV	X	DP Divide Into
SIGN	X	DP Sign Test
TAD	X	TP Add

Vector Arithmetic Operations

Mnemonic		Description
VAD	X	Vector Add
VSU	X	Vector Subtract
BVSU	X	Vector Subtract From
DOT	X	Vector Dot Product
VXSC	X	Vector Times Scalar
V/SC	X	Vector Divided by Scalar
VXV	X	Vector Cross Product
VPROJ	X	Vector Projection
VXM	X	Matrix Pre-Multiplication by Vector
MXV	X	Matrix Post-Multiplication by Vector

Scalar Functions

Mnemonic	Description
SQRT	DP Square Root
SIN	DP Sin
COS	DP Cosine
ARCSIN	DP Arcsin
ARCCOS	DP Arccosine
DSQ	DP Square
ROUND	Round to DP
DCOMP	TP Complement
ABS	TP Absolute Value

Vector Functions

Mnemonic	Description
UNIT	Unit Vector Function
ABVAL	Vector Length
VSQ	Square of Vector Length
VCOMP	Vector Complement
VDEF	Vector Define

Short Shift Instructions

Mnemonic	Description
SR1	Scalar Shift Right 1 Bit
SR2	Scalar Shift Right 2 Bits
SR3	Scalar Shift Right 3 Bits
SR4	Scalar Shift Right 4 Bits
SL1	Scalar Shift Left 1 Bit
SL2	Scalar Shift Left 2 Bits
SL3	Scalar Shift Left 3 Bits
SL4	Scalar Shift Left 4 Bits
SR1R	Scalar Shift Right 1 Bit and Round
SR2R	Scalar Shift Right 2 Bits and Round
SR3R	Scalar Shift Right 3 Bits and Round
SR4R	Scalar Shift Right 4 Bits and Round
SL1R	Scalar Shift Left 1 Bit and Round
SL2R	Scalar Shift Left 2 Bits and Round
SL3R	Scalar Shift Left 3 Bits and Round
SL4R	Scalar Shift Left 4 Bits and Round
VSR1	Vector Shift Right 1 Bit and Round
VSR2	Vector Shift Right 2 Bits and Round
VSR3	Vector Shift Right 3 Bits and Round
VSR4	Vector Shift Right 4 Bits and Round
VSR5	Vector Shift Right 5 Bits and Round
VSR6	Vector Shift Right 6 Bits and Round
VSR7	Vector Shift Right 7 Bits and Round
VSR8	Vector Shift Right 8 Bits and Round
VSL1	Vector Shift Left 1 Bit and Round
VSL2	Vector Shift Left 2 Bits and Round
VSL3	Vector Shift Left 3 Bits and Round
VSL4	Vector Shift Left 4 Bits and Round
VSL5	Vector Shift Left 5 Bits and Round
VSL6	Vector Shift Left 6 Bits and Round
VSL7	Vector Shift Left 7 Bits and Round
VSL8	Vector Shift Left 8 Bits and Round

General Shift Instructions

Mnemonic		Description
SL	X	General Scalar Shift Left
SRR	X	General Scalar Shift Right and Round
SLR	X	General Scalar Shift Left and Round
VSR	X	General Vector Shift Right
VSL	X	General Vector Shift Left

Normalization

Mnemonic		Description
NORM	X	Scalar Normalize

Branching, Sequence Changing, and Subroutine Linkage Instructions

Mnemonic		Description
GOTO	X	Go To
CALL	X	Call a Subroutine
CGOTO	X	
	Y	Computed Go To
CCALL	X	
	Y	Computed Subroutine Call
RVQ		Return via QPRET
STQ		Store QPRET
BPL	X	Branch on MPAC Plus
BZE	X	Branch on MPAC Zero
BMN	X	Branch on MPAC Minus
BHIZ	X	Branch on High Order Zero in MPAC
BOV	X	Branch on MPAC Overflow
BOVB	X	Branch on Overflow to Basic Instructions
RTB	X	Return to Basic Instructions
EXIT	X	Exit from Interpreter

Switch Instructions

Mnemonic		Description
SET	X	Set Switch
CLEAR	X	Clear Switch
INVERT	X	Invert Switch
SETGO	X	Set Switch and Go To
CLRGO	X	Clear Switch and Go To
INVGO	X	Invert Switch and Go To

Switch Test Instructions

Mnemonic		Description
BON	X	
	Y	Branch if Switch is On
BOFF	X	
	Y	Branch if Switch is Off
BONSET	X	
	Y	Branch if Switch is On, and Set Switch
BOFSET	X	
	Y	Branch if Switch is Off, and Set Switch
BONCLR	X	
	Y	Branch if Switch is On, and Clear Switch
BOFCLR	X	
	Y	Branch if Switch is Off, and Clear Switch
BONINV	X	
	Y	Branch if Switch is On, and Invert Switch
BOFINV	X	
	Y	Branch if Switch is Off, and Invert Switch

Index Register Instructions

Mnemonic		Description
AXT,1		
AXT,2	X	Address to Index True
AXC,1		
AXC,2	X	Address to Index Complemented
LXA,1		
LXA,2	X	Load Index from Erasable
LXC,1		
LXC,2	X	Load Index from Erasable Complemented
SXA,1		
SXA,2	X	Store Index into Erasable
XCHX,1		
XCHX,2	X	Exchange Index with Erasable
INCR,1		
INCR,2	X	Increment Index
XAD,1		
XAD,2	X	Index Register Add
XSU,1		
XSU,2	X	Index Register Subtract
TIX,1		
TIX,2	X	Transfer on Index

Miscellaneous Instructions

Mnemonic		Description
SSP	X	Set Single Precision
STADR		Push Up Stack on Store Code

APPENDIX G: COMMAND MODULE PROGRAMS (MAJOR MODES)

Command Module AGC Programs

Prelaunch	00	AGC Idling
and	01	Prelaunch or Service-Initialization
Service	02	Prelaunch or Service-Gyrocompassing
	03	Prelaunch or Service-Optical Verification of Gyrocompassing
	06	AGC Power Down
	07	System Test
Boost	11	Earth Orbit Insertion Monitor
	15	TLI Initiate/Cutoff
Coast	20	Universal Tracking
	21	Ground Track Determination
	22	Orbital Navigation
	23	Cislunar Midcourse Navigation
	24	Rate Aided Optics Tracking
	27	AGC Update
	29	Time-of-Longitude
Pre-Thrusting	30	External Delta V
	31	CSM Height Adjustment Maneuver (HAM)
	32	CSM Coelliptic Sequence Initiation (CSI)
	33	CSM Constant Delta Altitude (CDH)
	34	CSM Transfer Phase Initiation (TPI) Targeting
	35	CSM Transfer Phase Midcourse (TPM) Targeting
	36	CSM Plane Change (PC) Targeting
	37	Return to Earth
Thrusting	40	SPS
	41	RCS
	47	Thrust Monitor
Alignment	51	IMU Orientation Determination
	52	IMU Realign
	53	Backup IMU Orientation Determination
	54	Backup IMU Realign
Entry	61	Entry–Preparation
	62	Entry-CM/SM Separation and Preentry Maneuver
	63	Entry-Initialization

	64	Entry-Post 0.05G
	65	Entry-Upcontrol
	66	Entry-Ballistic
	67	Entry-Final Phase
Pre-Thrusting	72	LM Coelliptic Sequence Initiation (CSI)
Other	73	LM Constant Delta Altitude (CDH)
Vehicle	74	LM Transfer Phase Initiation (TPI) Targeting
	75	LM Transfer Phase (Midcourse) Targeting
	76	LM Target Delta V
	77	CSM Target Delta V
	79	Rendezvous Final Phase

APPENDIX H: COMMAND MODULE ROUTINES

Command Module AGC Routines

Routine	Routine Title
00	Final Automatic Request Terminate
01	Erasable and Channel Modification Routine
02	IMU Status Check
03	Digital Autopilot Data Load
05	S-Band Antenna
07	MINKEY Controller
21	Rendezvous Tracking Sighting Mark
22	Rendezvous Tracking Data Processing
23	Backup Rendezvous Tracking Sighting Mark
30	Orbit Parameter Display
31	Rendezvous Parameter Display Routine No. 1
33	AGC/LGC Clock Synchronization
34	Rendezvous Parameter Display Routine No. 2
35	Lunar Landmark Selection
36	Rendezvous Out-of-Plane Display Routine
40	SPS Thrust Fail
41	State Vector Integration (MID to AVE)
50	Coarse Align
52	Automatic Optics Positioning
53	Sighting Mark
54	Sighting Mark Display
55	Gyro Torquing
56	Alternate LOS Sighting Mark
57	Optics Calibration
60	Attitude Maneuver
61	Tracking Attitude
62	Crew-Defined Maneuver
63	Rendezvous Finat Attitude
67	Universal Tracking Rotation

APPENDIX I: COMMAND MODULE VERBS

Command Module AGC Verbs

Regular Verbs

01	Display Octal Component 1 in R1
02	Display Octal Component 2 in R1
03	Display Octal Component 3 in R1
04	Display Octal Components 1, 2 in R1, R2
05	Display Octal Components 1,2,3 in R1, R2, R3
06	Display decimal in R1 or R1, R2 or R1, R2, R3
07	Display DP decimal in R1, R2 {test only)
08	Spare
09	Spare
10	Spare
11	Monitor Octal Component 1 in R1
12	Monitor Octal Component 2 in R1
13	Monitor Octal Component 3 in R1
14	Monitor Octal Components 1, 2 in R1, R2
15	Monitor Octal Components 1, 2, 3 in R1, R2, R3
16	Monitor Decimal in R1 or R1, R2 or R1, R2, R3
17	Monitor DP decimal in R1, R2 (test only)
18	Spare
19	Spare
20	Spare
21	Load Component 1 into R1
22	Load Component 2 into R2
23	Load Component 3 into R3
24	Load Components 1, 2 into R1, R2
25	Load Components 1, 2, 3 into Rl, R2, R3
26	Spare
27	Display Fixed Memory
28	Spare
29	Spare
30	Request EXECUTIVE
31	Request WAITLIST
32	Recycle program
33	Proceed without DSKY inputs
34	Terminate function
35	Test lights
36	Request FRESH START
37	Change program (major mode)
38	Spare
39	Spare

Extended Verbs

40	Zero ICDU
41	Coarse align CDU's (specify N20 or N91)
42	Pulse torque gyros
43	Load IMU attitude error needles (test only)
44	Set Surface flag
45	Reset Surface flag
46	Establish G & N autopilot control
47	Move LM state vector into CM state vector
48	Request DAP Data Load routine (R03)
49	Start automatic attitude maneuver
50	Please perform
51	Please mark
52	Mark on offset landing site
53	Please perform COAS mark
54	Request Rendezvous Backup Sighting Mark routine (R23)
55	Increment AGC time (decimal)
56	Terminate tracking (P20)
57	Request display of full track flag (FULTKFLAG)
58	Reset Stick flag and set V50 N18 flag
59	Please calibrate
60	Set astronaut total attitude (N17) to present attitude
61	Display DAP following attitude errors (Mode 1)
62	Display total attitude errors with respect to Noun 22 (Mode 2)
63	Display total astronaut attitude error with respect to Noun 17 (Mode 3)
64	Request S-Band Antenna routine (R05)
65	Optical verification of prelaunch alignment
66	Vehicles are attached. Move this vehicle state vector to other vehicle.
67	Start W-matrix RMS error display
68	Spare
69	Cause RESTART
70	Update liftoff time (P27)
71	Start AGC update; block address (P27)
72	Start AGC update; single address (P27)
73	Start AGC update; AGC time (P27)
74	Initialize erasable dump via DOWNLINK
75	Backup liftoff
76	Spare
77	Spare
78	Update prelaunch azimuth
79	Spare
80	Enable LM state vector update
81	Enable CSM state vector update
82	Request Orbit Parameter display (R30)
83	Request Rendezvous Parameter display No. 1 (R31)
84	Spare
85	Request Rendezvous Parameter display No. 2 (R34)
86	Reject Rendezvous Backup Sighting Mark
87	Set VHP Range flag

88	Reset VHF Range flag
89	Request Rendezvous Final Attitude maneuver (R63)
90	Request Out of Plane Rendezvous display (R36)
91	Display BANKSUM
92	Spare
93	Enable W matrix initialization
94	Perform Cislunar Attitude maneuver
95	Spare
96	Terminate integration and go to P00
97	Please perform engine-fail (R40)
98	Spare
99	Please Enable Engine Ignition

APPENDIX J: COMMAND MODULE NOUNS

Command Module AGC Nouns

Number	Register Description	Data Format
00	Not in use	
01	Specify address (fractional)	.XXXXX fractional
		.XXXXX fractional
		.XXXXX fractional
02	Specify address (whole)	XXXXX. integer
		XXXXX. integer
		XXXXX. integer
03	Specify address (degree)	XXX.XX deg
		XXX.XX deg
		XXX.XX deg
04	Spare	
05	Angular error/difference	XXX.XX deg
06	Option code ID	Octal
	Option code	Octal
07	FLAGWORD operator	
	ECADR	Octal
	BIT ID	Octal
	Action	Octal
08	Alarm data	
	ADRES	Octal
	BBANK	Octal
	ERCOUNT	Octal
09	Alarm codes	
	First	Octal
	Second	Octal
	Last	Octal
10	Channel to be specified	Octal

11	TIG of CSI	00XXX. h
		000XX. min
		0XX.XX s
12	Option code (extended verbs only)	Octal
13	TIG of CDH	00XXX. h
		000XX. min
		0XX.XX s
14	Specified inertial velocity at TLI cutoff (V C/O)	XXXXX. ft/s
15	Increment address	Octal
16	Time of event (extended verbs only)	00XXX. h
		000XX. min
		0XX.XX s
17	Astronaut total attitude	R XXX.XX deg
		P XXX.XX deg
		Y XXX.XX deg
18	Desired automaneuver FDAI ball angles	R XXX.XX deg
		P XXX.XX deg
		Y XXX.XX deg
19	Spare	
20	Present ICDU angles	R (OG) XXX.XX deg
		P (IG) XXX.XX deg
		Y (MG) XXX.XX deg
21	PIPA'S	X XXXXX. pulses
		Y XXXXX. pulses
		Z XXXXX. pulses
22	Desired ICDU angles	R (OG) XXX.XX deg
		P (IG) XXX.XX deg
		Y (MG) XXX.XX deg
23	Spare	
24	Delta time for AGC clock	00XXX. h
		000XX. min
		0XX.XX s
25	CHECKLIST (used with V50)	XXXXX.
26	PRIO/DELAY, ADRES, BBCON	Octal
		Octal
		Octal
27	Self-test on/off switch	XXXXX.
28	Spare	
29	XSM launch azimuth	XXX.XX deg
30	Target codes	XXXXX.
		XXXXX.
		XXXXX.
31	Time of W matrix initialization	00XXX. h
		000XX. min
		0XX.XX s
32	Time from perigee	00XXX. h
		000XX. min
		0XX.XX s

33	Time of ignition (GETI)	00XXX. h
		000XX. min
		0XX.XX s
34	Time of event	00XXX. h
		000XX. min
		0XX.XX s
35	Time from event	00XXX. h
		000XX. min
		0XX.XX s
36	Time of AGC clock	00XXX. h
		000XX. min
		0XX.XX s
37	Time of ignition (TPI)	00XXX. h
		000XX. min
		0XX.XX s
38	Time of state being integrated	00XXX. h
		000XX. min
		0XX.XX s
39	Delta time for transfer	00XXX. h
		000XX. mm
		0XX.XX s
40	Time from ignition/cutoff (TFI/TFC)	XXbXX min/s
	VG	XXXX.X ft/s
	Delta V (accumulated)	XXXX.X ft/s
41	Target	
	Azimuth	XXX.XX deg
	Elevation	XX.XXX deg
42	Apocenter altitude	XXXX.X nmi
	Pericenter altitude	XXXX.X nmi
	Delta V (required)	XXXX.X ft/s
43	Latitude	XXX.XX deg (+ north)
	Longitude	XXX.XX deg (+ east)
	Altitude	XXXX.X nmi
44	Apocenter attitude	XXXX.X nmi
	Pericenter altitude	XXXX.X nmi
	TFF	XXbXX min/s
45	Marks (VHF/optics)	XXbXX marks
	Time from ignition of next burn	XXbXX min/s
	Middle gimbal angle	XXX.XX deg
46	DAP configuration	Octal
		Octal
47	CSM weight	XXXXX. lbs
	LM weight	XXXXX. lbs
48	Gimbal pitch trim	XXX.XX deg
	Gimbal yaw trim	XXX.XX deg
49	Delta R	XXXX.X nmi
	Delia V	XXXX.X ft/s
	VHF or optics code	0000X
50	Splash error	XXXX.X nmi

	Perigee	XXXX.X nmi
	TFF	XXbXX min/s
51	S-band antenna angles	
	RHO	XXX.XX deg
	GAMMA	XXX.XX deg
52	Central angle of active vehicle	XXX.XX deg
53	Range	XXX.XX nmi
	Range rate	XXXX.X ft/s
	Phi	XXX.XX deg
54	Range	XXX.XX nmi
	Range rate	XXXX.X ft/s
	Theta	XXX.XX deg
55	Perigee code	0000X.
	Elevation angle	XXX.XX deg
	Central angle of passive vehicle	XXX.XX deg
56	Reentry angle	XXX.XX deg
	Delta V	XXXXX. ft/s
57	Spare	
58	Pericenter altitude (post TPI)	XXXX.X nmi
	Delta V (TPI)	XXXX.X ft/s
	Delta V (TPF)	XXXX.X ft/s
59	Delta V LOS 1	XXXX.X ft/s
	Delta V LOS 2	XXXX.X ft/s
	Delta V LOS 3	XXXX.X ft/s
60	GMAX	XXX.XX g
	VPRED	XXXXX. ft/s
	GAMMA EI	XXX.XX deg (+ above)
61	Impact	
	Latitude	XXX.XX deg (+ north)
	Longitude	XXX.XX deg (+ east)
	Heads up/down	+/− 00001
62	Inertial velocity magnitude	XXXXX. ft/s
	Altitude rate	XXXXX. ft/s
	Altitude above pad radius	XXXX.X nmi
63	Range from EI altitude to splash	XXXX.X nmi
	Predicted inertial velocity	XXXXX. ft/s
	Time from EI altitude	XXbXX min/s
64	Drag acceleration	XXX.XX g
	Inertial velocity	XXXXX. ft/s
	Range to splash	XXXX.X nmi (+ is overshoot)
65	Sampled AGC time	00XXX. h
	(fetched in interrupt)	000XX. min
		XX.XX s
66	Commanded bank angle	XXX.XX deg
	Crossrange error	XXXX.X nmi (+ south)

	Downrange error	XXXX.X nmi
		(+ overshoot)
67	Range to target	XXXX.X nmi
		(+ overshoot)
	Present latitude	XXX.XX deg
		(+ north)
	Present longitude	XXX.XX deg
		(+ east)
68	Commanded bank angle	XXX.XX deg
	Inertial velocity	XXXXX. ft/s
	Altitude rate	XXXXX. ft/s
69	Commanded bank angle	XXX.XX deg
	Drag level	XXX.XX g
	Exit velocity	XXXXX. ft/s
70	Celestial body code (before mark)	Octal
	Landmark data	Octal
	Horizon data	Octal
71	Celestial bobv code (after mark)	Octal
	Landmark data	Octal
	Horizon data	Octal
72	Spare	
73	Altitude	XXXXX. nmi
	Velocity	XXXXX. ft/s
	Flight path angle	XXX.XX deg
74	Commanded bank angle	XXX.XX deg
	Inertial velocity	XXXXX. ft/s
	Drag acceleration	XXX.XX g
75	Delta altitude (CDH)	XXXX.X nmi
	Delta time (CDH-CSI or TPI-CDH	XXbXX min/s
	Delta time (TPt CDH or TPI-NOMTPI)	XXbXX min/s
76	Spare	
77	Spare	
78	GAMMA	XXX.XX deg
	RHO	XXX.XX deg
	OMICRON	XXX.XX deg
79	P20 rotation rate	X.XXXX deg/s
	P20 deadband	XXX.XX deg
80	Time from ignition/cutoff	XXbXX min/s
	Velocity to be gained	XXXXX. ft/s
	Delta V (accumulated)	XXXXX. ft/s
81	Delta VX (LV)	XXXX.X ft/s
	Delta VY (LV)	XXXX.X ft/s
	Delta VZ (LV)	XXXX.X ft/s
82	Delta VX (LV)	XXXX.X ft/s
	Delta VY (LV)	XXXX.X ft/s
	Delta VZ (LV)	XXXX.X ft/s
83	Delta VX (body)	XXXX.X ft/s
	Delta VY (body)	XXXX.X ft/s
	Delta VZ (body)	XXXX.X ft/s

84	Delta VX (LV of other vehicle)	XXXX X ft/s
	Delta VY (LV of other vehicle)	XXXX.X ft/s
	Delta VZ (LV of other vehicle)	XXXX.X ft/s
85	VGX (body)	XXXX.X ft/s
	VGY (body)	XXXX.X ft/s
	VGZ (body)	XXXX.X ft/s
86	Delta VX (LV)	XXXXX. ft/s
	Delta VY (LV)	XXXXX. ft/s
	Delta VZ (LV)	XXXXX. ft/s
87	Mark data	
	Shaft angle	XXX.XX deg
	Trunnion angle	XX.XXX deg
88	Celestial body unit vector	X .XXXXX
		Y .XXXXX
		Z .XXXXX
89	Landmark latitude	XX.XXX deg
		(+ north)
	Landmark longitude/2	XX.XXX deg
		(+ east)
	Landmark altitude	XXX.XX nmi
90	Rendezvous out of plane parameters	
	Y (Active)	XXX.XX nmi
	YDOT (Active)	XXXX.X ft/s
	YDOT (Passive)	XXXX.X ft/s
91	Present optics angles	
	Shaft	XXX.XX deg
	Trunnion	XX.XXX deg
92	New optics angles	
	Shaft	XXX.XX deg
	Trunion	XX.XXX deg
93	Delta gyro angles	X XX.XXX deg
		Y XX.XXX deg
		Z XX.XXX deg
94	Alternate LOS	
	Shaft angle	XXX.XX deg
	Trunnion angle	XX.XXX deg
95	Time from ignition (cutoff)	XXbXX min/s
	VG (P15)	XXXXX. ft/s
	VI (P15)	XXXXX. ft/s
96	Y (CSM)	XXX.XX nmi
	YDOT(CSM)	XXXX.X ft/s
	YDOT(LM)	XXXX.X ft/s
97	System test inputs	XXXXX.
		XXXXX.
		XXXXX.
98	System test results and input	XXXXX.
		.XXXXX
		XXXXX.

99	RMS in position	XXXXX. ft
	RMS in velocity	XXXX.X ft/s
	RMS option code	XXXXX.

APPENDIX K: COMMAND MODULE PROGRAM ALARMS

Command Module Program Alarms

Code	Interpretation	Issued By
00110	No mark since last mark reject	SXTMARK
00113	No inbits (Channel 16)	SXTMARK
00114	Mark made but not desired	SXTMARK
00115	Optics torque request with switch not at CMC	Extended verb optics CDU
00116	Optics Switch altered before 15 second ZERO time elapsed	T4RUPT
00117	Optics torque request with optics not available	Extended verb optics CDU
00120	Optics torque request with optics not ZEROED	T4RUPT
00121	Optics CDU's no good at time of mark	SXTMARK
00205	Bad PIPA reading	SERVICER
00206	Zero encode not allowed with coarse align + gimbal lock	IMU mode switch
00207	ISS turn-on request not present for 90 seconds	T4RUPT
00210	IMU not operating	IMU mode switch, IMU-2 R02, P51
00211	Coarse align error-drive > 2 degrees	IMU mode switch
00212	PIPA fail but PIPA is not being used	IMU mode switch, T4RUPT
00213	IMU not operating with turn-on request	T4RUPT
00214	Program using IMU when turned off	T4RUPT
00217	Bad return from Stall routines	CURTAINS
00220	IMU not aligned (no REFSMMAT)	R02, P51
00401	Desired gimbal angles yield Gimbal Lock	Fine Align, IMU-2
00402	Second MINKEY pulse torque must be done	P52
00404	Target out of view (trunnion angle > 90 degrees)	R52
00405	Two stars not available	P52, P54
00406	Rendezvous navigation not operating	R21, R23
00421	W-Matrix overflow	INTEGRV
00600	Imaginary roots on first iteration	P32, P72
00601	Perigee altitude after CSI < 85 nmi earth orbit, 35,000 feet moon orbit	P32, P72
00602	Perigee altitude after CDH < 85 nmi earth orbit, 35,000 feet moon orbit	P32, P72
00603	CSI to CDH time < 10 minutes	P32, P33, P72, P73
00604	CDH to TPI time < 10 minutes	P32, P72
00605	Number of iterations exceeds loop maximum	P32, P72 P37
00606	DV exceeds maximum	P32, P72

Code	Interpretation	Issued By
00611	No TIG for given elevation angle	P34, P74
00612	State vector in wrong sphere of influence	P37
00613	Reentry angle out of limits	P37
00777	PIPA fail caused ISS warning	T4RUPT
01102	AGC self-test error	SELF-CHECK
01105	Downlink too fast	T4RUPT
01106	Uplink too fast	T4RUPT
01107	Phase table failure; assume erasable memory is destroyed	RESTART
01301	ARCSIN-ARCCOS argument too large	INTERPRETER
01407	VG increasing	S40.8
01426	IMU unsatisfactory	P61, P62
01427	IMU reversed	P61, P62
01520	V37 request not permitted at this time	V37
01600	Overflow in Drift Test	Optical Prealignment Calibration
01601	Bad IMU torque	Optical Prealignment Calibration
01703	Insufficient time for integration	R41
03777	ICDU fail caused ihe ISS warning	T4RUPT
04777	ICDU, PIPA fails caused the ISS warning	T4RUPT
07777	IMU fail caused the ISS earning	T4RUPT
10777	IMU, PIPA fails caused the ISS warning	T4RUPT
13777	IMU, ICDU fails caused the ISS warning	T4RUPT
14777	IMU, ICDU, PIPA fails caused the ISS warning	T4RUPT
20430	Integration abort due to subsurface state vector	All calls to Integration
20607	No solution from Time-Theta or Time-Radius	TIMETHET, TIMERAD
20610	Lambda less than unity	P37
21204	Negative or zero WAITLIST call	WAITLIST
21206	Second job attempts to go to sleep via Keyboard and Display program	PINBALL
21210	Two programs using device at same time	IMU Mode Switch
21302	SORT called with negative argument	INTERPRETER
21501	Keyboard and Display alarm during internal use (NVSUB)	PINBALL
21502	Illegal flashing display	GOPLAY
21521	P01 illegally selected	P01
31104	Delay routine busy	EXECUTIVE
31201	Executive overflow-no VAC areas available	EXECUTIVE
31202	Executive overflow-no core sets available	EXECUTIVE
31203	WAITLIST overflow-too many tasks	WAITLIST
31211	Illegal interrupt of Extended Verb	SXTMARK, P23

APPENDIX L: LUNAR MODULE PROGRAMS (MAJOR MODES)

Lunar Module AGC Programs

Service	00	LGC Idling
	06	LGC Power Down
Ascent	12	Powered Ascent
Coast	20	Rendezvous Navigation
	21	Ground Track Determination
	22	Lunar Surface Navigation
	25	Preferred Tracking Attitude
	27	LGC Update
Prethrusting	30	External Delta V
	32	Coelliptic Sequence Initiation (CSI)
	33	Constant Delta Altitude (CDH)
	34	Transfer Phase Initiation (TPI)
	35	Transfer Phase Midcourse (TPM)
Thrusting	40	DPS
	41	RCS Thrusting
	42	APS
	47	Thrust Monitor
Alignments	51	IMU Orientation Determination
	52	IMU Realign
	57	Lunar Surface Align
Descent	63	Braking Phase
	64	Approach Phase
	66	Landing Phase (ROD)
	68	Landing Confirmation
Aborts	70	DPS Abort
and	71	APS Abort
Backups	72	CSM Coelliptic Sequence Initiation (CSI) Targeting
	73	CSM Constant Delta Altitude (CDH) Targeting
	74	CSM Transfer Phase Initiation (TPI) Targeting
	75	State Vector Update (CSM)
	77	State Vector Update (LM)

APPENDIX M: LUNAR MODULE ROUTINES

Lunar Module AGC Routines

Routine	Routine Title
00	Final Automatic Request Terminate
01	Erasable Modification
02	IMU Status Check
03	DAP Data Load
04	Rendezvous Radar/Landing Radar Self-Test
05	S-Band Antenna
09	R10/R11/R12 Service
10	Landing Analog Displays
11	Abort Discretes Monitor
12	Descent State Vector Update
13	Landing Auto Modes Monitor
20	Landing Radar/Rendezvous Radar Read
21	Rendevzous Radar Designate
22	Rendezvous Radar Data Read
23	Rendezvous Radar Manual Acquisition
24	Rendezvous Radar Search
25	Rendezvous Radar Monitor
26	Lunar Surface RR Designate
30	Orbit Parameter Display
31	Rendezvous Parameter Display
33	LGC/AGC Clock Synchronization
36	Out-of-Plane Rendezvous Display
40	DPS/APS Thrust Fail
41	State Vector Integration (MIDTOAVE)
47	ACS Initialization
50	Coarse Align
51	In-Flight Fine Align
52	Auto Optics Positioning
53	AOT Mark
54	Sighting Data Display
55	Gyrotorquing
56	Terminate Tracking
57	MARKRUPT
53	Celestial Body Definition
59	Lunar Surface Sighting Mark
60	Attitude Maneuver
61	Preferred Tracking Attitude
62	Crew-Defined Maneuver
65	Fine Preferred Tracking Attitude
76	Extended Verb Interlock
77	LR Spurious Test

APPENDIX N: LUNAR MODULE VERBS

Lunar Module AGC Verbs

Regular Verbs

00	Not in use
01	Display Octal Component 1 in R1
02	Display Octal Component 2 in R1
03	Display Octal Component 3 in R1
04	Display Octal Components 1, 2 in R1, R2
05	Display Octal Components 1, 2, 3 in R1, R2, R3
06	Display decimal in R1 or R1, R2 or R1, R2, R3
07	Display DP decimal in R1, R2 (test only)
08–10	Spare
11	Monitor Octal Component 1 in R 1
12	Monitor Octal Component 2 in R1
13	Monitor Octal Component 3 in R1
14	Monitor Octal Components 1, 2 in R1, R2
15	Monitor Octal Components 1,2, 3 tn R1.R2, R3
16	Monitor decimal in R1 or R1, R2 or R1, R2, R3
17	Monitor DP decimal in R1, R2 (test only)
18–20	Spare
21	Load Component 1 into R1
22	Load Component 2 into R2
23	Load Component 3 into R3
24	Load Components 1, 2 into R1, R2
25	Load Components 1, 2, 3 into R1, R2, R3
26	Spare
27	Display Fixed Memory
28–29	Spare
30	Request EXECUTIVE
31	Request WAITLIST
32	Recycle program
33	Proceed without DSKY inputs
34	Terminate function
35	Test lights
36	Request FRESH START
37	Change program (major mode)
38–39	Spare

Extended Verbs

40	Zero CDU's (specify N20 or N72)
41	Coarse align CDU's (specify N20 or N72)
42	Fine align IMU
43	Load IMU attitude error needles
44	Terminate RR continuous designate (V41N72 Option 2)
45–46	Spare

47	Initialize AGS (R47)
48	Request DAP Data Load routine (R03)
49	Request Crew Defined Maneuver routine (R62)
50	Please perform
51	Spare
52	Mark X reticle
53	Mark Y reticle
54	Mark X or Y reticle
56	Terminate tracking (P20 and P25)
57	Permit Landing Radar updates
58	Inhibit Landing Radar updates
59	Command LR to Position 2
60	Display vehicle attitude rates on FDAI error needles
61	Display DAP following attitude errors
62	Display total attitude errors with respect to N22
63	Sample radar once per second (R04)
64	Request S-Band Antenna routine (R051
65	Disable U and V jet firings during DPS burns
66	Vehicles are attached. Move this vehicle state vector to other vehicle
67	Display W matrix
68	Cause Lunar Terrain model to be bypassed
69	Cause RESTART
70	Start LGC update, liftoff time (P27)
71	Start LGC update, block address (P27)
72	Start LGC update, single address (P27)
73	Start LGC update, LGC time (P27)
74	Initialize erasable dump via DOWNLINK
75	Enable U and V jet firings during DPS bums
77	Rate Command and Attitude Hold mode
78	Start LR spurious test (R77)
79	Stop LR spurious test
80	Enable LM state vector update
81	Enable CSM state vector update
82	Request Orbit Parameter display (R30)
83	Request Rendezvous Parameter display (R31)
84	Spare
85	Display Rendezvous Radar LOS azimuth and elevation
86–88	Spare
89	Request Rendezvous Final Attitude maneuver (R63)
90	Request Out of Plane Rendezvous display (R36)
91	Display BANKSUM
92	Start IMU performance tests {ground use)
93	Enable W matrix initialization
94	Spare
95	No update of either state vector allowed (P20 or P22)
96	Interrupt integration and go to P00
97	Perform Engine Fail procedure (R40)
98	Spare
99	Please Enable Engine Ignition

APPENDIX O: LUNAR MODULE NOUNS

Lunar Module AGC Nouns

Number	Register Description	Data Format
00	Not in use	
01	Specify address (fractional)	.XXXXX fractional
		.XXXXX fractional
		.XXXXX fractional
02	Specify address (whole)	XXXXX. integer
		XXXXX. integer
		XXXXX. integer
03	Specify address (degree)	XXX.XX deg
		XXX.XX deg
		XXX.XX deg
04	Angular error/difference	XXX.XX deg
05	Angular error/difference	XXX.XX deg
06	Option code ID	Octal
	Option code	Octal
	Data code	Octal
07	Channel/E Memory operator	
	ECADR	Octal
	BIT ID	Octal
	Action	Octal
08	Alarm data	
	ADRES	Octal
	BBANK	Octal
	ERCOUNT	Octal
09	Alarm codes	
	First	Octal
	Second	Octal
	Last	Octal
10	Channel to be specified	Octal
11	TIG of CSI/T (APO APSIS)	00XXX. h
		000XX. min
		0XX.XX s
12	Option code (extended verbs only)	Octal
		Octal
13	TIG of CDH	00XXX. h
		000XX. min
		0XX.XX s
14	CHECKLIST (extended verbs only)	XXXXX.
15	Increment address	Octal
16	Time of event (extended verbs only)	00XXX. h
		000XX. min
		0XX.XX s
17	Spare	

Number	Register Description	Data Format
18	Desired automaneuver FDAI ball angles	R XXX.XX deg P XXX.XX deg Y XXX.XX deg
19	Spare	
20	Present ICDU angles	OG XXX.XX deg IG XXX.XX deg MG XXX.XX deg
21	PIPA's	X XXXXX. pulses Y XXXXX. pulses Z XXXXX. pulses
22	Desired ICDU angles	OG XXX.XX deg IG XXX.XX deg MG XXX.XX deg
23	Spare	
24	Delta time for LGC clock	00XXX. h 000XX. min 0XX.XX s
25	CHECKLIST (used with V50)	XXXXX.
26	PRIO/DELAY ADRES BBCON	Octal Octal Octal
27	Self-test on/off switch	XXXXX.
28–31	Spare	
32	Time from perigee	00XXX. h 000XX. min 0XX.XX s
33	Time of ignition	00XXX. h 000XX. min 0XX.XX s
34	Time of event	00XXX. h 000XX. min 0XX.XX s
35	Time from event	00XXX. h 000XX. min 0XX.XX s
36	Time of LGC clock	00XXX. h 000XX. min 0XX.XX s
37	Time of ignition (TPI)	00XXX. h 000XX. min 0XX.XX s
38	Time of state being integrated	00XXX. h 000XX. min 0XX.XX s
39	Spare	
40	Time from ignition/cutoff	XXbXX min/s

Number	Register Description	Data Format
	VG	XXXX.X ft/s
	Delta V (accumulated)	XXXX.X ft/s
41	Target (V92 only)	XXX.XX deg
	Azimuth	XX.XXX deg
	Elevation	
42	Apocenter attitude	XXXX.X nmi
	Pericenter altitude	XXXX.X nmi
	Delta V (required)	XXXX.X ft/s
43	Latitude	XXX.XX deg (+ north)
	Longitude	XXX.XX deg (+ east)
	Altitude	XXXX.X nmi
44	Apocenter altitude	XXXX.X nmi
	Pericenter altitude	XXXX.X nmi
	TFF	XXbXX min/s
45	Marks	XXXXX.
	Time from ignition of next/last burn	XXbXX min/s
	Middle gimbal angle	XXX.XX deg
46	DAP configuration	Octal
	Switch function fail code	Octal
47	LM weight	XXXXX. lbs
	CSM weight	XXXXX. lbs
48	Gimbal pitch trim	XXX.XX deg
	Gimbal roll trim	XXX.XX deg
49	Delta R	XXXX.X nmi
	Delta V	XXXX.X ft/s
	Radar data source code	0000X
50	Spare	
51	S-band antenna angles	
	Pitch (Alpha)	XXX.XX deg
	Yaw (Beta)	XXX.XX deg
52	Central angle of active vehicle	XXX.XX deg
53	Spare	
54	Range	XXX.XX nmi
	Range rate	XXXX.X ft/s
	Theta	XXX.XX deg
55	Number of apsidal crossings	XXXXX.
	Elevation angle	XXX.XX deg
	Central angle of passive vehicle	XXX.XX deg
56	RR LOS	
	Azimuth	XXX.XX deg
	Elevation	XXX.XX deg
57	Spare	
58	Pericenter altitude (post TPI)	XXXX.X nmi
	Delta V (TPI)	XXXX.X ft/s
	Delta V (TPF)	XXXX.X ft/s
59	Delta V LOS 1	XXXX.X ft/s

Number	Register Description	Data Format
	Delta V LOS 2	XXXX X ft/s
	Delta V LOS 3	XXXX.X ft/s
60	Forward velocity	XXXX.X ft/s
	Altitude rate	XXXX.X ft/s
	Computed altitude	XXXXX. ft/s
61	Time to go in braking phase	XXbXX min/s
	Time from ignition	XXbXX min/s
	Crossrange distance	XXXXX nmi
62	Absolute value of velocity	XXXX.X ft/s
	Time from ignition	XXbXX min/s
	Delta V (accumulated)	XXXX.X ft/s
63	Delta Altitude (+LR > LGC)	XXXXX. ft
	Altitude rate	XXXX.X ft/s
	Computed altitude	XXXXX. ft
64	Time left for redesignations (TR)/LPD	XXbXX seconds/deg
	Altitude rate	XXXX.X ft/s
	Computed altitude	XXXXX. ft
65	Sampled LGC lime (fetched in interrupt)	00XXX. h
		000XX. min
		0XX.XX s
66	LR slant range (R2)	XXXXX. ft
	LR position (R3)	00001/00002
67	LR VX	XXXXX. ft/s
	LR VY	XXXXX. ft/s
	LR VZ	XXXXX. ft/s
68	Ground range to landing site	XXXX.X nmi
	Time to go in braking phase	XXbXX min/s
	Absolute value of velocity	XXXX.X ft/s
69	Landing site correction	Z XXXXX. ft
	Landing site correction	Y XXXXX. ft
	Landing site correction	X XXXXX. ft
70	AOT detent code/star code (before mark)	Octal
		Octal
		Octal
71	AOT detent code/star (after mark)	Octal
	Mark X/Cursor Counter (Max = 5)	Octal
	Mark Y/Spiral Counter (Max = 5)	Octal
72	RR trunnion angle	XXX.XX deg
	(360 degrees – CDU trunnion angle)	XXX.XX deg
	RR shaft angle	
73	Desired RR trunnion angle	XXX.XX deg
	(360 degrees – CDU trunnion angle)	XXX.XX deg
	Desired RR shaft angle	
74	Time from ignition	XXbXX min/s
	Yaw after vehicle rise	XXX.XX deg
	Pitch after vehicle rise	XXX.XX deg

Number	Register Description	Data Format
75	Delta altitude (CDH)	XXXX.X nmi
	Delta time (CDH-CSI or TPI-CDH) (Modular 60)	XXbXX min/s
	Delta time (TPI-CDH or TPI-NOMTPI) (Modutar 60)	XXbXX min/s
70	Desired downrange velocity	XXXX.X ft/s
	Desired radial velocity	XXXX.X ft/s
	Crossrange distance	XXXX.X nmi
77	Time to engine cutoff	XXbXX min/s
	Velocity normal to CSM plane (LM) (+ Rt)	XXXX.X ft/s
	Absolute value of velocity	XXXX.X ft/s
78	RR range	XXX.XX nmi
	RR range rate	XXXXX. ft/s
79	Cursor angle	XXX.XX deg
	Spiral angle	XXX.XX deg
	Position code	+ 0000X
80	Data indicator	XXXXX.
	Omega	XXX.XX deg
81	Delta VX (LV) (+ Fwd)	XXXX.X ft/s
	Delta VY (LV) (+ Rt)	XXXX.X ft/s
	Delta VZ (LV) (+ Dn)	XXXX.X ft/s
82	Delta VX (LV) (+ Fwd)	XXXX.X ft/s
	Delta VY (LV) (+ Rt)	XXXX.X ft/s
	Delta VZ (LV) (+ Dn)	XXXX.X ft/s
83	Delta VX (body) (+ Up)	XXXX.X ft/s
	Delta VY (body) (+ Rt)	XXXX.X ft/s
	Delta VZ (body) (+ Fwd)	XXXX.X ft/s
84	Delta VX (LV of other vehicle) ((+ R x V) x R)	XXXX.X ft/s
	Delta VY (LV of other vehicle) (+ (V x R))	XXXX.X ft/s
	Delta VZ (LV of other vehicle) (+ (-R))	XXXX.X ft/s
85	VGX (body) (+ Up)	XXXX.X ft/s
	VGY (body) (+ Rt)	XXXX.X ft/s
	VGZ (body) (+ Fwd)	XXXX.X ft/s
86	VGX (LV) (+ Fwd)	XXXX.X ft/s
	VGY (LV) (+ Rt)	XXXX.X ft/s
	VGZ (LV) (+ Dn)	XXXX.X ft/s
87	Backup optics LOS	
	Azimuth (+ Rt)	XXX.XX deg
	Elevation (+ Up)	XXX.XX deg
88	Component of celestial body unit vector	X .XXXXX
		Y .XXXXX
		Z .XXXXX
89	Landmark latitude (+ North)	XX.XXX deg

Number	Register Description	Data Format
	Landmark longitude/2 (+East)	XX.XXX deg
	Landmark altitude	XXX.XX nmi
90	Rendezvous out of plane parameters	
	Y	XXX.XX nmi
	Y dot	XXXX.X ft/s
	PSI	XXX.XX deg
91	Altitude	XXXXX. nmi
	Velocity	XXXXX. ft/s
	Flight path angle	XXX.XX deg
92	Percent of full thrust (10,500 lb)	00XXX %
	Altitude rate	XXXX.X ft/s
	Computed altitude	XXXXX. ft
93	Delta gyro angles	X XX.XXX deg
		Y XX.XXX deg
		Z XX.XXX deg
94	VGX (LM) (+Up)	XXXX.X ft/s
	Altitude Rate	XXXX.X ft/s
	Computed Altitude	XXXXX. ft
95	Spare	XXXXX.
96	Spare	XXXXX.
97	System test inputs	XXXXX.
98	System test results and input	XXXXX.
		XXXXX.
		XXXXX.
99	RMS in position	XXXXX. ft
	RMS in velocity	XXXX.X ft/s
	RMS in bias	XXXXX. milliradians

APPENDIX P: LUNAR MODULE PROGRAM ALARMS

Lunar Module AGC Program Alarms

Code	Interpretation	Issued By
00111	Mark missing	R53
00112	Mark or mark reject not being accepted or ROD input and Average G off	R57
00113	No inbits	R57
00114	Mark made but not desired	R57
00115	No marks in last pair to reject	R57
00206	Zero Encode not allowed with Coarse Align plus Gimbal Lock	IMU mode switching V40, N20
00207	ISS turn-on request not present for 90 seconds	T4RUPT
00210	IMU not operating	IMU mode switching, R02
00211	Coarse Align error	IMU mode switching, P51, P57, R50

Code	Interpretation	Issued By
00212	PIPA fail but PIPA is not being used	IMU mode switching. T4RUPT
00213	IMU not operating with turn-on request	T4RUPT
00214	Program using IMU when turned off	T4RUPT
00217	Bad return from IMUSTALL	P51, P57, R50
00220	Bad REFSMMAT	R02, R47
00401	Desired gimbal angles yields Gimbal Lock	In-flight alignment, IMU-2, FINDCDUW
00402	FINDCDUW routine not controlling attitude because of inadequate pointing vectors	FINDCDUW
00404	Two Stars not available in any detent	R59
00405	Two stars not available	R51
00421	W-matrix overflow	INTEGRV
00501	Radar antenna out of limits	R23
00502	Bad radar gimbal angle input	V41N72
00503	Radar antenna designate fail	R21, V41 N72
00510	Radar auto discrete not present	R25, V40N72
00511	LR not in Position 2 or repositioning	R12
00514	RR goes out of Auto mode while in use	P20, P22
00515	RR CDU Fail discrete present	R25
00520	RADARUPT not expected at this time	Radar read, P20, P22, R12
00522	LR position change	R12
00523	LR antenna did not achieve position 2	R12, V59
00525	Delta Theta greater than 3 degrees	R22
00526	Range greater than 400 nmi	P20, P22
00527	LOS not in Mode 2 coverage in P22 or vehicle maneuver required in P20	R24
00530	LOS not in Mode 2 coverage while on lunar surface after 600 seconds	R21
00600	Imaginary roots on first iteration	P32, P72
00601	Perigee altitude after CSI < 85 nmi earth orbit <35,000 feet moon orbit	P32, P72
00602	Perigee altitude after CDH < 85 nmi earth orbit <35,000 feet moon orbit	P32, P72
00603	CSI to CDH time less than 10 minutes	P32, P33, P72, P73
00604	CDH to TPI time less than 10 minutes	P32, P72
00605	Number of iterations exceeds P32/P72 loop maximum (> 15)	P32, P72
00606	DV exceeds maximum	P32, P72
00611	No TIG for given E angle	P33, P34, P73, P74
00701	llegal option code selected	P57
00777	PIPA Fail caused ISS warning	T4RUPT
01102	DOWNLINK too fast	T4RUPT
01106	UPLINK too fast	T4RUPT

Code	Interpretation	Issued By
01107	Phase table failure. Assume erasable memory is suspect.	RESTART
01301	ARCSIN-ARCCOS input angle too large	INTERPRETER
01406	Bad return from ROOTSPRS during descent guidance	P63, P64
01407	VG increasing (Delta V accumulated at 90 degrees from desired thrust vector)	P40, P42 P63, P64, P66
01410	Unintentional overflow in guidance	
01412	Descent ignition algorithm not converging	P63
01466	Throttle servicing insufficient	P66
01520	V37 request not permitted at this time	R00
01600	Overflow in drift test	Ground Test
01601	Bad IMU torque	Ground Test
01703	Too close to ignition; slip time of ignition	R41
01706	Incorrect program selected for vehicle configuration	P40, P42
02001	Jet failures have disabled Y-Z translation	DAP
02002	Jet failures have disabled X translation	DAP
02003	Jet failures have disabled P rotations	DAP
02004	Jet failures have disabled U-V rotations	DAP
03777	ICDU fail caused the ISS warning	T4RUPT
04777	ICDU, PIPA fails caused the ISS warning	T4RUPT
07777	IMU fail caused the ISS warning	T4RUPT
10777	IMU, PIPA fails caused the ISS warning	T4RUPT
13777	IMU, ICDU fails caused the ISS warning	T4RUPT
14777	IMU, ICDU, PIPA fails caused the ISS warning	T4RUPT
20105	AOT mark system in use	R53
20430	Acceleration overflow in integration	Orbital integrator
20607	No solution from TIME-THETA or TIME-RADIUS	TIMETHET, TIMERAD
21103	Unused CCS branch executed	ABORT
21204	WAITLIST, VARDELAY, FIXDELAY, DELAYJOB, or LONGCALL called with zero or negative delta time	WAITLIST
21302	SORT called with negative argument	INTERPRETER
21406	Bad return from ROOTSPS during descent preignition	Ignition algorithm
21501	Keyboard and Display alarm during internal use (NVSUB)	PINBALL
31104	Delay routine busy	EXECUTIVE
31201	Executive overflow-no VAC areas	EXECUTIVE
31202	Executive overflow-no core sets	EXECUTIVE
31203	WAITLIST overftow-too many tasks	WAITLIST
31206	Second job attempts to go to sleep via Keyboard and Display program	PINBALL
31207	No VAC area for marks	R53

Code	Interpretation	Issued By
31210	Two programs using device at same time	IMU mode switch
31211	Illegal interrupt of extended verb	R53
31502	Two priority displays waiting	GOPLAY
32000	DAP still in progress at next T5RUPT	DAP

APPENDIX Q: COMMAND MODULE AND LUNAR MODULE DOWNLISTS

Command Module Downlists

Powered List
    P40 SPS Thrust
    P41 RCS Thrust
    P47 Thrust Monitor
    P61 Entry Preparation Program

Coast and Align List
    P00 CMC Idling
    P01 Prelaunch Initialization
    P02 Gyro Compassing
    P03 Optical Verification of azimuth
    P06 CMC Power Down
    P07 System Test
    P11 Earth Orbit Injection (EOI) Monitor
    P51 IMU Orientation Determination
    P52 IMU Realignment Program
    P53 Backup IMU Orientation Determination
    P54 Backup IMU Realignment

Rendezvous and Prethrust List
    P17 CSM TPI Search
    P20 Rendezvous Navigation
    P21 Ground Track Determination 23 Cislunar Navigation
    P30 External AV Maneuver Guidance
    P31 Lambert Aim Point Maneuver Guidance
    P34 Transfer Phase Initiation (TPI) Guidance
    P35 Transfer Phase Midcourse (TPM) Guidance
    P37 Return to Earth Maneuver Guidance
    P38 Stable Orbit Rendezvous Guidance
    P39 Stable Orbit Rendezvous Midcourse Guidance
    P74 LM Transfer Phase Initiation (TPI) Targeting
    P75 LM Transfer Phase Midcourse (TPM) Targeting
    P77 LM TPI Search Program
    P78 LM Stable Orbit Rendezvous
    P79 LM Stable Orbit Rendezvous Midcourse Targeting

Entry and Update List
  P27 CMC Update
  P62 CM/SM Separation and Pre-entry Maneuver
  P63 Entry Initialization
  P64 Post 0.05 G Entry Mode
  P65 Up Control Entry Mode
  P66 Ballistic Entry Mode
  P67 Final Entry Mode

Program 22 List
  P22 Orbital Navigation

Lunar Module Downlists

Orbital Maneuvers List
  P40 DPS Thrust
  P41 RCS Thrust
  P42 APS Thrust
  P47 Thrust Monitor

Coast and Align List
  P00 LGC Idling
  P06 LGC Power Down
  P51 IMU Orientation Determination
  P52 IMU Realignment

Rendezvous and Prethrust List
  P20 Rendezvous Navigation
  P21 Ground Track Determination
  P25 Preferred Tracking Attitude
  P30 External AV Maneuver Guidance
  P32 Coelliptic Sequence Initiation (CSI)
  P33 Constant Differential Altitude (CDH)
  P34 Transfer Phase Initiation (TPI)
  P35 Transfer Phase Midcourse (TPM)
  P72 CSM CSI Targeting
  P73 CSM CDH Targeting
  P74 CSM TPI Targeting
  P75 CSM TPM Targeting
  P76 Target DELTA V

Descent and Ascent List
  P12 Powered Ascent Guidance
  P63 Braking Phase Guidance
  P64 Approach Phase Guidance
  P66 Rate of Descent (ROD) Landing Phase Guidance
  P68 Confirm Lunar Landing
  P70 DPS Abort Guidance
  P71 APS Abort Guidance

Lunar Surface Align List
  P22 RR Lunar Surface Navigation
  P57 Lunar Surface Alignment

AGS Initialization and Update List
  P27 LGC Update
  R47 AGS Initialization Routine

APPENDIX R: AGC NAVIGATION STAR CATALOG

Code	Star	Visual Magnitude	Right Ascension		Declination	
			Hr	Min	Deg	Min
0	Planet					
1	Alpheratz (Alpha Andromedae)	2.1	0	6	28	53
2	Diphda (Beta Ceti)	2.2	0	42	−18	11
3	Navi (Gamma Cassiopeiae)	2.2	0	54	60	27
4	Achernar (Alpha Eridani)	0.6	1	36	−57	25
5	Polaris (Alpha Ursae Minoris)	2.1	1	58	89	6
6	Armour (Theta Eridani)	3.4	2	57	−40	26
7	Menkar (Alpha Ceti)	2.8	3	0	3	56
10	Mirfak (Alpha Persei)	1.9	3	22	49	44
11	Aldebaran (Alpha Tauri)	1.1	4	34	16	26
12	Rigel (Beta Orionis)	0.3	5	12	−8	15
13	Capella (Alpine Aurigae)	0.2	5	13	45	57
14	Canopus (Alpha Carinae)	−0.9	6	23	−52	40
15	Sirius (Alpha Canis Majoris)	−1.6	6	44	−16	40
16	Procyon (Alpha Canis Minoris)	0.5	7	37	5	19
17	Regor (Gamma Velorum)	1.9	8	8	−47	14
20	Donce (Iota Ursae Majoris)	3.1	8	50	48	30
21	Alphard (Alpha Hydrae)	2.2	9	26	−8	30
22	Regulus (Alpha Leonia)	1.3	10	6	12	9
23	Denebola (Beta Leonis)	2.2	11	47	14	46
24	Gienah (Gamma Corvi)	2.8	12	13	−17	20
25	Acrux (Alpha Crucis)	1.6	12	24	−62	49
26	Spica (Alpha Virginis)	1.2	13	23	−10	58
27	Alkaid (Eta Ursae Majoris)	1.9	13	46	49	30
30	Menkent (Theta Centauri)	2.3	14	4	−36	11
31	Arcturus (Alpha Bootis)	0.2	14	14	19	22
32	Alphecca (Alpha Coronae Borealis)	2.3	15	33	26	50
33	Antares (Alpha Scorpii)	1.2	16	27	−26	21
34	Atria Austrinus (Alpha Trianguli)	1.9	16	43	−68	56
35	Resalhague (Alpha Ophiuchi)	2.1	17	33	12	35
36	Vega (Alpha Lyrae)	0.1	18	36	38	45
37	Nunki (Sigma Sagittarii)	2.1	18	53	−26	20

Code	Star	Visual Magnitude	Right Ascension		Declination	
40	Altair (Alpha Aquilae)	0.9	19	49	8	46
41	Dabih (Beta Capricorni)	3.2	20	19	−14	54
42	Peacock (Alpha Pavonis)	2.1	20	23	−56	51
43	Deneb (Alpha Cygni)	1.3	20	40	45	9
44	Enif (Epsilon Pegasi)	2.5	21	42	9	42
45	Fomalhaut (Alpha Piscis Austrinus)	1.3	22	56	−29	49

APPENDIX S: CONFIGURING THE CSM AND LM DAP (ROUTINE 03)

Reflecting the wide range of mission requirements and configurations, the DAP initialization requires a wide variety of inputs. The DAP configuration program, Routine 03, is started using extended Verb 48. While the specifics differ between the CSM and LM, both DAPs require the following as input: Spacecraft weight and configuration, maneuvering rate, deadband, and thruster configuration. Using the LM for the example, Routine 03 first responds with Noun 46, to specify the spacecraft configuration. In Register 1, five different parameters are specified, each consisting of a single octal number.

LM DAP Setting

Noun 46
R1 codes: ABCDE

A – Data Code
>   1: Ascent stage only
>   2: Ascent and Descent stage
>   3: Ascent and Descent stage, docked with CSM

B – Translation
>   0: 2 jet translation (fuel system A)
>   1: 2 jet translation (fuel system B)
>   2: 4 jet translation (fuel system A)
>   3: 4 jet translation (fuel system B) (Fuel system specification is not meaningful)

C – Scaling (Rate Command)
>   0: Fine scaling (ACA 4 degrees/sec)
>   1: Normal scaling (ACA 20 degrees/second)

D – Deadband
>   0: 0.3 Degrees
>   1: 1 Degree
>   2: 5 Degrees

E – KALCMANU Rate
>   0: .2 degrees/second

1: .5 degrees/second
2: 2 degrees/second
3: 10 degrees/second

Verb 21 is used to change the data in Register 1 to reflect the required configuration. After register 1 has been entered, the PRO key is pressed and Routine 03 then requests the weight of the LM and CSM (if docked) in Verb 47. After the weights are entered, the PRO key is pressed again. Finally, if the Descent Stage is still attached to the LM (specified in the first digit of Register 1 in Verb 46), the gimbal angles of the descent engine are entered through Verb 48. With the final pressing of the PRO key, the DAP begins its initialization and assumes control of the spacecraft.

CSM DAP Setting

Noun 46
Register 1 Codes: ABCDE

A – Configuration
        0: No DAP is requested
        1: CSM Alone
        2: CSM and LM
        3: S-IVB, CSM and LM (S-IVB control)
        6: CSM and LM ascent stage

B – XTAC: X-Translations using Quads AC
        0: Do not use AC
        1: Use AC

C – XTBD: X Translations using Quads BD
        0: Do not use BD
        1: Use BD

D – DB: Angular Deadband for Attitude Hold and Automatic Maneuvers
        0: +/− 0.5 degrees
        1: +/− 5.0 degrees

E – RATE: Rate Specification for RHC in Hold or AUTO Mode, and for KALCMANU-Supervised Automatic Maneuvers
        0: 0.05 degrees/sec
        1: 0.2 degrees/sec
        2: 0.5 degrees/sec
        3: 4.0 degrees/sec

Register 2 Codes: ABCDE

A – AC Roll: Roll-Jet Selection
        1: Use AC Roll
        0: Use BD Roll

B – QUAD A
    1: Quad is operational
    0: Quad has failed
C – QUAD B
    1: Quad is operational
    0: Quad has failed
D – QUAD C
    1: Quad is operational
    0: Quad has failed
E – QUAD D
    1: Quad is operational
    0: Quad has failed

Noun 47

Register 1: CSM Weight
Register 2: LM Weight

Noun 48

Register 1: SPS Pitch gimbal trim offset in 1/100 degree
Register 2: SPS Yaw gimbal trim offset in 1/100 degree

Glossary of terms and abbreviations

AC	Alternating Current
ACA	Attitude Control Assembly
AELD	Ascent Engine Latching Device
AGC	Apollo Guidance Computer
AGS	Abort Guidance System
AOT	Alignment Optical Telescope
APS	Ascent Propulsion System
BBANK	Both Banks
CDH	Constant Delta Height
CDR	Commander
CDU	Coupling Data Unit
CM	Command Module
CMC	Command Module Computer
CMP	Command Module Pilot
COAS	Crewman Optical Alignment Sight
CPU	Central Processing Unit
CSI	Coelliptic Sequence Initiation
CSM	Command and Service Module
CWEA	Caution and Warning Electronics Assembly
DAP	Digital Autopilot
DECA	Descent Engine Control Assembly
DEDA	Data Entry and Display Assembly
DPS	Descent Propulsion System
DSKY	Display and Keyboard
EBANK	Erasable Storage Bank
ECS	Environmental Control System
EMP	Erasable Memory Programs
EMS	Entry Monitor System
FBANK	Fixed Storage Bank
FDAI	Flight Director Attitude indicator
FEB	Fixed Extension Bank
GPS	Global Positioning System

HAM	Height Adjustment Maneuver
I/O	Input-Output
IAW	Interpreter Address Word
IIW	Interpreter Instruction Word
IMU	Inertial Measurement Unit
LGC	Lunar Module Guidance Computer
LLOS	Landmark Line of Sight
LM	Lunar Module
LMP	Lunar Module Pilot
LPD	Landing Point Designator
LVDC	Launch Vehicle Digital Computer
MET	Mission Elapsed Time
MSFN	Manned Space Flight Network
OHC	Optics Hand Controller
ORDEAL	Orbital Rate Display, Earth and Lunar
OS	Operating System
PAD	Pre-Advisory Data
PC	Plane Change
PDI	Powered Descent Initiation
PGNS	Primary Guidance and Navigation System
PIPA	Pulsed Integrating Pendulous Accelerometer
PTC	Passive Thermal Control
PU	Propellant Utilization
RCS	Reaction Control System
REFSMMAT	Reference to a Stable Member Matrix
RHC	Rotational Hand Controller
ROD	Rate of Descent
RSI	Roll Stability Indicator
RTCC	Real-Time Computer Center
SCS	Stablization and Control System
SLOS	Star Line of Sight
SPS	Service Propulsion System
TLI	Translunar Injection
TPF	Terminal Phase Midcourse
TPI	Terminal Phase Initiation
TPM	Terminal Phase Midcourse
TTHC	Thrust/Translational Hand Controller
VAC	Vector Accumulator
VHF	Very High Frequency

Bibliography

Alonso, R.L., Blair-Smith, H. and Hopkins, A.L. 'Some Aspects of the Logical Design of a
Control Computer: A case study', *IEEE Transactions*, December 1963.

Bellcomm Inc. 'Colossus on C-Prime: Case 310', Washington, D.C.: Bellcomm Inc.,
November 26, 1968.

Bellcomm Inc. 'Description of Apollo Entry Guidance', TM-66-2012-2. Washington, D.C.:
Bellcomm, Inc., August 4, 1966.

Blair-Smith, Hugh. 'AGC Memo #9: Block II Instructions', Cambridge, Mass.: MIT
Instrumentation Laboratory, July 1, 1966.

Bowler, D.J. 'Auxiliary Memory System Final Report on Phase 1', E-2254. Cambridge, Mass.:
MIT Instrumentation Laboratory, April 1968.

Copps, Edward M. 'Recovery from Transient Failures of the Apollo Guidance Computer', E-
2307. Cambridge, Mass.: MIT Instrumentation Laboratory, August 1968.

Crisp, R. and Keene, D. 'Apollo Command and Service Module Reaction Control by the
Digital Autopilot', E-1964. Cambridge, Mass.: MIT Instrumentation Laboratory, May
1966.

Frank, A.J., Knotts, E.F. and Johnson, B.C. 'An Entry Monitor System for Maneuverable
Vehicles', *Journal of Spacecraft and Rockets*, 3, no. 8 (August 1966): 1229–1234.

General Motors. 'Apollo 15', Milwaukee, Wisconsin: General Motors AC Delco Electronics,
June 1971.

General Motors. 'Apollo Guidance and Navigation System Lunar Module Student Study
Guide, System Course 3100', Milwaukee, Wisconsin: General Motors AC Delco
Electronics, January 15, 1967.

Green, Alan I. 'Keyboard and Display Program and Operation', E-1905. Cambridge, Mass.:
MIT Instrumentation Laboratory, January 1966.

Green, Alan I. and Filene, Robert J. 'Keyboard and Display Program and Operation 2', E-
2129. Cambridge, Mass.: MIT Instrumentation Laboratory, June 1967.

Green, Alan I. and Rocchio, Joseph J. 'Keyboard and Display System Program for AGC
(Program Sunrise)', E-1574. Cambridge, Mass.: MIT Instrumentation Laboratory,
August 1964.

Grumman. 'LEM CES and AGS', Bethpage, NY: Grumman Aircraft Engineering Company,
September 1966.

Hall, Eldon C. 'Case History of the Apollo Guidance Computer', E-1970. Cambridge, Mass.:
MIT Instrumentation Laboratory, June 1966.

Hall, Eldon C. 'MIT's Role in Project Apollo Volume III, The Computer Subsystem', R-700.
Cambridge, Mass.: MIT Instrumentation Laboratory, August 1972.

Hall, Eldon C. *Journey to the Moon: The History of the Apollo Guidance Computer*. Reston, Va. AIAA Press, 1996.

Hopkins, Albert, Alonso, Ramon and Blair-Smith, Hugh. 'Logical Description for the Apollo Guidance Computer (AGC 4)', R-393. Cambridge, Mass.: MIT Instrumentation Laboratory, March 1963.

Hopkins, Albert L., Alonso, Ramon L. and Blair-Smith, Hugh. 'Preliminary Mod 3C Programmers Manual', Cambridge, Mass.: MIT Instrumentation Laboratory, November 1961, E-1077.

Hyle, C.T. 'Launch Abort Philosophy for Manned Space Flights', 68-FM-104. Houston, Texas: National Aeronautics and Space Administration, April 29, 1968.

Johnson, Madeline S. and Giller, Donald R. 'MIT's Role in Project Apollo Volume V, The Software Effort', R-700. Cambridge, Mass.: MIT Instrumentation Laboratory, March 1971.

Klumpp, Allan R. 'A Manually Retargeted Automatic Landing System for the Lunar Module', *Journal of Spacecraft and Rockets*, 5, no. 2 (February 1968): 129–138.

Lawton, T.J. and Muntz, C.A. 'Organization of Computation and Control in the Apollo Guidance Computer', E-1758. Cambridge, Mass.: MIT Instrumentation Laboratory, April 1965.

Lugo, Jose R.P. 'Lunar Module Attitude Controller Assembly Input Processing', Presentation at MAPLD 2004, Washington D.C., June 2004.

Martin, Fredrick H. and Battin, Richard H. 'Computer-Controlled Steering of the Apollo Spacecraft', E-2159. Cambridge, Mass.: MIT Instrumentation Laboratory, August 1967.

McGee, Leonard A. and Schmidt, Stanley F. 'Discovery of the Kalman Filter as a Practical Tool for Aerospace and Industry', NASA Technical Memorandum 86847. Moffett Field, California: National Aeronautics and Space Administration, November 1985.

Mindell, David A. *Digital Apollo: Human and Machine in Spaceflight*, Cambridge, Mass, The MIT Press, 2008.

MIT. 'Digital Development Memo #267, Block II Keyboard and Display Program (RETRED44)', Cambridge, Mass.: MIT Instrumentation Laboratory, July 28, 1965.

MIT. 'Guidance System Operational Plan for Manned Earth Orbit and Lunar Missions Using Program Colossus 1 (Rev. 237) and Colossus 1A (Rev. 249): Section 1, Prelaunch (Rev. 1)', R-577. Cambridge, Mass.: MIT Instrumentation Laboratory, December 1968.

MIT. 'Guidance System Operational Plan for Manned Earth Orbit and Lunar Missions Using Program Colossus 1 (Rev. 237) and Colossus 1A (Rev. 249): Section 4, GNCS Operations (Rev. 6)', R-577. Cambridge, Mass.: MIT Instrumentation Laboratory, December 1968.

MIT. 'Guidance System Operational Plan for Manned Earth Orbit and Lunar Missions Using Program Colossus 2E: Section 3, Digital Autopilot (Rev. 11)', R-577. Cambridge, Mass.: MIT Instrumentation Laboratory, June 1970.

MIT. 'Guidance System Operational Plan for Manned Earth Orbit and Lunar Missions Using Program Colossus 3, Section 5, Guidance Equations (Rev. 14)', R-577. Cambridge, Mass.: MIT Instrumentation Laboratory, March 1971.

MIT. 'Guidance System Operational Plan for Manned Earth Orbit and Lunar Missions Using Program Colossus 3, Section 7, Erasable Memory Programs (Rev. 1)', R-577. Cambridge, Mass.: MIT Instrumentation Laboratory, October 1972.

MIT. 'Guidance System Operational Plan for Manned Earth Orbit and Lunar Missions Using Program Colossus 3: Section 2, Data Links (Rev. 14)', R-577. Cambridge, Mass.: MIT Instrumentation Laboratory, February 1971.

MIT. 'Guidance System Operational Plan for Manned Earth Orbit and Lunar Missions Using

Program Luminary 1C (LM131 Rev 1.): Section 2, Data Links (Rev. 8)', R-567. Cambridge, Mass.: MIT Instrumentation Laboratory, March 1970.

MIT. 'Guidance System Operational Plan for Manned Earth Orbit and Lunar Missions Using Program Luminary 1C (LM131 Rev 1.): Section 3, Digital Autopilot (Rev. 4)', R-567. Cambridge, Mass.: MIT Instrumentation Laboratory, March 1970.

MIT. 'Guidance System Operational Plan for Manned Earth Orbit and Lunar Missions Using Program Luminary 1C (LM131): Section 4, PGNCS Operational Modes (Rev. 7)', R-567. Cambridge, Mass.: MIT Instrumentation Laboratory, December 1969.

MIT. 'Guidance System Operational Plan for Manned Earth Orbit and Lunar Missions Using Program Luminary 1C (LM131): Section 5, Guidance Equations (Rev. 8)', R-567. Cambridge, Mass.: MIT Instrumentation Laboratory, April 1970.

MIT. 'Space Navigation Guidance and Control, Vol 1 of 2', R-500. Cambridge, Mass.: MIT Instrumentation Laboratory, June 1965.

MIT. 'Space Navigation Guidance and Control, Vol 2 of 2', R-500. Cambridge, Mass.: MIT Instrumentation Laboratory, June 1965.

MIT. 'The Complete Sunrise Being a Description of Program Sunrise (Sunrise 33 – NASA dwg #1021102)', R-467. Cambridge, Mass.: MIT Instrumentation Laboratory, September 1964.

MIT. 'Users Guide to Skylark GN&CS Major Modes and Routines', R-738. Cambridge, Mass.: MIT Instrumentation Laboratory, January, 1973.

Moore, William E. 'The Generalized Forward Iterator', 66-FM-55. Houston, Texas: NASA Manned Spacecraft Center, June 15, 1966.

Muntz, Charles A. 'Users Guide to the Block II AGC/LGC Interpreter', R-489. Cambridge, Mass.: MIT Instrumentation Laboratory, April 1965.

NASA. 'Apollo 11 Flight Plan', Houston, Texas: NASA Manned Spacecraft Center, July 1, 1969.

NASA. 'Apollo 12 LM Rendezvous / Abort Book', Part No. SKB32100081-393. Houston, Texas: NASA Manned Spacecraft Center, October 27, 1969.

NASA. 'Apollo 12 LM Timeline Book', Part No. SKB32100081-388. Houston, Texas: NASA Manned Spacecraft Center, September 1969.

NASA. 'Apollo 14 LM Activation Checklist'. Houston, Texas: NASA Manned Spacecraft Center, September 15, 1970.

NASA. 'Apollo 14 LM Timeline Book'. Houston, Texas: NASA Manned Spacecraft Center, September 15, 1970.

NASA. 'Apollo 14 Mission Report', MSC-04112. Houston, Texas: NASA Manned Spacecraft Center, May 1971.

NASA. 'Apollo 15 CSM G&C Checklist'. Houston, Texas: NASA Manned Spacecraft Center, March 22, 1971.

NASA. 'Apollo 15 LM Data Card Book'. Houston, Texas: NASA Manned Spacecraft Center, June 10, 1971.

NASA. 'Apollo 16 CSM G&C Checklist', SKB3100122-310. Houston, Texas: NASA Manned Spacecraft Center, March 15, 1972.

NASA. 'Apollo CSM and LM Onboard Navigation System Constraints', 68-FM-106. Houston, Texas: NASA Manned Spacecraft Center, May 1, 1968.

NASA. 'Apollo Experience Report: Evolution of the Rendezvous Maneuver Plan for Lunar Landing Missions', TN D-7388. Houston, Texas: NASA Manned Spacecraft Center, August 1973.

NASA. 'Apollo Experience Report: Guidance and Control Systems', TN D-8249. Houston, Texas: NASA Manned Spacecraft Center, June 1976.

NASA. 'Apollo Experience Report: Guidance and Control Systems: CSM Service Propulsion System Gimbal Actuators', TN D-7969. Houston, Texas: NASA Manned Spacecraft Center, July 1975.

NASA. 'Apollo Experience Report: Guidance and Control Systems: Digital Autopilot Design Development', TN D-7289. Houston, Texas: NASA Manned Spacecraft Center, June 1973.

NASA. 'Apollo Experience Report: Guidance and Control Systems: Lunar Module Stabilization and Control System', TN D-8086. Houston, Texas: NASA Manned Spacecraft Center, November 1975.

NASA. 'Apollo Experience Report: Guidance and Control Systems: Orbital rate Drive Electronics for the Apollo Command Module and Apollo Lunar Module', TN D-7784. Houston, Texas: NASA Manned Spacecraft Center, September 1974.

NASA. 'Apollo Experience Report: Mission Planning for Apollo Entry', TN D-6725. Houston, Texas: NASA Manned Spacecraft Center, March 1972.

NASA. 'Apollo Experience Report: Onboard Navigational and Alignment Software', TN D-6741. Houston, Texas: NASA Manned Spacecraft Center, March 1972.

NASA. 'Apollo Experience Report: Primary Guidance, Navigation and Control System Development', TN D-8227. Houston, Texas: NASA Manned Spacecraft Center, May 1976.

NASA. 'Apollo LM Cue Cards'. Houston, Texas: NASA Manned Spacecraft Center, May 10, 1971.

NASA. 'Apollo Operations Handbook Block II Spacecraft Volume I Spacecraft Description: SM2A-03 BLOCK II-(1)'. Houston, Texas: NASA Manned Spacecraft Center, October 1969.

NASA. 'Apollo Operations Handbook Lunar Module LM6 and Subsequent Volume I Subsystems Data: LMA790-3-LM6 and Subsequent'. Houston, Texas: NASA Manned Spacecraft Center, September 1969.

NASA. 'Description and Performance of the Saturn Launch Vehicle's Navigation, Guidance and Control System', TN D-5869. Houston, Texas: National Aeronautics and Space Administration, July 1970.

NASA. 'LM Descent/Phasing Summary Document Mission F', MSC-CF-P-69-5. Houston, Texas: NASA Manned Spacecraft Center, January 1969.

NASA. 'Lunar Descent and Ascent Trajectories', 70-FM-80. Houston, Texas: NASA Manned Spacecraft Center, April 21, 1970.

NASA. 'Lunar Module Pilot Control Considerations', TN D-4131. Houston, Texas: NASA Manned Spacecraft Center, February 1968.

NASA. 'Program Notes for Colossus 3 and Luminary 1E J-Series Missions Flight Programs Rev 1', MSC 05225. Houston, Texas: NASA Manned Spacecraft Center, September 1972.

NASA. 'Project Apollo Coordinate System Standards', SE 008-001-1. Houston, Texas: NASA Manned Spacecraft Center, June 1965.

Raytheon. 'Auxiliary Memory for Apollo Guidance Computer Contract NAS 9-5994'. Sudbury, Mass.: Raytheon Company Space and Information Systems, March 11, 1968.

Savage, Bernard I. and Drake, Alice. 'AGC4 Basic Training Manual Vol I of II', E-2052. Cambridge, Mass., MIT Instrumentation Laboratory, January 1967.

Sears, Norman E. (Editor). 'Primary G&N System Lunar Orbit Operations Vol II of II', R-446. Cambridge, Mass.: MIT Instrumentation Laboratory, April 1964.

Smally, Edwin D. 'Block II AGC Self-Check and Show-Banksum', E-2065. Cambridge, Mass.: MIT Instrumentation Laboratory, December 1966.

Stengel, Robert F. 'Manual Attitude Control of the Lunar Module', *Journal of Spacecraft and Rockets*, 7, no. 8 (August 1970): 941–947.

Tomayko, James E. *Computers in Spaceflight: The NASA Experience*. Linthicum Heights, Md: NASA History Office, 1988.

Trageser, Milton B. and Hoag, David G. 'Apollo Spacecraft Guidance System', R-495. Cambridge, Mass.: MIT Instrumentation Laboratory, June 1965.

TRW Systems Group. 'Lunar Module Abort Guidance System (LM/AGS) Design Survey', CR-86169. TRW Systems Group, September 1968.

Windall, William S. 'Lunar Module Digital Autopilot', *Journal of Spacecraft and Rockets*, 8, no. 1 (January 1971): 56–61.

Young, Kenneth A. and Alexander, James D. 'Apollo Lunar Rendezvous', *Journal of Spacecraft and Rockets*, 7, no. 9 (September 1970): 56–61.

Illustration credits

Index